Lecture Notes in Physics

Volume 1013

The series Lecture Notes in Physics (LNP), founded in 1969, reports new developments in physics research and teaching - quickly and informally, but with a high quality and the explicit aim to summarize and communicate current knowledge in an accessible way. Books published in this series are conceived as bridging material between advanced graduate textbooks and the forefront of research and to serve three purposes:

- to be a compact and modern up-to-date source of reference on a well-defined topic;
- to serve as an accessible introduction to the field to postgraduate students and non-specialist researchers from related areas;
- to be a source of advanced teaching material for specialized seminars, courses and schools.

Both monographs and multi-author volumes will be considered for publication. Edited volumes should however consist of a very limited number of contributions only. Proceedings will not be considered for LNP.

Volumes published in LNP are disseminated both in print and in electronic formats, the electronic archive being available at springerlink.com. The series content is indexed, abstracted and referenced by many abstracting and information services, bibliographic networks, subscription agencies, library networks, and consortia.

Proposals should be sent to a member of the Editorial Board, or directly to the responsible editor at Springer:

Dr Lisa Scalone
lisa.scalone@springernature.com

Subhendra Mohanty

Gravitational Waves from a Quantum Field Theory Perspective

 Springer

Subhendra Mohanty
Theory Division
Physical Research Laboratory
Ahmedabad, India

ISSN 0075-8450   ISSN 1616-6361   (electronic)
Lecture Notes in Physics
ISBN 978-3-031-23769-0   ISBN 978-3-031-23770-6   (eBook)
https://doi.org/10.1007/978-3-031-23770-6

This Springer imprint is published by the registered company Springer Nature Switzerland AG
The registered company address is: Gewerbestrasse 11, 6330 Cham, Switzerland

Dedicated to Anushmita and Srubabati.

Preface

The observations of gravitational wave signal from coalescing black holes or neutron stars have opened a new window for astronomy. This has also opened the avenue for studying the nature of gravity itself. One of the important questions in physics is to determine whether Einstein's General Relativity is a theory of geometry or whether it can be studied as a quantum field theory like electromagnetism, weak and strong interactions.

In this book, we use the methods of quantum field theory to calculate the rates of gravitational radiation from different sources like binaries, quasi-normal mode oscillations of black holes, phase transitions in early universe, gravitational waves from cosmic defects and finally gravitational waves generated during inflation. In QFT, the radiation rates are calculated perturbatively using the Fermi Golden Rule. We also show how the gravitational radiation rate can be calculated using path integral methods.

We do a detailed computation of gravitational waves from elliptical orbits with applications in binary pulsar timings. The field theory method can also used to derive the radiation rate of ultralight scalars and vector bosons from compact binaries. We calculate gravitational wave radiation from hyperbolic orbits and show that we have a memory signal—a gravitational wave signal where there is a permanent change in the spacetime-metric at the detector after the passage of the gravitational wave.

We do a general analysis of memory signals and show that they can be related to the soft-graviton theorems. Soft-graviton theorems relate the amplitude of gravitational radiation to scattering of bodies by the inclusion of a universal kinematical factor. The usefulness of field theory techniques is brought out best in deriving the memory signals using soft-graviton amplitudes.

In addition to the memory terms from soft-gravitons, there is another non-linear memory signal which is generated by the stress tensor of the primary gravitons. The importance of the secondary or the non-linear memory will demonstrate that the gravitons have self-couplings as expected from the non-linear nature of Einstein's theory.

The path integral methods are useful for calculating the back-reaction which we demonstrate by using the Schwinger-Keldysh method for deriving the Burke-Thorne back-reaction force on sources of gravitational radiation. The calculation of the binary radiation in a gravitational wave source in a graviton thermal bath but not necessarily in thermal equilibrium can be carried out using the Schwinger-Keldysh method with thermal propagators.

At the end of the binary coalescence, the black hole which is formed oscillates with a complex quasinormal mode frequency. These give rise to exponentially damped sinusoidal signals. The oscillations are at the photon radius $3GM$ and can arise from other possible final state objects like worm-holes.

The spins of the binaries process in the course of mergers, and this precession is registered in the gravitational wave signal. We derive the equations for the spin dynamics of extended objects and apply it to the problem of binary mergers.

An interesting area where further enquiry is warranted is the question of the refractive index of gravitational waves. It is known that gravitational waves are damped when propagating through matter with non-zero shear viscosity. It is also known that gravitational waves are unaffected by cold dark matter, i.e. they propagate with the velocity of light through pressure-less matter. Study of gravitational wave propagation through diverse forms of matter could be useful in determining the nature of dark matter/energy using gravitational wave observations from distant sources.

We introduce Grand Unified Theories (GUTs) and discuss their consequences in the generation of stochastic gravitational waves. Depending on the GUT symmetry-breaking patterns, there may be cosmic strings and domain walls formed during phase transitions which have cosmological consequences. These can give rise to stochastic gravitational waves. Phase transitions at intermediate scales 10^6 GeV give rise to gravitational waves which may be observed at LIGO frequencies.

An important source of gravitational waves is inflation. The scalar perturbations generated during inflation give rise to anisotropies in the CMB which have been observed by COBE, WMAP and PLANCK satellite-based experiments. The tensor perturbations are expected to be also generated during inflation and give rise to B-mode polarisations in the CMB. The B-mode polarisation generated by primordial gravitational waves has not yet been observed. The cosmological correlations in the inflationary era are more naturally formulated using the "in-in" formalism. Study of cosmological correlators like the bi-spectrum can prove the assumption that the observed density perturbations in the CMB and matter power spectrum had their origin as zero-point fluctuations of the inflaton and graviton fields.

The different chapters are self-contained and the reader can dive into the topic of his choice without needing to read all the earlier chapters. The treatment is more pedagogical compared a review article. The main results discussed are worked out in the text. We hope that the readers will find the subjects covered interesting and the exposition clear, and working through the derivations will help them find novel ideas for research in this field.

I thank my students Akhilesh Nautiyal, Suratna Das, Soumya Rao, Moumita Das, Tanushree Kaushik, Gaurav Tomar, Girish Chakravarti, Bhavesh Chauhan, Ashish Narang and Priyank Parashari for prodding me on to explore me new areas over the years.

I thank my collaborators in this field Aragam Prasanna, Gaetano Lambiase, Eduard Masso, Anjan Joshipura, Ketan Patel, Soumya Jana, Sukannya Bhattacharya, Tanmay Poddar, Arindam Mazumdar, Kaushik Bhattacharya, Prakrut Chaubal, Gaurav Goswami, Prafulla Panda, Sarira Sahu, Anshu Gupta, Joydeep Chakrabortty, Arpan Hait and Suraj Prakash for the collaborative work, a lot of which appears in this book. I thank Diptarka Das, Debtosh Chowdhury, Nilay Kundu, Sukanta Panda and Rahul Srivastava for fruitful discussions on many topics covered in this book.

I thank my colleagues at PRL and my family for their support.

I thank B. Ananthanaryan, IISc Bangalore, for introducing me to the Springer eco-system.

And last but not the least, I thank my Springer Nature Editor, Lisa Scalone, for patiently guiding me through this book, our second project together.

Ahmedabad, India   Subhendra Mohanty
October 2022

Contents

Units and Acronyms

We use are the natural units $c = \hbar = 1$, and the signature of the metric is $(-, +, +, +)$. To compare with experiments we restore the final results in CGS units and when discussing quantum aspects we restore $\hbar$. Newtons constant $G = 1/M_p^2$ where the Planck mass $M_p = 1.22 \times 10^{-19}$ GeV.

GW	Gravitational Waves
EFT	Effective Field Theory
ISCO	Innermost stable circular orbit
LIGO	The Laser Interferometer Gravitational-Wave Observatory
VIRGO	The VIRGO gravitational wave observatory
KAGRA	The Kamioka Gravitational Wave Detector
LISA	The Laser Interferometer Space Antenna
IPTA	International Pulsar Timing Array
BBO	Big Bang Observatory
PN	Post Newtonian
NLO	Next to leading order
FLRW	Friedmann-Lemaitre-Robertson-Walker
FOPT	First-Order Phase Transition
SOPT	Second-Order Phase Transition
CW	Coleman-Weinberg
BH	Black Hole
PBH	Primordial Black Hole
NS	Neutron Star
WD	White Dwarf
DM	Dark Matter
DE	Dark Energy
CDM	Cold Dark Matter
HDM	Hot Dark Matter
WDM	Warm Dark Matter
CMB	Cosmic Microwave Background
LSS	Large-Scale Structure
QNM	Quasi-Normal Mode Oscillations

Introduction

1

Abstract

Gravitational waves have opened a new window in astronomy and becoming important tools for testing theories of astrophysics and cosmology. They can also be used to test the nature of general relativity (GR) in particular about whether GR is a geometrical theory or a quantum field theory like the other interactions of physics. Studying GR as a quantum field theory (QFT) helps in uncovering new aspects even in the classical phenomenon. Gravitational waves radiated by binaries give rise to a backreaction and dissipation in the orbital dynamics which are studied using the techniques of non-equilibrium field theory. Zero-point fluctuations of the metric give rise to the primordial gravitational waves generated during inflation. Gravitational waves also arise in phase transitions and from cosmological defects like strings and monopoles. All these aspects of gravitational waves are studied in a unified formalism using the techniques of QFT.

1.1 Overview

In the quantum field theory of gravity, the spacetime metric is expanded as a perturbation around a background $g_{\mu\nu} = \bar{g}_{\mu\nu} + \kappa h_{\mu\nu}$ and the fluctuation $h_{\mu\nu}$ is treated as the graviton field. The graviton is created or annihilated in interaction vertices and it has a propagator which can be off-shell (in-loops, for instance). Like in the usual perturbative treatment of quantum field theory, the scattering amplitudes with gravitons in the propagator or external legs are expanded in powers of the gravitational coupling $\kappa = \sqrt{32\pi G}$, see [1–4].

As the gravitational coupling is $\kappa \propto 1/M_P$ (where $M_P \equiv 1/\sqrt{G} = 1.9 \times 10^{19}$GeV is the Planck mass), in a process with n-graviton vertex for example amplitudes will be proportional to $(E/M_P)^n$ and will become non-unitary when

© The Author(s), under exclusive license to Springer Nature Switzerland AG 2023
S. Mohanty, *Gravitational Waves from a Quantum Field Theory Perspective*,
Lecture Notes in Physics 1013, https://doi.org/10.1007/978-3-031-23770-6_1

the energy transferred E of the scattering particles exceeds the Planck mass. This is one of the consequences of the fact that the quantum theory of $h_{\mu\nu}$ is not a renormalisable field theory. For practical calculations (like bending of light by a star or the Newtonian potential between massive particles) where the energy transferred in miniscule compared to the Planck scale, the violation of unitarity at Planck energies is not a problem for making predictions using the QFT of the gravitons. The tree level scattering processes give the same result as weak field classical gravity. The amplitude calculation techniques of QFT can sometimes be easier than the corresponding classical GR calculation of the same process [5, 6].

The quantum nature of gravitons occurs in diagrams with gravitons in the loops (there are exceptions to this rule [7–9]). Loop corrections give rise to higher dimensional operators which are divergent and the theory can be renormalised by introducing higher dimensional operators at each order to absorb the infinities. Starting with Einstein's GR with $\mathcal{L}_0 = (2/\kappa^2)R$, the one loop corrections give rise to divergent operators of dimension four [10]

$$\mathcal{L}_1 = \frac{1}{8\pi^2\epsilon}\left(\frac{1}{120}\bar{R}^2 + \frac{7}{20}\bar{R}_{\mu\nu}\bar{R}^{\mu\nu}\right), \tag{1.1}$$

where $\epsilon = 4 - d$ and the bars denote evaluation in the background metric. One must therefore introduce the higher dimension operators with unknown coefficients $c_1\bar{R}^2 + c_2\bar{R}_{\mu\nu}\bar{R}^{\mu\nu}$ and absorb the one loop divergent terms (1.1) in the renormalised couplings c_1 and c_2.

Loop corrections also give rise to infrared divergent terms and these corrections depend only on the parameters of the effective theory like the masses of the particles and the coupling constant κ. The loop corrections in this case do not depend upon the unknown ultra-violet completion of the theory [11], and using these one can make computations new terms in the effective action from the loops which will have well-determined coefficients. These arise from the non-analytic infrared divergent terms of the type $1/\sqrt{-q^2}$ and $\log q^2$ (q being the momentum transferred). These cannot be written as a series in powers of q^2, so these do not give the higher dimensional operators. Fourier transforming to position space, one can identify these pieces as giving the classical and quantum corrections to the tree level classical Newtonian potential. The FT of $1/|\mathbf{q}|$ is $1/(2\pi r^2)$ and the FT of $\log \mathbf{q}^2$ is $-1/(2\pi r^3)$ and these are the classical and quantum corrections, respectively, to the Newtonian potential from loop diagrams.

The Newtonian potential with one loop correction is [12]

$$V_1(r) = \frac{-Gm_1m_2}{r}\left[1 + 3\frac{G(m_1 + m_2)}{r} + \frac{41}{10\pi}\frac{G\hbar}{r^2}\right]. \tag{1.2}$$

Here the second term that is also there in classical GR [13] arises from the $1/1/\sqrt{-q^2}$ terms in the loops and the order $\hbar$ quantum correction arises from the $\log q^2$ terms.

Similarly graviton loops give rise to long range corrections to the Reissner–Nordström and Kerr–Newman metrics from the $1/\sqrt{-q^2}$ and $\log q^2$ terms in the loops [14].

The development that has the greatest impact on the field of gravitational waves is the effective field theory of macroscopic bodies [15–20] where the stars are treated quantum mechanically as point particles on the world line and the gravitational fields are treated as quantum fields.

Field theory calculations have reproduced the Peter Mathews formula [21] and this technique has been used to calculate the radiation of other possible light particles like scalars [21,22] and vector bosons [23] from highly eccentric compact binaries. Observations from the Hulse–Taylor binary continue to refine the error bars [24] and subsequent discoveries of other double pulsars [25] or pulsars in a binary where the companion is neutron stars [26] or white dwarf [27] have proved to be fertile laboratories for testing theories of gravitation and particle physics.

The waveform of gravitational waves can also be determined from the probability amplitude of a graviton emission by a source (like a coalescing binary). The amplitude of graviton emission at the source

$$\mathcal{A}_\lambda(k_0, \mathbf{n}k_0) = -\frac{\kappa^2}{2}\epsilon_{\mu\nu}^{*\lambda}(\mathbf{n})\tilde{T}^{\mu\nu}(k_0, \mathbf{n}k_0) \qquad (1.3)$$

can be used for calculating the gravitational wave metric observed at the detector as

$$h_{\alpha\beta}(\mathbf{x}, t) = \frac{1}{4\pi}\frac{1}{r}\int \frac{dk_0}{(2\pi)}\sum_{\lambda=1}^{2}\epsilon_{\alpha\beta}^\lambda(\mathbf{n})\mathcal{A}_\lambda(k_0, \mathbf{n}k_0)e^{-ik_0(t-r)} \qquad . \qquad (1.4)$$

This relation is useful as it relates the quantum amplitude (which can be a multi-loop vertex) can give us the classical metric perturbation at the detector.

Soft-graviton theorems can be important for computing the graviton emission probability from a complex scattering process. According to the soft-theorems [28–31], if a scattering process of n-particle is $\mathcal{A}_n$, then the amplitude of a graviton emission from one of the legs of the n-particles can be written as the n-particle amplitude multiplied by a kinematic factor

$$\mathcal{A}_{n+1}(p_a, q, \lambda)$$

$$= \frac{\kappa}{2}\epsilon_\lambda^{*\mu\nu}(q)\sum_{a=1}^{n}\left(\frac{p_{a\mu}p_{a\nu}}{p_a \cdot q} + \frac{p_{a\mu}q_\beta J_a^{\beta\nu}}{p_a \cdot q} - \frac{1}{2}\frac{q_\alpha q_\beta J_a^{\alpha\mu}J_a^{\beta\nu}}{p_a \cdot q}\right)\mathcal{A}_n(p_a), \qquad (1.5)$$

where p_a is the momentum and $J_a^{\alpha\beta} = x_a^\alpha p_a^\beta - x_a^\beta p_a^\alpha + S_a^{\alpha\beta}$ the total angular momentum of particle 'a'.

An interesting effect that is predicted for gravitational waves is the memory effect. Gravitational waves from oscillatory sources produce an oscillatory displacement between the mirrors of the detectors. If the source is non-oscillatory and unbounded like binaries in a hyperbolic orbit, then there is a gravitational impulse on the detector which produces a permanent displacement between the mirrors even after the wave has passed by. This is called the linear memory. The permanent displacement can also arise from bound sources like coalescing binaries when the primary gravitational waves act as the source of the secondary gravitational waves with the memory effect. The memory effect has been predicted since long [32–34] and the memory effect signal from coalescing binaries has been computed [35–37]. What is interesting from a theoretical perspective is that the memory effect has been derived from the soft-graviton theorems [38]. Using the connection between memory effects and soft-graviton theorems, it is possible to work out the effect of spin of the bodies and quantum corrections to the soft theorems on the memory signal from binaries and test these in observations.

With the observations of black hole and neutron star mergers by the LIGO and VIRGO collaborations [39–41], a new window to the universe has been opened in the field of gravitational wave astronomy. Interpreting the observations and making predictions of signal require immense theoretical support. Filtering the events from the background and extracting the parameters of the binaries (like their masses and spins) require the generation of theoretical templates of gravitational wave signals from coalescent binaries [42,43]. The techniques for calculating waveforms of binary mergers [44] range from the Hamiltonian formalism where the motion of stars and gravitational fields are treated via their ADM-Hamiltonian evolution [45], post-Newtonian corrections using the effective-one-body (EOB) method [46], numerical simulations [47] and effective field theory [15–20] (Fig. 1.1).

Gravitational wave signal observations at LIGO and VIRGO and future detectors, especially in multi-messenger observations (with concurrent gamma rays or neutrino observations from the same source like in the GW170817 event [49]), open up new horizons for testing theories of gravity and particle physics beyond the standard models [50–52] and a field theoretic treatment of the generation of gravitational waves is particularly amenable for applications to new theories.

At present LIGO has the sensitivity of observing neutron star/solar mass black hole mergers in the range of 80 Mpc. With upgrades, this range is projected to increase to 330 Mpc by 2025 [53], and in addition the present LIGO, VIRGO and KAGRA network will be joined by LIGO-India [54], with five detectors operating the sensitivity for sky localisation and parameter extraction which will improve greatly [53].

Long before the first black hole merger events were first observed by LIGO in 2015, gravitational waves were tested indirectly through the Hulse–Taylor binary pulsar 1913+16 [55] discovered in 1974. According to GR, the binary orbit time period would decay and the observations of the secular period decay [56] matched to within 1% the 1PN calculation by Peters and Mathews [57]. The Peters and Mathew calculation is well suited for binary pulsar calculations as it is exact to all orders in the orbit eccentricity and most binaries in the early stages have highly eccentric orbits.

Coalescing binaries of total mass M separated by distance R radiate gravitational waves at a frequency $f \sim 10\,\text{kHz}(M_\odot/M)(R_s/R)^{3/2}$. Terrestrial detectors like

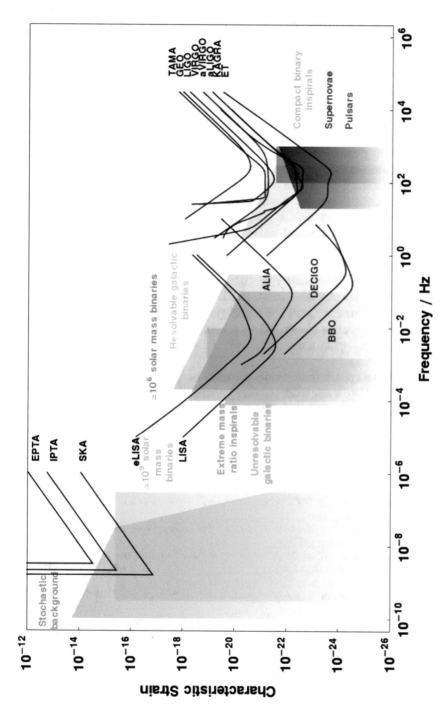

Fig. 1.1 The frequency spectrum of gravitational waves has a span from $f \sim 10^{-10}$ Hz generated in phase transitions in the early universe which can be observed as stochastic GW by pulsar timings (International Pulsar Timing Arrays (IPTA) collaboration) to $f \sim 10$ kHz generated by inspiralling binaries and observed at the LIGO, VIRGO and KAGRA detectors. Reprinted from [48]. Figure credit: ©The Author(s) 2015. Reproduced under CC-BY-3.0 license

LIGO, VIRGO and KAGRA are most sensitive in the frequency range of 1 Hz–10 kHz and they are well suited for observing mergers of black holes of masses in the range $(1 - 100) M_\odot$.

There is a background gravitational radiation from the unresolved signals from astrophysical binaries, or from phase transitions in the early universe. These will be observed as stochastic gravitational waves [58,59]. The stochastic signal can be filtered from a much larger instrumental noise by taking the cross correlation of independent detectors. In the cross correlation, the noise that is independent in each detector will cancel and the signal will add up. Space-based detectors like LISA [60] will be sensitive to gravitational waves in the frequency range of 10^{-4}–1 Hz. Other than extending the mass range of the observed black hole mergers, LISA will be sensitive to gravitational waves generated by phase transitions in the early universe phase. A first order phase transition will release the free energy when the vacuum goes from the higher energy state to the ground state as gravitational waves. The peak frequency of the gravitational waves generated depends on the temperature at which phase transition takes place T_* as $f \sim 10^{-7}(T_*/(\lambda_* H_* \text{GeV})$ Hz. First order phase transitions at the electroweak scale $T_* = 10^2$ GeV will produce gravitational waves with frequency $f \sim 10^{-3}$ Hz, which is in the range of LISA's detection capability. On the other hand, quark hadron phase transition which is believed to have occurred at T_*=100 MeV will generate gravitational waves at peak frequency 10^{-7} Hz which is outside LISA's peak sensitivity, but these can be seen by monitoring the timing of a group of pulsar as is done by the NANOGrav [61] and IPTA [62] collaborations.

The terrestrial and pulsar timing array networks have the ability to measure the two extra vector and two scalar modes of the metric [63,64]. These extra polarisation modes can be dynamical degrees of freedom in theories like $f(R)$ gravity [65] and their observations can be important probes on theories beyond Einstein's gravity.

References

1. R.P. Feynman, F.B. Morinigo, W.G. Wagner, B. Hatfield, *Feynman Lectures on Gravitation*. (Addison-Wesley, Reading, 1995)
2. S. Weinberg, *Gravitation and Cosmology: Principles and Applications of the General Theory of Relativity* (Wiley, Hoboken, 1972)
3. M.J.G. Veltman, Quantum theory of gravitation, in *Methods in Field Theory*. Les Houches, Session XXVIII, ed. by R. Balian, J. Zinn-Justin (North Holland, Netherlands, 1976), pp. 265-328
4. J.F. Donoghue, Introduction to the effective field theory description of gravity (1995). [arXiv:gr-qc/9512024 [gr-qc]]
5. N.E.J. Bjerrum-Bohr, P.H. Damgaard, G. Festuccia, L. Planté, P. Vanhove, General relativity from scattering amplitudes. Phys. Rev. Lett. **121**(17), 171601 (2018). [arXiv:1806.04920 [hep-th]]
6. C. Cheung, I.Z. Rothstein, M.P. Solon, From scattering amplitudes to classical potentials in the post-minkowskian expansion. Phys. Rev. Lett. **121**(25), 251101 (2018). [arXiv:1808.02489 [hep-th]]

7. B.R. Holstein, J.F. Donoghue, Classical physics and quantum loops. Phys. Rev. Lett. **93**, 201602 (2004). [arXiv:hep-th/0405239 [hep-th]]
8. D.A. Kosower, B. Maybee, D. O'Connell, Amplitudes, observables, and classical scattering. J. High Energy Phys. **02**, 137 (2019). [arXiv:1811.10950 [hep-th]]
9. N.E.J. Bjerrum-Bohr, P.H. Damgaard, L. Planté, P. Vanhove, Classical gravity from loop amplitudes. Phys. Rev. D **104**(2), 026009 (2021). [arXiv:2104.04510 [hep-th]]
10. G. 't Hooft, M.J.G. Veltman, One loop divergencies in the theory of gravitation. Ann. Inst. H. Poincare Phys. Theor. A **20**, 69–94 (1974)
11. J.F. Donoghue, General relativity as an effective field theory: the leading quantum corrections. Phys. Rev. D **50**, 3874-3888 (1994). https://doi.org/10.1103/PhysRevD.50.3874. [arXiv:gr-qc/9405057 [gr-qc]]
12. N.E.J. Bjerrum-Bohr, J.F. Donoghue, B.R. Holstein, Quantum gravitational corrections to the nonrelativistic scattering potential of two masses. Phys. Rev. D **67**, 084033 (2003). [erratum: Phys. Rev. D **71**, 069903 (2005)] [arXiv:hep-th/0211072 [hep-th]]
13. Y. Iwasaki, Quantum theory of gravitation vs. classical theory. Prog. Theor. Phys. **46**, 1587 (1971)
14. J.F. Donoghue, B.R. Holstein, B. Garbrecht, T. Konstandin, Quantum corrections to the Reissner-Nordström and Kerr-Newman metrics. Phys. Lett. B **529**, 132–142 (2002). [erratum: Phys. Lett. B **612**, 311–312 (2005)]. [arXiv:hep-th/0112237 [hep-th]]
15. W.D. Goldberger, I.Z. Rothstein, An effective field theory of gravity for extended objects. Phys. Rev. D **73**, 104029 (2006). [arXiv:hep-th/0409156 [hep-th]]
16. W.D. Goldberger, Les Houches lectures on effective field theories and gravitational radiation (2007). [arXiv:hep-ph/0701129 [hep-ph]]
17. S. Foffa, R. Sturani, Effective field theory methods to model compact binaries. Class. Quant. Grav. **31**(4), 043001 (2014). [arXiv:1309.3474 [gr-qc]]
18. R.A. Porto, The effective field theorist's approach to gravitational dynamics. Phys. Rept. **633**, 1–104 (2016). https://doi.org/10.1016/j.physrep.2016.04.003. [arXiv:1601.04914 [hep-th]]
19. M. Levi, Effective field theories of post-newtonian gravity: a comprehensive review. Rept. Prog. Phys. **83**(7), 075901 (2020). [arXiv:1807.01699 [hep-th]]
20. D.A. Kosower, B. Maybee, D. O'Connell, Amplitudes, observables, and classical scattering. J. High Energy Phys. **02**, 137 (2019). https://doi.org/10.1007/JHEP02(2019)137. [arXiv:1811.10950 [hep-th]]
21. S. Mohanty, P. Kumar Panda, Particle physics bounds from the Hulse-Taylor binary. Phys. Rev. D **53**, 5723 (1996)
22. T. Kumar Poddar, S. Mohanty, S. Jana, Constraints on ultralight axions from compact binary systems. Phys. Rev. D **101**(8), 083007 (2020). [arXiv:1906.00666 [hep-ph]]
23. T. Kumar Poddar, S. Mohanty, S. Jana, Vector gauge boson radiation from compact binary systems in a gauged $L_\mu - L_\tau$ scenario. Phys. Rev. D **100**(12), 123023 (2019). [arXiv:1908.09732 [hep-ph]]
24. J.M. Weisberg, Y. Huang, Relativistic measurements from timing the binary pulsar PSR B1913+16. Astrophys. J. **829**(1), 55 (2016)
25. M. Kramer, et al., Tests of general relativity from timing the double pulsar. Science **314**, 97 (2006)
26. J. Antoniadis, et al., A massive pulsar in a compact relativistic binary. Science **340**, 6131 (2013)
27. P.C.C. Freire, et al., The relativistic pulsar-white dwarf binary PSR J1738+0333 II. The most stringent test of scalar-tensor gravity. Mon. Not. Roy. Astron. Soc. **423**, 3328 (2012)
28. S. Weinberg, Infrared photons and gravitons. Phys. Rev. **140**, B516 (1965)
29. F. Cachazo, A. Strominger, Evidence for a New Soft Graviton Theorem (2014). [arXiv:1404.4091 [hep-th]]
30. A. Laddha, A. Sen, Sub-subleading soft graviton theorem in generic theories of quantum gravity. J. High Energy Phys. **10**, 065 (2017). [arXiv:1706.00759 [hep-th]]
31. B. Sahoo, A. Sen, Classical and quantum results on logarithmic terms in the soft theorem in four dimensions. J. High Energy Phys. **02**, 086 (2019). [arXiv:1808.03288 [hep-th]]

32. Y.B. Zel'dovich, A.G. Polnarev, Radiation of gravitational waves by a cluster of superdense stars", Soviet Astr.**18**, 17 (1974). https://adsabs.harvard.edu/full/1974SvA....18...17Z

33. V.B. Braginskii, K.S. Thorne, Gravitational-wave bursts with memory and experimental prospects. Nature **327** 123–125 (1987)

34. D. Christodoulou, Nonlinear nature of gravitation and gravitational wave experiments. Phys. Rev. Lett. **67**, 1486–1489 (1991)

35. A.G. Wiseman, C.M. Will, Christodoulou's nonlinear gravitational wave memory: evaluation in the quadrupole approximation. Phys. Rev. **D44**(10), R2945–R2949 (1991)

36. M. Favata, The gravitational-wave memory effect. Class. Quant. Grav. **27**, 084036 (2010). [arXiv:1003.3486 [gr-qc]]

37. M. Ebersold, Y. Boetzel, G. Faye, C.K. Mishra, B.R. Iyer, P. Jetzer, Gravitational-wave amplitudes for compact binaries in eccentric orbits at the third post-Newtonian order: memory contributions. Phys. Rev. D **100**(8), 084043 (2019). [arXiv:1906.06263 [gr-qc]]

38. A. Strominger, A. Zhiboedov, Gravitational memory, BMS supertranslations and soft theorems. J. High Energy Phys. **01**, 086 (2016). [arXiv:1411.5745 [hep-th]]

39. B.P. Abbott, et al. [LIGO Scientific and Virgo], GWTC-1: a gravitational-wave transient catalog of compact binary mergers observed by LIGO and Virgo during the first and second observing runs. Phys. Rev. X **9**(3), 031040 (2019). https://doi.org/10.1103/PhysRevX.9.031040. [arXiv:1811.12907 [astro-ph.HE]]

40. R. Abbott, et al. [LIGO Scientific and Virgo], GWTC-2: compact binary coalescences observed by LIGO and virgo during the first half of the third observing run. Phys. Rev. X **11**, 021053 (2021). https://doi.org/10.1103/PhysRevX.11.021053. [arXiv:2010.14527 [gr-qc]]

41. R. Abbott, et al. [LIGO Scientific, VIRGO and KAGRA], GWTC-3: Compact Binary Coalescences Observed by LIGO and Virgo During the Second Part of the Third Observing Run (2021). [arXiv:2111.03606 [gr-qc]]

42. B. Allen, W.G. Anderson, P.R. Brady, D.A. Brown, J.D.E. Creighton, FINDCHIRP: an algorithm for detection of gravitational waves from inspiraling compact binaries. Phys. Rev. D **85**, 122006 (2012). [arXiv:gr-qc/0509116 [gr-qc]]

43. L. Blanchet, Gravitational radiation from post-newtonian sources and inspiralling compact binaries. Living Rev. Rel. **17**, 2 (2014). https://doi.org/10.12942/lrr-2014-2. [arXiv:1310.1528 [gr-qc]]

44. A. Buonanno, B.S. Sathyaprakash, Sources of Gravitational Waves: Theory and Observations (2014). [arXiv:1410.7832 [gr-qc]]

45. G. Schäfer, P. Jaranowski, Hamiltonian formulation of general relativity and post-Newtonian dynamics of compact binaries. Living Rev. Rel. **21**(1), 7 (2018). https://doi.org/10.1007/s41114-018-0016-5. [arXiv:1805.07240 [gr-qc]]

46. T. Damour, Coalescence of two spinning black holes: an effective one-body approach. Phys. Rev. D **64**, 124013 (2001) https://doi.org/10.1103/PhysRevD.64.124013. [arXiv:gr-qc/0103018 [gr-qc]]

47. F. Pretorius, Binary Black Hole Coalescence (2007). [arXiv:0710.1338 [gr-qc]]

48. C.J. Moore, R.H. Cole, C.P.L. Berry, Gravitational-wave sensitivity curves. Class. Quant. Grav. **32**(1), 015014 (2015). https://doi.org/10.1088/0264-9381/32/1/015014. [arXiv:1408.0740 [gr-qc]]

49. B.P. Abbott, et al. [LIGO Scientific and Virgo], GW170817: Observation of gravitational waves from a binary neutron star inspiral. Phys. Rev. Lett. **119**(16), 161101 (2017). [arXiv:1710.05832 [gr-qc]]

50. B.P. Abbott, et al. [LIGO Scientific and Virgo], Tests of general relativity with GW150914. Phys. Rev. Lett. **116**(22), 221101 (2016). [erratum: Phys. Rev. Lett. **121**(12), 129902 (2018)]. https://doi.org/10.1103/PhysRevLett.116.221101. [arXiv:1602.03841 [gr-qc]]

51. N. Yunes, K. Yagi, F. Pretorius, Theoretical physics implications of the binary Black-Hole mergers GW150914 and GW151226. Phys. Rev. D **94**(8), 084002 (2016). [arXiv:1603.08955 [gr-qc]]

52. N. Christensen, R. Meyer, Parameter estimation with gravitational waves. Rev. Mod. Phys. **94**(2), 025001 (2022). [arXiv:2204.04449 [gr-qc]]

53. B.P. Abbott, et al. [KAGRA, LIGO Scientific, Virgo and VIRGO], Prospects for observing and localizing gravitational-wave transients with advanced LIGO, advanced Virgo and KAGRA. Living Rev. Rel. **21**(1), 3 (2018). [arXiv:1304.0670 [gr-qc]]

54. B. Iyer, et al., LIGO-India. Tech. Rep. LIGO-M1100296, LIGO (2011). https://dcc.ligo.org/LIGO-M1100296/public/main

55. R.A. Hulse, J.H. Taylor, Discovery of a pulsar in a binary system. Astrophys.J.Lett. **195**, L51–L53 (1975)

56. J.M. Weisberg, D.J. Nice, J.H. Taylor, Timing measurements of the relativistic binary pulsar PSR B1913+16. Astrophys. J. **722**, 1030–1034 (2010). [arXiv:1011.0718 [astro-ph.GA]]

57. P.C. Peters, J. Mathews, Gravitational radiation from point masses in a Keplerian orbit. Phys. Rev. **131**, 435–439 (1963)

58. C. Caprini, D.G. Figueroa, Cosmological backgrounds of gravitational waves. Class. Quant. Grav. **35**(16), 163001 (2018). https://doi.org/10.1088/1361-6382/aac608 [arXiv:1801.04268 [astro-ph.CO]]

59. J.D. Romano, N.J. Cornish, Detection methods for stochastic gravitational-wave backgrounds: a unified treatment. Living Rev. Rel. **20**(1), 2 (2017). https://doi.org/10.1007/s41114-017-0004-1. [arXiv:1608.06889 [gr-qc]]

60. Laser Interferometer Space Antenna (LISA), https://lisa.nasa.gov/

61. Z. Arzoumanian, et al. [NANOGrav], The NANOGrav 12.5 yr data set: search for an isotropic stochastic gravitational-wave background. Astrophys. J. Lett. **905**(2), L34 (2020). https://doi.org/10.3847/2041-8213/abd401. [arXiv:2009.04496 [astro-ph.HE]]

62. J. Antoniadis, et al. [IPTA]. The international pulsar timing array second data release: search for an isotropic gravitational wave background. Monthly Notices Roy. Astron. Soc. **510**(4), 4873–4887 (2022)

63. B.P. Abbott, et al. [LIGO Scientific and Virgo], Search for tensor, vector, and scalar polarizations in the stochastic gravitational-wave background. Phys. Rev. Lett. **120**(20), 201102 (2018). [arXiv:1802.10194 [gr-qc]]

64. Z. Arzoumanian, et al. [NANOGrav], The NANOGrav 12.5-year data set: search for non-einsteinian polarization modes in the gravitational-wave background. Astrophys. J. Lett. **923**(2), L22 (2021). [arXiv:2109.14706 [gr-qc]]

65. H. Rizwana Kausar, L. Philippoz, P. Jetzer, Gravitational wave polarization modes in $f(R)$ theories. Phys. Rev. D **93**(12), 124071 (2016). [arXiv:1606.07000 [gr-qc]]

Prologue: Gravitational Waves in Classical General Relativity

2

Abstract

We discuss classical theory of general relativity and we review aspects of gravitational wave generation and detection in the standard GR framework. This chapter serves as the background to compare and contrast quantum field theory method which follows in the rest of the book for the study of gravitational waves.

2.1 General Relativity

In Einstein's general relativity, dynamics of test particles are determined by a metric tensor $g_{\mu\nu}(x^\alpha)$ which defines the line element

$$ds^2 = g_{\mu\nu}dx^\mu dx^\nu \tag{2.1}$$

in the curved spacetime. Test particle orbits follow the geodesic equation

$$\frac{d^2 x^\mu}{d\tau^2} + \Gamma^\mu_{\alpha\beta}\frac{dx^\alpha}{d\tau}\frac{dx^\beta}{d\tau} = 0, \tag{2.2}$$

where for massive particles $d\tau = \sqrt{-ds^2}$ the proper time interval and for massless particles $d\tau = d\lambda$, where λ is any affine parameter along the geodesic. An affine parameter is defined by the property that the tangent vector $u^\mu = dx^\mu/d\lambda$ is covariantly constant along the geodesic

$$u^\mu \nabla_\mu u^\nu = 0. \tag{2.3}$$

© The Author(s), under exclusive license to Springer Nature Switzerland AG 2023
S. Mohanty, *Gravitational Waves from a Quantum Field Theory Perspective*,
Lecture Notes in Physics 1013, https://doi.org/10.1007/978-3-031-23770-6_2

The covariant derivative is defined by

$$\nabla_\mu u^\nu = \partial_\mu u^\nu + \Gamma^\nu_{\mu\alpha} u^\alpha. \tag{2.4}$$

The Christoffel connection components are related to the metric tensor as

$$\Gamma^\beta_{\mu\nu} = \frac{1}{2} g^{\alpha\beta} \left(\partial_\nu g_{\alpha\mu} + \partial_\mu g_{\alpha\nu} - \partial_\alpha g_{\mu\nu} \right). \tag{2.5}$$

If there are two particles $x^\mu_A(\tau)$ and $x^\mu_B(\tau)$ in a geodesic orbits, the separation $\xi^\mu = x^\mu_A(\tau) - x^\mu_B(\tau)$ between the two particles obeys the geodesic deviation equation

$$\frac{D^2 \xi^\mu}{D\tau^2} = -R^\mu{}_{\nu\rho\sigma} \xi^\rho \frac{dx^\nu}{d\tau} \frac{dx^\rho}{d\tau}, \tag{2.6}$$

where $D/D\tau$ is the covariant derivative along the curve parameterised by τ given by

$$\frac{D}{D\tau} = u^\alpha \nabla_\alpha. \tag{2.7}$$

The Riemann tensor $R^\mu{}_{\nu\rho\sigma}$ represents the tidal force and is given in terms of the metric tensor by

$$R^\mu{}_{\nu\rho\sigma} = \partial_\rho \Gamma^\mu_{\nu\sigma} - \partial_\sigma \Gamma^\mu_{\nu\rho} + \Gamma^\mu_{\alpha\rho} \Gamma^\alpha_{\nu\sigma} - \Gamma^\mu_{\alpha\sigma} \Gamma^\alpha_{\nu\rho}. \tag{2.8}$$

Contractions of the Riemann tensor give the Ricci tensor and Ricci scalar, respectively,

$$R_{\mu\nu} = R^\alpha{}_{\mu\alpha\nu}, \quad R = g^{\mu\nu} R_{\mu\nu}. \tag{2.9}$$

Einstein's equation relates the metric tensor to the stress tensor of the matter distribution

$$G_{\mu\nu} \equiv R_{\mu\nu} - \frac{1}{2} R = 8\pi G T_{\mu\nu}. \tag{2.10}$$

Contracting this with the metric $g_{\mu\nu}$ we have

$$R = -4\pi G T \tag{2.11}$$

where we used $g^{\mu\nu} g_{\mu\nu} = 4$ and defined the trace of the stress tensor $T \equiv T^\mu{}_\mu$. Using (2.11), we can write Einstein's equation (2.10) also in the form

$$R_{\mu\nu} = 8\pi G \left(T_{\mu\nu} - \frac{1}{2} g_{\mu\nu} T \right). \tag{2.12}$$

In regions outside mass distributions, $R_{\mu\nu} = 0$, and however the Riemann tensor need not be zero in these regions.

The Einstein tensor $G_{\mu\nu}$ obeys the Bianchi identity

$$\nabla_\mu G^{\mu\nu} = 0. \tag{2.13}$$

From this identity, Einstein's equation implies

$$\nabla_\mu T^{\mu\nu} = 0, \tag{2.14}$$

and the stress tensor is covariantly conserved.

2.2 Linearised Gravity

The linearised gravitational $h_{\mu\nu}$ field is the expansion of the metric around the Minkowski background

$$g_{\mu\nu}(x) = \eta_{\mu\nu} + h_{\mu\nu}(x), \tag{2.15}$$

and we consider the weak field situations $|h_{\mu\nu}| \ll 1$ and keep $h_{\mu\nu}$ to the leading order in the field expansions. The raising and lowering of indices is done by the Minkowski space metric $\eta_{\mu\nu} = \eta^{\mu\nu} = diagonal(-1, 1, 1, 1)$.

The Christoffel connection and Riemann tensor to the linear order in $h_{\mu\nu}$ are

$$\Gamma^\lambda_{\mu\nu} = \frac{1}{2}\eta^{\lambda\rho}\left(\partial_\mu h_{\rho\nu} + \partial_\nu h_{\rho\mu} - \partial_\rho h_{\mu\nu}\right),$$

$$R_{\mu\nu\alpha\beta} = \frac{1}{2}\left(\partial_\nu\partial_\alpha h_{\mu\beta} + \partial_\mu\partial_\beta h_{\nu\alpha} - \partial_\nu\partial_\beta h_{\mu\alpha} - \partial_\mu\partial_\alpha h_{\nu\beta}\right),$$

$$\tag{2.16}$$

and the Ricci tensor and Ricci scalar to the linear order are

$$R_{\mu\nu} = \frac{1}{2}\left(-\Box h_{\mu\nu} + \partial_\lambda\partial_\mu h^\lambda{}_\nu + \partial_\lambda\partial_\nu h^\lambda{}_\mu - \partial_\mu\partial_\nu h^\lambda{}_\lambda\right),$$

$$R = -\Box h^\mu_\mu + \partial^\alpha\partial^\beta h_{\alpha\beta}. \tag{2.17}$$

where $\Box \equiv \eta^{\alpha\beta}\partial_\alpha\partial_\beta$. Einstein's equation $R_{\mu\nu} = 16\pi G(T_{\mu\nu} - \frac{1}{2}g_{\mu\nu}T^\alpha{}_\alpha)$ to the linear order in $h_{\mu\nu}$ reads

$$\Box h_{\mu\nu} - \partial_\lambda\partial_\mu h^\lambda{}_\nu - \partial_\lambda\partial_\nu h^\lambda{}_\mu + \partial_\mu\partial_\nu h^\lambda{}_\lambda = -16\pi G\bar{T}_{\mu\nu}, \tag{2.18}$$

where

$$\bar{T}_{\mu\nu} \equiv T_{\mu\nu} - \frac{1}{2} T^{\alpha}{}_{\alpha} \eta_{\mu\nu} .$$ (2.19)

We can use the general covariance symmetry of general relativity

$$h'_{\mu\nu}(x') = h_{\mu\nu}(x) - \left(\partial_{\mu} \xi_{\nu}(x) + \partial_{\nu} \xi_{\mu}(x) \right)$$ (2.20)

to choose a gauge where (2.18) will be in the simplest form. We choose the de-Donder gauge condition

$$g^{\mu\nu} \Gamma^{\lambda}_{\mu\nu} = 0$$

$$\Rightarrow \quad \partial_{\mu} h^{\mu}{}_{\nu} = \frac{1}{2} \partial_{\nu} h^{\mu}{}_{\mu} ,$$ (2.21)

and the e.o.m. for the gravitational waves (2.18) reduces to the form

$$\Box h_{\mu\nu} = -16\pi G \bar{T}_{\mu\nu}.$$ (2.22)

2.3   Gravitational Waves in Minkowski Background

The harmonic gauge condition (2.21) (four equations) reduces the 10 degrees of freedom of $h^{\mu\nu}$ to 6. If the gravitational wave is propagating in empty space, then its e.o.m. is $\Box h_{\mu\nu} = 0$ and we can choose four harmonic functions ξ^{μ} (which obey $\Box \xi^{\mu} = 0$) to make four more gauge transformations (2.20) and still retain the form of e.o.m. $\Box h'_{\mu\nu} = 0$. This reduces the propagating degrees of gravitational waves to 2, which implies that gravitational waves propagating in free space have two independent polarisations. When interpreted as gravitons, these are the two independent helicity states of a massless spin-2 particle.

The plane wave solutions of gravitational waves in free space are of the form

$$h_{\mu\nu}(x) = \mathcal{A} \epsilon_{\mu\nu} e^{ik \cdot x} + \mathcal{A}^{*} \epsilon^{*}_{\mu\nu} e^{-ik \cdot x} ,$$ (2.23)

where $\epsilon_{\mu\nu}$ is a symmetric tensor ($\epsilon_{\mu\nu} = \epsilon_{\nu\mu}$) that represents the polarization states of the gravitons. The plane wave satisfies the equation of motion (2.22) in free space $\Box h_{\mu\nu} = 0$ if

$$k_{\alpha} k^{\alpha} = 0 ,$$ (2.24)

while the de-Donder gauge condition (2.21) is satisfied if

$$k_{\mu} \epsilon^{\mu}{}_{\nu} = \frac{1}{2} k_{\nu} \epsilon^{\mu}{}_{\mu} .$$ (2.25)

This condition reduces the free parameters in $\epsilon_{\mu\nu}$ from 10 to 6. Gravitational waves propagating in empty space have 2 degrees of polarization.

In the de-Donder gauge, we use the harmonic functions ξ^μ to impose the transversality conditions

$$k^\mu \epsilon_{\mu\nu} = k^\nu \epsilon_{\nu\mu} = 0. \tag{2.26}$$

The transversality condition (2.26) reduces the free graviton degrees of freedom from 6 to 2. The transversality condition must be consistent with the harmonic gauge condition (2.25), which implies $\epsilon^\mu{}_\mu = 0$. Therefore the free graviton polarization tensor is transverse and traceless and represents two polarization states. The polarization states of a graviton in free space propagating with momentum $k = (\omega, 0, 0, \omega)$ have the following two transverse–traceless polarization states:

$$\epsilon^{(+)}_{\mu\nu} = \begin{pmatrix} 0 & 0 & 0 & 0 \\ 0 & 1 & 0 & 0 \\ 0 & 0 & -1 & 0 \\ 0 & 0 & 0 & 0 \end{pmatrix}, \qquad \epsilon^{(\times)}_{\mu\nu} = \begin{pmatrix} 0 & 0 & 0 & 0 \\ 0 & 0 & 1 & 0 \\ 0 & 1 & 0 & 0 \\ 0 & 0 & 0 & 0 \end{pmatrix}.$$

The free graviton propagating in a general direction $\hat{n}$ is given by the solution of $\Box h_{ij} = 0$

$$h_{ij}(x) = \epsilon^+_{ij} h_+ e^{i\omega(\hat{n}\cdot x - t)} + \epsilon^\times_{ij} h_\times e^{-i\omega(\hat{n}\cdot x - t)}. \tag{2.27}$$

Of the 6 degrees of freedom of the gravitational wave h_{ij} (2.27), only the two transverse and traceless modes h^{TT}_{ij} which have the property $h^{iTT}_i = 0$ and $\hat{n}^i h^{TT}_{ij} = \hat{n}^j h^{TT}_{ij} = 0$ are the propagating degrees of freedom and the remaining are removed by a choice of gauge.

The projector that projects out the gauge modes of the general tensor h_{ij} and retains only the transverse and traceless part

$$h^{TT}_{ij} = \Lambda_{ij,kl} h_{kl}. \tag{2.28}$$

The TT projector for a wave propagating in the direction $\hat{\mathbf{n}}$ can be constructed from the transverse projection operator $P_{ij} = \delta_{ij} - \hat{n}_i \hat{n}_j$ as follows [1,2]:

$$\Lambda_{ij,kl}(\hat{\mathbf{n}}) \equiv P_{ik} P_{jl} - \frac{1}{2} P_{ij} P_{kl}. \tag{2.29}$$

The TT projector has the properties

$$\Lambda_{ij,kl} = \Lambda_{ji,kl} = \Lambda_{kl,ij}, \qquad n^i \Lambda_{ij,kl} = \delta^{ij} \Lambda_{ij,kl} = 0, \qquad \Lambda_{ij,kl} \Lambda_{kl,rs} = \Lambda_{ij,rs}.$$

$$\tag{2.30}$$

For a gravitational wave propagating in direction given by the unit vector $\mathbf{n}$, the two independent polarisations are defined w.r.t. the mutually orthogonal unit vectors $\mathbf{n}$, $\mathbf{p}$ and $\mathbf{q}$ are

$$h_+ = \left(p_i p_j - q_i q_j \right) h_{ij}^{TT} , \tag{2.31}$$

$$h_\times = \left(p_i q_j + q_i p_j \right) h_{ij}^{TT} . \tag{2.32}$$

In spherical $(\mathbf{e}_r, \mathbf{e}_\theta, \mathbf{e}_\phi)$ coordinates for gravitational waves in the radial direction $\mathbf{n} = \mathbf{e}_r$, the $+$ and $\times$ polarisations can be written in Cartesian coordinates as

$$\epsilon_{ij}^+(\mathbf{n}) = (\mathbf{e}_{\theta i}\mathbf{e}_{\theta j} - \mathbf{e}_{\phi i}\mathbf{e}_{\phi j})$$

$$\epsilon_{ij}^\times(\mathbf{n}) = (\mathbf{e}_{\theta i}\mathbf{e}_{\phi j} + \mathbf{e}_{\phi i}\mathbf{e}_{\theta j}), \tag{2.33}$$

where in the Cartesian coordinates

$$\mathbf{n} = (\sin\theta\cos\phi, \sin\theta\sin\phi, \cos\theta) ,$$

$$\mathbf{e}_\theta = (\cos\theta\cos\phi, \cos\theta\sin\phi, -\sin\theta) ,$$

$$\mathbf{e}_\phi = (-\sin\phi, \cos\phi, 0) . \tag{2.34}$$

On rotating the coordinates around $\mathbf{n}$ by angle ψ, the polarisation tensors transform as

$$\begin{pmatrix} \mathbf{e}_\theta \\ \mathbf{e}_\phi \end{pmatrix} \rightarrow \begin{pmatrix} \cos 2\psi & -\sin 2\psi \\ \sin 2\psi & \cos 2\psi \end{pmatrix} \begin{pmatrix} \mathbf{e}_\theta \\ \mathbf{e}_\phi \end{pmatrix} .$$

The angle ψ in a given frame $(\mathbf{n}, \mathbf{e}_\theta, \mathbf{e}_\phi)$ is called the polarisation angle.

One can define a complex amplitude

$$h(t, \mathbf{x}) = \left(\epsilon_{ij}^+ - i\epsilon_{ij}^\times \right) h_{ij}^{TT}(t, \mathbf{x})$$

$$= h_+(t, \mathbf{x}) - ih_\times(t, \mathbf{x}). \tag{2.35}$$

This has the property

$$|h|^2 = h_+^2 + h_\times^2 = \frac{1}{2}h_{ij}^{TT}h^{ijTT}. \tag{2.36}$$

Under rotation around the axis $\mathbf{n}$ by the angle ψ, the complex amplitude h transforms as

$$h \rightarrow e^{-2i\psi}h . \tag{2.37}$$

For spherical waves propagating in the radial direction, the complex amplitude $h(r, \theta, \phi, t)$ can be expanded in terms of the spin-2 weighted spherical harmonics $_{-2}Y_{lm}(\theta, \phi)$ as

$$h(r, \theta, \phi, t) = h_+(r, \theta, \phi, t) - ih_\times(r, \theta, \phi, t) = \sum_{l=2}^{\infty} \sum_{m=-l}^{l} h_{lm}(t - r) \; _{-2}Y_{lm}(\theta, \phi).$$

$$(2.38)$$

This representation is useful in calculations of the gravitational waves during the 'ring down' phase of the post merger black hole.

2.4 Interaction of Gravitational Waves with Test Masses

Gravitational waves are measured by LIGO, VIRGO, KAGRA and other similar detectors by laser interferometry. When a gravitational waves $h_{\mu\nu}(t)$ passes through the interferometer, the distance between the mirrors changes as a function of time.

Denote the coordinates of two mirrors A and B by the coordinates $\mathbf{x}_A$ and $\mathbf{x}_B$. The spatial separation between the bodies is denoted by $\xi^i \equiv x_A^i - x_B^i$. When a gravitational wave passes through the detector, the e.o.m. of ξ^k will be described by the geodesic deviation equation

$$\frac{d^2 \xi^i}{d\tau^2} = -R^i{}_{0j0} \xi^j.$$

$$(2.39)$$

The Riemann tensor for the gravitational metric $g_{\mu\nu} = \eta_{\mu\nu} + h_{\mu\nu}$ with $h_{\mu\nu}$ of gravitational waves in the TT gauge has non-zero components

$$R_{i0j0} = R_{0i0j} = -R_{i00j} = -R_{0ij0} = -\frac{1}{2} h_{ij,00}^{TT}.$$

$$(2.40)$$

For $h_{\mu\nu}$ given by h_{ij}^{TT}, we have $d\tau = dt$, and from (2.39) and (2.40) we have

$$\frac{d^2 \xi^i}{dt^2} = \frac{1}{2} \ddot{h}_{ij}^{TT} \xi^i.$$

$$(2.41)$$

For a gravitational wave propagating along the z-axis, there are two polarisation modes $h_+(t, z) = h_{xx}^{TT} = -h_{yy}^{TT}$ and $h_\times(t, z) = h_{xy}^{TT}$. The wave vector of the GW is $k^\mu = (\omega, 0, 0, \omega)$. The metric in the local Lorentz frame during the passage of the GW is

$$ds^2 = -dt^2 + dz^2 + [1 + h_+(\cos \omega(t - z))] dx^2 + [1 - h_+(\cos \omega(t - z))] dy^2$$
$$+ 2h_\times(\cos \omega(t - z)) dx dy.$$

$$(2.42)$$

2.5   Polarisations of Gravitons in Theories Beyond Einstein's Gravity

In Einstein's gravity, there are only two independent polarisations which propagate. In more general theories like $f(R)$ [3] and Horndeski theories [4,5] of gravity, other degrees of freedom of the metric like the scalar or vector mode of the metric may become dynamical. Here we list the six independent polarisations of the graviton h_{ij}, which may become dynamical in theories beyond Einstein's gravity.

A graviton plane wave in the z direction can be described by the matrix

$$h_{ij}(z,t) = \sum_{A=1}^{6} h_A(z,t) e_{ij}^A(\hat{z}), \tag{2.43}$$

where the 6 independent polarisation tensors can be written as

$$e_{+}(\hat{z}) = \begin{pmatrix} 1 & 0 & 0 \\ 0 & -1 & 0 \\ 0 & 0 & 0 \end{pmatrix}, \quad e_{\times}(\hat{z}) = \begin{pmatrix} 0 & 1 & 0 \\ 1 & 0 & 0 \\ 0 & 0 & 0 \end{pmatrix}, \quad e_{b}(\hat{z}) = \begin{pmatrix} 1 & 0 & 0 \\ 0 & 1 & 0 \\ 0 & 0 & 0 \end{pmatrix},$$

$$e_{l}(\hat{z}) = \sqrt{2} \begin{pmatrix} 0 & 0 & 0 \\ 0 & 0 & 0 \\ 0 & 0 & 1 \end{pmatrix}, \quad e_{x}(\hat{z}) = \begin{pmatrix} 0 & 0 & 1 \\ 0 & 0 & 0 \\ 1 & 0 & 0 \end{pmatrix}, \quad e_{y}(\hat{z}) = \begin{pmatrix} 0 & 0 & 0 \\ 0 & 0 & 1 \\ 0 & 1 & 0 \end{pmatrix}. \tag{2.44}$$

The polarisation tensors are normalised such that $e_{ij}^A e_{ji}^B = 2\delta^{AB}$. The modes $\{h_+, h_\times, h_b, h_x, h_y, h_l\}$ are called the 'plus, cross, breathing, x, y and longitudinal' modes, respectively. We can write h_{ij} in terms of the polarizations modes as

$$h_{xx} = (h_+ + h_b), \quad h_{yy} = (-h_+ + h_b), \quad h_{xy} = h_\times,$$

$$h_{zz} = \sqrt{2} h_l, \quad h_{xz} = h_x, \quad h_{yz} = h_y \quad \text{and} \quad h \equiv \eta^{ij} h_{ij} = -(2h_b + \sqrt{2} h_l). \tag{2.45}$$

To understand the effect of the polarisation modes on test masses, we start with the geodesic deviation equation for test masses separated by ξ_i in a gravitational wave in the $\hat{z}$ direction

$$\frac{d^2 \xi_i}{dt^2} = -R_{titj} \xi^j, \tag{2.46}$$

and we express the components R_{0i0j} of the Riemann tensor responsible for geodesic deviation in terms of the polarisation tensors e_{ij}^A as

$$R_{titj}(t, \hat{z}) \equiv \sum_{A=1}^{6} p_A(t, \hat{z}) e_{ij}^A = \begin{pmatrix} \frac{1}{2}(p_4 + p_6) & p_5 & p_2 \\ p_5 & \frac{1}{2}(-p_4 + p_6) & p_3 \\ p_2 & p_3 & p_1 \end{pmatrix}. \quad (2.47)$$

The polarisation amplitudes $p_A(t, \hat{z})$ in terms of the gravitational wave amplitude are

$$p_1^{(l)}(t, \hat{z}) = R_{tztz} = -\frac{1}{2} \left(\partial_t^2 h_{zz} - 2\partial_z \partial_t h_{tz} + \partial_z^2 h_{tt} \right)$$

$$p_2^{(x)}(t, \hat{z}) = R_{tztx} = -\frac{1}{2} \left(\partial_t^2 h_{xz} - \partial_z \partial_t h_{tx} \right)$$

$$p_3^{(y)}(t, \hat{z}) = R_{tzty} = -\frac{1}{2} \left(\partial_t^2 h_{yz} - \partial_z \partial_t h_{ty} \right)$$

$$p_4^{(+)}(t, \hat{z}) = R_{txtx} - R_{tyty} = -\frac{1}{2} \left(\partial_t^2 (h_{xx} - h_{yy}) \right)$$

$$p_5^{(x)}(t, \hat{z}) = R_{txty} = -\frac{1}{2} \left(\partial_t^2 h_{xy} \right)$$

$$p_6^{(b)}(t, \hat{z}) = R_{txtx} + R_{tyty} = -\frac{1}{2} \left(\partial_t^2 (h_{xx} + h_{yy}) \right). \quad (2.48)$$

Due to the gauge freedom not all the 10 components of h_{ij} will be independent. In (Fig. 2.1) we show the six independent polarisation modes corresponding to the test particle motion which follows from Eq. (2.48).

2.6 Generation of Gravitational Waves

Gravitational waves couple to the stress tensor of matter (and of gravitons). This means that a time-dependent stress tensor can be a source for gravitational waves.

The e.o.m. (2.22) for the TT components of $h_{\mu\nu}$ can also be written as

$$\Box h_{ij}^{TT} = -16\pi G \Lambda_{ij,kl} \bar{T}_{kl} = -16\pi G \Lambda_{ij,kl} T_{kl}, \quad (2.49)$$

as $\Lambda_{ij,kl} \eta_{kl} = 0$. This can be solved by Green's function method. The Green's function $G(t, \mathbf{x}; t', \mathbf{x}')$ is the solution of the delta function source equation

$$\Box G(t, \mathbf{x}; t', \mathbf{x}') = \delta^3(\mathbf{x} - \mathbf{x}') \delta(t - t'). \quad (2.50)$$

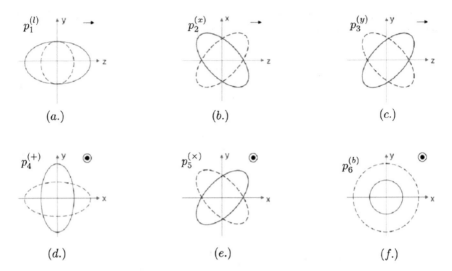

Fig. 2.1 Pattern of test particles motion in the x–y plane for a gravitational wave along z-axis. The polarisation modes shown are (**a**) $p_1^{(l)}$ longitudinal mode, (**b**) $p_2^{(x)}$ x-mode, (**c**) $p_3^{(y)}$ y-mode, (**d**) $p_4^{(+)}$ plus-mode, (**e**) $p_5^{(\times)}$ cross-mode and (**f**) $p_6^{(b)}$ breathing-mode. Of the six possible polarisations only the $+$ and $\times$ polarisations are the propagating degrees of freedom in Einstein's gravity

Green's function solution of (2.50) is

$$G(t, \mathbf{x}; t', \mathbf{x}') = -\frac{\delta(t' - [t - |\mathbf{x} - \mathbf{x}'|])}{4\pi \, |\mathbf{x} - \mathbf{x}'|}. \tag{2.51}$$

The signal emitted at time t' from a source located at $\mathbf{x}'$ reaches the detector at $\mathbf{x}$ at retarded time $t = [t' + |\mathbf{x} - \mathbf{x}'|]$.

With the Green's function (2.51), we can solve for the gravitational wave $h_{ij}^{TT}(t, \mathbf{x})$ generated by the source $T_{kl}(t', \mathbf{x}')$ as

$$\tilde{h}_{ij}(t, \mathbf{x})^{TT} = \Lambda_{ij,kl} \int dt' d^3\mathbf{x}' \, G(t, \mathbf{x}; t', \mathbf{x}') T_{kl}(t', \mathbf{x}')$$

$$= 4G \, \Lambda_{ij,kl} \int d^3\mathbf{x}' \, \frac{T_{kl}(t - |\mathbf{x} - \mathbf{x}'|, \mathbf{x}')}{|\mathbf{x} - \mathbf{x}'|}. \tag{2.52}$$

We choose the origin of the coordinate system at the source. The location of the detector is $\mathbf{x} = \mathbf{n}r$. When the distance of the detector $|\mathbf{x}| = r$ is large compared to the size of the source R_s, we can expand

$$|\mathbf{x} - \mathbf{x}'| = r - \mathbf{n} \cdot \mathbf{x}' + O\left(\frac{R_s^2}{r}\right), \tag{2.53}$$

and we can write (2.52) (to the leading order in $1/r$) as

$$h_{ij}^{\text{TT}}(t, \mathbf{x}) = \frac{4G}{r} \Lambda_{ij,kl} \int d^3x' \, T_{kl}\left(t - r + \mathbf{n} \cdot \mathbf{x}', \mathbf{x}'\right). \qquad (2.54)$$

The source is time dependent. For non-relativistic sources, we can solve (2.54) by a Taylor expansion of $T^{kl}(t - r + \mathbf{n} \cdot \mathbf{x}', \mathbf{x}')$ around the point $\mathbf{n} \cdot \mathbf{x}' = t - r = 0$,

$$
\begin{aligned}
T^{kl}\left(t - r + \mathbf{n} \cdot \mathbf{x}', \mathbf{x}'\right) = {} & T^{kl}\left(t - r, \mathbf{x}'\right) + \mathbf{n} \cdot \mathbf{x}' \, \dot{T}^{kl}\left(t - r, \mathbf{x}'\right) \\
& + \frac{1}{2}(\mathbf{n} \cdot \mathbf{x}')^2 \, \ddot{T}^{kl}\left(t - r, \mathbf{x}'\right) + \cdots \Big|_{(t-r,\mathbf{x}')}, \quad (2.55)
\end{aligned}
$$

where overdots on T^{kl} represent derivatives w.r.t. time t. This expansion is valid in the low velocity regime where the wavelength of GW is much larger than the size of the source, $\lambda_{gw}/(2\pi) \sim (c/v)d \gg d$.

We can relate the moments of T^{ij} and T^{0j} components of the stress tensor with the moments of the mass distributions of the source by using the conservation equation $\partial_\mu T^{\mu\nu} = 0$. We define the moments of the T^{00} which are

$$M \equiv \int d^3x' \, T^{00}, \quad M^i \equiv \int d^3x' \, T^{00} x'^i, \quad M^{ij} \equiv \int d^3x' \, T^{00} x'^i x'^j.$$

$$(2.56)$$

We define the moments of the anisotropic pressure T^{ij} the source as

$$S^{kl} \equiv \int d^3x' T^{kl}, \quad S^{kl,i} \equiv \int d^3x' T^{kl} x'^i, \quad S^{kl,ij} \equiv \int d^3x' T^{kl} x'^i x'^j.$$

$$(2.57)$$

We also define the moments of the energy flux T^{0i} as

$$P^i \equiv \int d^3x' \, T^{0i}, \quad P^{ij} \equiv \int d^3x' \, T^{0i} x'^j, \quad P^{ijk} \equiv \int d^3x' \, T^{0i} x'^j x'^k.$$

$$(2.58)$$

The expression for the gravitational waveform at the detector (2.54) can therefore be expressed as a function of the time derivatives of the moments of the stress tensor evaluated at time $t - r$

$$h_{ij}^{TT}(t, \mathbf{x}) = \frac{4G}{r} \Lambda_{ij,kl} \left[S^{kl} + \hat{n}_r \dot{S}^{kl,r} + \frac{1}{2}\hat{n}_r \hat{n}_s \ddot{S}^{kl,rs} + \cdots \right]_{t-r}. \qquad (2.59)$$

The terms in the series expansion in (2.59) are the time derivatives of multipole moments of mass distribution with the leading term which are suppressed by powers of $\frac{d}{dt}|\mathbf{x}'| \sim v'/c$ the velocity of the source constituents.

The leading term in (2.55) is the quadrupole approximation of the source. In the quadrupole approximation, the gravitational wave signal is given by

$$h_{ij}^{TT}(t, \mathbf{x}) = \frac{4G}{r} \Lambda_{ij,kl} \int d^3x' \, T^{kl}(t - r, \mathbf{x}') .$$
(2.60)

Using the conservation equation $\partial_0 T^{00} + \partial_i T^{i0} = 0$, we see that

$$\dot{M} = \int d^3\mathbf{x}' \, \partial_0 T^{00} = -\int d^3\mathbf{x}' \, \partial_i T^{i0} = -\int dS_i \, T^{i0} = 0,$$
(2.61)

where we have applied the Stokes theorem and assumed that T^{0i} vanishes on the surface $|\mathbf{x}'| \to \infty$. We can similarly show that

$$\dot{M}^{jk} = \int d^x \mathbf{x}' \, \partial_0 T^{00} x'^j x'^k = -\int d^3\mathbf{x}' \, \partial_i T^{i0} x'^j x'^k$$

$$= \int d^3\mathbf{x}' \, T^{i0} \left(x'^j \delta_i^k + \delta_i^j x'^k \right) .$$
(2.62)

Taking one more time derivative of (2.62) and using relation $\partial_0 T^{0i} + \partial_l T^{li} = 0$, we obtain

$$\ddot{M}^{jk} = \int d^3\mathbf{x}' \, \partial_0 T^{0i} \left(x'^j \delta_i^k + \delta_i^j x'^k \right) = -\int d^3\mathbf{x}' \, \partial_l T^{li} \left(x'^j \delta_i^k + \delta_i^j x'^k \right)$$

$$= 2 \int d^3\mathbf{x}' \, T^{jk}(t - r, \mathbf{x}') .$$
(2.63)

The moments of the stress tensor can in this way be related to time derivatives of mass distribution of the source. Using these relations, we can write the gravitational waveform (2.54) in terms of the time derivatives of the mass distribution of the source as

$$h_{ij}^{TT}(t, \mathbf{x}) = \frac{4G}{r} \Lambda_{ij,kl} \left[\frac{1}{2} \ddot{M}^{kl} \right]_{t-r} .$$
(2.64)

We can split the tensor $\ddot{M}^{kl}$ as a sum of the traceless part and the trace

$$\ddot{M}^{kl} = \left(\ddot{M}^{kl} - \frac{1}{3} \eta^{kl} \ddot{M}^r{}_r \right) + \frac{1}{3} \eta^{kl} \ddot{M}^r{}_r .$$
(2.65)

The TT projection operator $\Lambda_{ij,kl}$ operating on the second term gives zero, and retaining only the traceless quadrupole distribution

$$\ddot{Q} \equiv \ddot{M}^{kl} - \frac{1}{3}\eta^{kl}\ddot{M}^r{}_r = \frac{d^2}{dt^2}\left[\int d^4\mathbf{x}'\rho(t',\mathbf{x}')\left(x'^k x'^l - \frac{1}{3}\eta^{kl}x'^i x'_i \right) \right]_{t'=t-r}$$

(2.66)

(where $\rho = T^{00}$), we can write the expression (2.64) as

$$h_{ij}^{TT}(t, r\mathbf{n}) = \frac{2G}{r}\Lambda_{ij,kl}(\mathbf{n})\,\ddot{Q}^{kl}(t-r,\mathbf{n}).$$

(2.67)

The projection operators $\Lambda_{ij,kl}(\mathbf{n})$ are a function of the direction of the source $\mathbf{n}$.

2.7   Energy Flux of Gravitational Waves

We can calculate the energy radiated in gravitational waves due to the motion of the sources as follows. The stress tensor of gravitational waves is derived by taking perturbations of Einstein's equation up to second order in $h_{\mu\nu}$. The Landau–Lifshitz stress tensor or the gravitational waves is defined as

$$\tau_{\mu\nu} = \frac{1}{8\pi G}\langle R^{(2)}_{\mu\nu} - \frac{1}{2}\bar{g}_{\mu\nu}R^{(2)}\rangle,$$

(2.68)

where the angular brackets indicate time-averaging over the oscillatory terms. The Ricci tensor at the second order in $h_{\mu\nu}$ in the TT gauge $\partial^\mu h_{\mu\nu} = 0$, $h^\alpha{}_\alpha = 0$ and $\Box h_{\alpha\beta} = 0$ is

$$R^{(2)}_{\mu\nu} = \frac{1}{4}\partial_\mu h_{\alpha\beta}\partial_\nu h^{\alpha\beta}.$$

(2.69)

The Landau–Lifshitz pseudo-tensor of the gravitational waves in the TT gauge is given by

$$t_{\mu\nu} = \frac{1}{32\pi G}\left(\partial_\mu h_{ij}^{TT}\right)\left(\partial_\mu h^{ij\,TT}\right).$$

(2.70)

The energy density of the gravitational waves is

$$\rho_{GW} = t_{00} = \frac{1}{32\pi G}\left(\partial_t h_{ij}^{TT}\right)\left(\partial_t h^{ij\,TT}\right).$$

(2.71)

The energy flux out of a given volume around the source is

$$\frac{dE_{gw}}{dt} = \int_V d^3\mathbf{x}\, \partial_t t^{00} = -\int_V d^3\mathbf{x}\, \partial_i t^{i0} = -\int_S dS_i t^{i0}$$

$$= -\frac{1}{32\pi G} \int_S dS_i \left(\partial^i h_{lm}^{TT}\right)\left(\partial^i h^{lm\,TT}\right), \tag{2.72}$$

where we have used the conservation of gravitational wave stress tensor $\partial_\mu t^{\mu\nu} = 0$ and the Stokes theorem to convert the volume integral into a surface integral. Taking the sphere of radius r as the surface with the source at the origin the flux of gravitational wave crossing the surface is

$$\frac{dE_{gw}}{dt} = -\frac{1}{32\pi G} \int_S r^2 d\Omega \left(\partial^r h_{lm}^{TT}\right)\left(\partial^r h^{lm\,TT}\right)$$

$$= \frac{1}{32\pi G} \int_S r^2 d\Omega \left(\partial^t h_{lm}^{TT}\right)\left(\partial^t h^{lm\,TT}\right), \tag{2.73}$$

where in the last step we replaced ∂^r by $-\partial^t$ as a radially propagating gravitational waves h_{lm} is a function of $(t - r)$.

The luminosity of the source is the energy flux crossing a 2-sphere of radius r surrounding the source

$$\frac{dE_{gw}}{dtd\Omega} = \frac{r^2}{32\pi G} \left\langle \partial_t h_{ij}^{TT} \partial_t h^{ij\,TT} \right\rangle, \tag{2.74}$$

where the angular brackets denote time-averaging over times longer than the period of the GW. Expressing the waveform h_{ij}^{TT} in terms of the source from (2.67), we obtain

$$\frac{dE_{gw}}{dtd\Omega} = \frac{G}{8\pi} \left\langle \overset{...}{Q}_{ij}^{TT} \overset{...}{Q}^{ij\,TT} \right\rangle \Big|_{t-r}. \tag{2.75}$$

The formula for the rate of gravitational energy radiated as the triple time derivative of the quadrupole moment of the body in retarded time was given by Einstein in 1918.

If $\mathbf{n}$ is the direction of the graviton with respect to the source coordinates, the projected components of the product of quadrupoles in (2.75) are explicitly

$$\overset{...}{Q}_{ij}^{TT} \overset{...}{Q}_{ij}^{TT} = \Lambda_{ij,kl}\, \overset{...}{Q}_{kl}\, \Lambda_{ij,rs}\, \overset{...}{Q}_{rs} = \Lambda_{kl,rs}\, \overset{...}{Q}_{kl}\, \overset{...}{Q}_{rs}$$

$$= \overset{...}{Q}_{kl}\, \overset{...}{Q}^{kl} - 2\overset{...}{Q}_{kl}\, \overset{...}{Q}^{ks} n^l n_s + \frac{1}{2}\overset{...}{Q}_{kl}\, \overset{...}{Q}_{rs} n^k n^l n^r n^s. \tag{2.76}$$

In the quadrupole approximation, Q_{ij} are independent of the angles and we can do the angular integrations

$$\int d\Omega n^i n^j = \frac{4\pi}{3}\delta_{ij},$$

$$\int d\Omega n^i n^j n^l n^m = \frac{4\pi}{15}\left(\delta_{ij}\delta_{lm} + \delta_{il}\delta_{jm} + \delta_{im}\delta_{jm}\right) \tag{2.77}$$

with integrals of odd powers of n^i vanishing. The angular integral over the quadrupole gives us

$$\int d\Omega \, \Lambda_{ij,kl} \, \dddot{Q}_{ij} \, \dddot{Q}^{kl} = \frac{8\pi}{5}\left(\dddot{Q}_{ij}\dddot{Q}^{ji} - \frac{1}{3}\left|\dddot{Q}^i{}_i\right|^2\right). \tag{2.78}$$

The expression for the energy radiated (2.75) after angular integrations becomes

$$\frac{dE_{gw}}{dt} = \frac{G}{5}\left\langle \dddot{Q}_{ij}\dddot{Q}^{ji} - \frac{1}{3}\left(\dddot{Q}^i{}_i\right)^2\right\rangle_{(t-r)}, \tag{2.79}$$

where the angular brackets stand for time-averaging.

2.8 Waveforms from Compact Binaries

Consider a binary system of compact stars with masses m_1 and m_2 in a circular Kepler orbit separated by distance R. Choosing the origin of the coordinate at the center of mass, the two stars revolve around the CM at radii $r_1 = m_2 R/(m_1 + m_2)$ and $r_2 = -m_1 R/(m_1 + m_2)$, respectively, with angular velocity given by the Kepler law

$$\Omega^2 = \frac{G(m_1 + m_2)}{R^3}. \tag{2.80}$$

Choose the plane of orbit of the stars to be the $x - y$ plane. The orbit of the one body reduced mass is $\mathbf{R} = R(\cos\Omega t, \sin\Omega t, 0)$. The components of the quadrupole moment of the binary system are therefore

$$M_{xx} = \frac{\mu R^2}{2}\left(1 + \cos(2\Omega t)\right), \quad M_{yy} = \frac{\mu R^2}{2}\left(1 - \cos(2\Omega t)\right),$$

$$M_{xy} = \frac{-\mu R^2}{2}\sin(2\Omega t), \tag{2.81}$$

where $\mu = m_1 m_2/(m_1 + m_2)$ is the reduced mass and we can see that the fundamental frequency of the quadruple oscillation which is the frequency of the gravitational waves generated is twice the orbital frequency of the binary, $\Omega_{gw} = 2\Omega$. The non-zero components of the traceless quadrupole moment $Q_{ij} = M_{ij} - (1/3)\delta_{ij}M_l^l$ are

$$Q_{xx} = \frac{\mu R^2}{2}\left(-\frac{1}{3} + \cos(2\Omega t)\right), \; Q_{yy} = \frac{\mu R^2}{2}\left(\frac{1}{3} - \cos(2\Omega t)\right),$$

$$Q_{xy} = \frac{\mu R^2}{2}\sin(2\Omega t). \tag{2.82}$$

Now consider the observer to be on the z-axis, $\mathbf{n} = (0, 0, 1)$. The non-zero components of $Q_{ij}^{TT} = \Lambda_{ij,kl}Q^{kl}$ are

$$Q_{xx}^{TT} = \frac{1}{2}\left(Q_{xx} - Q_{yy}\right) = \frac{\mu R^2}{2}\cos(2\Omega t),$$

$$Q_{yy}^{TT} = -\frac{1}{2}\left(Q_{xx} - Q_{yy}\right) = -\frac{\mu R^2}{2}\cos(2\Omega t),$$

$$Q_{xy}^{TT} = Q_{xy} = \frac{\mu R^2}{2}\sin(2\Omega t). \tag{2.83}$$

The gravitational waveform observed along the z-axis at a distance r will be

$$h_{xx}^{TT} = \frac{2G}{r}\ddot{Q}_{xx}^{TT} = -\frac{4G}{r}\mu^2 R^2\Omega^2\cos(2\Omega t_r)$$

$$h_{yy}^{TT} = \frac{2G}{r}\ddot{Q}_{yy}^{TT} = \frac{4G}{r}\mu^2 R^2\Omega^2\cos(2\Omega t_r)$$

$$h_{xy}^{TT} = \frac{2G}{r}\ddot{Q}_{xy}^{TT} = -\frac{4G}{r}\mu^2 R^2\Omega^2\sin(2\Omega t_r), \tag{2.84}$$

where $t_r = t - r$ is the retarded time and the gravitational wave angular frequency is $\Omega_{gw} = 2\Omega$. The waveform of $h_+ = h_{xx}^{TT} = -h_{yy}^{TT}$ and $h_\times = h_{xy}^{TT}$ is therefore

$$h_+(t) = \frac{G}{r}\mu^2 R^2\Omega_{gw}^2\cos(\Omega_{gw}t + \phi_0)$$

$$h_\times(t) = \frac{G}{r}\mu^2 R^2\Omega_{gw}^2\sin(\Omega_{gw}t + \phi_0), \tag{2.85}$$

and we can replace R in terms of Kepler frequency $\Omega = \Omega_{gw}/2$ using (2.80) and write $h+$ in (2.85) as

$$h_+(t) = \frac{G^{5/3}}{r}2^{1/3}\mu(m_1 + m_2)^{2/3}\Omega_{gw}^{2/3}\cos(\Omega_{gw} + \phi_0) \tag{2.86}$$

and similarly for $h_\times$. We can combine the mass factors by defining a 'chirp mass'

$$\mathcal{M}_c \equiv \mu^{3/5}(m_1 + m_2)^{2/5} \tag{2.87}$$

and write the expressions for the gravitational wave signals from binaries as

$$h_+(t) = h\,\cos(\Omega_{gw}t + \phi_0)\,,$$
$$h_\times(t) = h\,\sin(\Omega_{gw}t + \phi_0)\,, \tag{2.88}$$

where the amplitude of the GW of both the polarizations is

$$h = \frac{G^{5/3}}{r}\,2^{1/3}\mathcal{M}_c^{5/3}\,\Omega_{gw}^{2/3}\,. \tag{2.89}$$

Numerically the amplitude of gravitational waves from binary black holes in the last seconds before coalescence is

$$h \simeq 10^{-21}\left(\frac{\mathcal{M}_c}{10M_\odot}\right)^{5/3}\left(\frac{f}{100Hz}\right)^{2/3}\left(\frac{100Mpc}{r}\right)\,, \tag{2.90}$$

where $f = \Omega_{gw}/(2\pi)$ is the frequency of GW.

Inclined Binary Plane
Now we generalize the geometry such that the plane of binary orbit is tilted by an angle i w.r.t. the x–y plane as shown in Fig. 2.2. The quadrupole moment tensor Q'_{ij} in the tilted plane can be computed from the stress tensor of the orbit in x–z plane (2.81) by a rotation around the y-axis by an angle i

$$Q'_{ij} = R^y(i)_{ik}R^y(i)_{lj}Q_{kl} = \left(R^y(i)\,\hat{Q}\,R^y(i)^T\right)_{ij}. \tag{2.91}$$

Fig. 2.2 Gravitational waveform detected along the z-axis from a binary orbit with plane inclined at an angle i with respect to the x-y plane

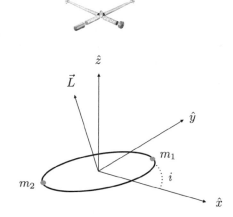

To obtain the TT projection of a wave along $\mathbf{n} = (0, 0, 1)$, we use the projection operator $\Lambda_{ij,kl}$ (2.29) calculated with $\mathbf{n} = (0, 0, 1)$. The TT components of the moment tensor are

$$
\Lambda_{ij,kl} Q'_{kl} = \begin{pmatrix} \frac{1}{2}(Q'_{xx} - Q'_{yy}) & Q'_{xy} & 0 \\ Q'_{yx} & -\frac{1}{2}(Q'_{xx} - T'_{yy}) & 0 \\ 0 & 0 & 0 \end{pmatrix}
$$

$$
= \begin{pmatrix} \frac{1}{2}(Q_{xx} \cos^2 i - Q_{yy}) & Q_{xy} \cos i & 0 \\ Q_{yx} \cos i & -\frac{1}{2}(Q_{xx} \cos^2 i - Q_{yy}) & 0 \\ 0 & 0 & 0 \end{pmatrix} \quad (2.92)
$$

The waveforms generated in this configuration of the source are of the form

$$
h_{ij}^{TT} = \frac{2G}{r} \Lambda_{ij,kl} \ddot{Q}'_{kl}, \quad (2.93)
$$

and the two modes of gravitational waves from an binary orbit inclined at an angle i with respect to the observer (along the z-axis) are

$$
h_+(t, z) = \frac{G^{5/3}}{r} 2^{1/3} \mathcal{M}_c^{5/3} \Omega_{gw}^{2/3} \left(\frac{1 + \cos^2 i}{2} \right) \cos(\Omega_{gw} t + \phi_0),
$$

$$
h_\times(t, z) = \frac{G^{5/3}}{r} 2^{1/3} \mathcal{M}_c^{5/3} \Omega_{gw}^{2/3} (\cos i) \sin(\Omega_{gw} t + \phi_0), \quad (2.94)
$$

where the amplitude for the $i = 0$ case is given in (2.89). If from the Earth the binary orbit plane is edge-on ($i = \pi/2$), then there will be no $h_\times$ polarisation gravitational waves from that source.

2.9   Energy Radiated by Compact Binary Stars

We consider a pair of binary stars of masses m_1 and m_2 orbiting each other with a separation R. Without loss of generality, we can assume the orbit binary in the x–y axis and we consider direction of the gravitational wave as $\mathbf{n} = (\sin\theta \cos\phi, \sin\theta \sin\phi, \cos\theta)$. We use the expression (2.79) for calculating the energy radiated

$$
\frac{dE_{gw}}{dt} = \frac{G}{5} \left\langle \dddot{Q}_{ij} \dddot{Q}^{ji} - \frac{1}{3} \left(\dddot{Q}^i_{\ i} \right)^2 \right\rangle. \quad (2.95)
$$

Using Q_{ij} for the binary orbit in the x–y plane given in (2.82), we find the non-zero components of $\dddot{Q}_{ij}$

$$\dddot{Q}_{xx} = 4\mu R^2 \Omega^3 \sin(2\Omega t), \quad \dddot{Q}_{yy} = -4\mu R^2 \Omega^3 \sin(2\Omega t),$$

$$\dddot{Q}_{xy} = -4\mu R^2 \Omega^3 \cos(2\Omega t). \tag{2.96}$$

Substituting (2.96) in (2.95), we obtain the rate of energy radiated in binaries given by

$$
\begin{aligned}
\frac{dE_{gw}}{dt} &= \frac{G}{5}\left\langle \dddot{Q}_{xx}^2 + \dddot{Q}_{yy}^2 + 2\dddot{Q}_{xy}^2 - \frac{1}{3}(\dddot{Q}_{xx} + \dddot{Q}_{yy})^2 \right\rangle \\
&= \frac{32G}{5}\mu^2 R^4 \Omega^6 \left\langle \sin^2 2\Omega t + \cos^2 2\Omega t \right\rangle \\
&= \frac{32G}{5}\mu^2 R^4 \Omega^6.
\end{aligned}
\tag{2.97}
$$

From the energy loss formula, we can calculate the change in time period ($P_b = 2\pi/\Omega$). From Kepler's law $\Omega^2 R^3 = G(m_1 + m_2)$, we have $\dot{R}/R = (2/3)(\dot{P}_b/P_b)$. The gravitational energy is $E = -Gm_1 m_2/2R$, which implies $\dot{R}/R = (\dot{E}/E)$. Using these two relations, we get $\dot{P}_b/P_b = -(3/2)(\dot{E}/E)$. The energy radiated as gravitational waves results in a decrease in the total energy of the binary, $dE_{gw}/dt = -dE/dt$. Using the formula (2.97) for energy radiated in gravitational waves, we obtain expression for the time period loss

$$
\begin{aligned}
\frac{dP_b}{dt} &= 6\pi G^{-3/2}(m_1 m_2)^{-1}(m_1 + m_2)^{-1/2} R^{5/2}\left(\frac{dE}{dt}\right) \\
&= -\frac{192\pi}{5}G^{5/3}\Omega^{5/3}\frac{m_1 m_2}{(m_1 + m_2)^{1/3}}.
\end{aligned}
\tag{2.98}
$$

The loss of time period is measured in binary pulsar observations and comparison of the observed value of $\dot{P}_b$ with the prediction (2.98) in the Hulse–Taylor binary pulsar gave the first confirmation of the existence of gravitational waves. We will discuss this in detail in Sect. 4.3 where we derive the rate of change of time period taking into account eccentricity of the binary orbit, which provides a sizable correction to (2.98). This eccentricity factor is of order ~ 10 for eccentricity $e \sim 0.6$.

2.10 Chirp Signal of Coalescing Binaries

Coalescing binaries in the last 15 min or so of their orbits are the main source of the signals seen at LIGO and VIRGO detectors whose detection efficiency peaks at $f_{GW} = 2\pi \Omega_{gw} \sim 100\,\text{Hz}$ range. During the coalescing phase of black hole or neutron star mergers, the orbits become circular and the orbital frequency increases as the separation decreases (maintaining the Kepler relation (2.80). The orbital energy of a binary pair with masses m_1 and m_2 separated by a distance R is $E = (Gm_1m_2/2r)$ and the angular frequency is $\Omega = (GM/R^3)^{1/2}$, where $M = m_1 + m_2$. The change of gravitational wave frequency $f = \Omega_{gw}/(2\pi) = \Omega/\pi$ of the GW due to energy loss by emission of gravitational is

$$\frac{df}{dt} = \frac{df}{dR}\frac{da}{dE}\frac{dE}{dt}$$

$$= 6\pi^{1/3} G^{-2/3} f^{1/3} (m_1m_2)^{-1}(m_1 + m_2)^{1/3}\left(\frac{dE}{dt}\right). \qquad (2.99)$$

Using the expression for energy loss by gravitational waves from (2.97), we obtain

$$\frac{df}{dt} = \frac{96}{5}\pi^{8/3} (G\mathcal{M}_c)^{5/3} f^{11/3}. \qquad (2.100)$$

We can integrate (2.100) to give us the time dependence of the frequency of the gravitational waves from inspiralling binaries

$$f(t) \simeq \left(\frac{5}{256}\right)^{3/8}\frac{1}{\pi} (G\mathcal{M}_c)^{-5/8}(t - t_c)^{-3/8}. \qquad (2.101)$$

We see that frequency increases with time till the coalesce time t_c which is regarded as the last stable circular orbit $R_c = 6GM$ ($M = m_1 + m_2$) at which point the frequency reaches the value $f_c = (6^{3/2}\pi GM)^{-1}$. After $R \sim 6GM$ in a BH–BH binary, the two black holes will coalesce within one orbit. For NS–NS or NS–BH inspirals, the NS will be tidally disrupted after the binary separation reaches $R = 6GM$. In the inspiralling phase, the GRW amplitude (2.89) rises with frequency till the coalesce time $t = t_c$ and this is the characteristic 'chirp signal' of the inspiralling binaries. In Fig. 2.3, we show the observation of the gravitational wave signal of binary black hole merger event GW150914 [6]. The chirp phase is followed by the merger to form the final black hole. The perturbations of this final black hole produce GRW which are seen in the signal. The black hole perturbations are damped by GW emission in the ring down phase. The effect on the gravitational wave signal due to emission of ultralight scalar or vector particles can be determined by adding to dE_{gw}/dt for gravitational emission in (2.99) the energy loss rate for scalars from Eq. (4.92) and for vectors from (4.110)

$$\frac{dE}{dt} = \frac{dE_{gw}}{dt} + \frac{dE_s}{dt} + \frac{dE_V}{dt}. \qquad (2.102)$$

The chirp frequency will rise faster when there is a coherent radiation of other particles besides gravitons as discussed in Sects. 4.7 and 4.8.

It may be possible that the masses of the stars change due to emission of neutrinos, axions, dark photons or by neutron–mirror neutron oscillations [7]. In

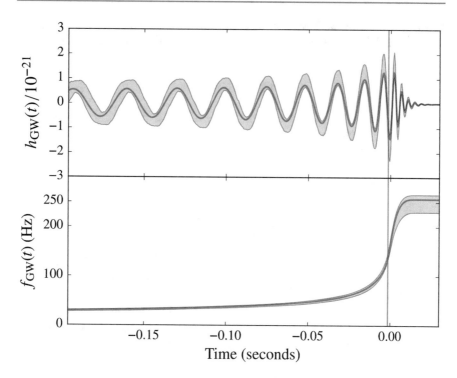

Fig. 2.3 Gravitational wave amplitude and frequency at 90% CL for the black hole binary in spiral event GW150914. Reprinted from [6]. Figure credit: ©American Physical Society. Reproduced under CC-BY-4.0 license

the mass of the neutron star changes, there will be a corresponding change in the orbital angular speed $\Omega = (R^3/GM)^{1/2}$, where $M = m_1 + m_2$. The change in the gravitational wave frequency would be

$$\frac{1}{f}\frac{df}{dt} = \frac{2}{M}\frac{dM}{dt}. \tag{2.103}$$

Since $dM/dt < 0$, the radiation of particles from the interior of the stars would decrease the rise of chirp frequency and can therefore be probed in gravitational wave signals.

2.11 Post-newtonian Corrections to Binary Orbit

The waveforms given in (2.94) were derived assuming a Newtonian orbit and did not take into account the effect of energy loss. There are post-Newtonian (PN) corrections to the waveforms which take into account the GR corrections to the

Newtonian orbit and the backreaction of the binding energy radiated to gravitational waves [8–11]. The PN corrections are an expansion in the small parameter

$$\gamma = \frac{GM}{c^2 R} \sim \frac{v^2}{c^2}, \tag{2.104}$$

where $M = m_1 + m_2$ is the total mass of the binary, R the typical separation and v the orbital velocity. An n-th order PN corrections would have terms with factors $(GM/c^2 R)^{n_1} (v^2/c^2)^{n_2/2}$ for with $n = n_1 + n_2$.

The post-Newtonian orbital parameters are given as expansions in the dimensionless parameters

$$x = \left(\frac{GM\omega}{c^3} \right) \sim O\left(\frac{1}{c^2} \right), \tag{2.105}$$

where $\omega = (GM/R^3)^{1/2}$ is the Newtonian angular velocity. The other expansion parameters are the symmetric mass ratio $\eta = \frac{\mu}{M}$ of the reduced mass $\mu = m_1 m_2/M$ and the total mass M.

The Newtonian equations of motion of the orbital separation $y^i = y_1^i - y_2^i$ and relative velocity $v^i = v_1^i - v_2^i$ of the binary can be corrected by taking into account the dissipative effect of the energy loss by gravitational radiation

$$\frac{dv^i}{dt} = -\omega^2 y^i - \frac{32}{5} \frac{G^3 M^2 \eta}{c^5 R^4} v^i + O\left(\frac{1}{c^7} \right). \tag{2.106}$$

The dissipative term is the radiation reaction where the gravitational radiation was computed for Newtonian orbits. As a result of the dissipative radiation reaction force, the separation between the binaries decreases and the angular frequency increases as

$$\dot{R} = -\frac{64}{5} \frac{G^3 M^3 \eta}{R^3 c^5} + O\left(\frac{1}{c^7} \right),$$

$$\dot{\omega} = -\frac{96}{5} \frac{GM\eta}{R^3} \gamma^{5/2} + O\left(\frac{1}{c^7} \right). \tag{2.107}$$

The corrections to the orbital parameters at the 2PN order are

$$\omega^2 = \frac{GM}{R^3} \left\{ 1 + (-3 + \eta)\gamma + \left(6 + \frac{41}{4}\eta + \eta^2 \right) \gamma^2 + O\left(\frac{1}{c^5} \right) \right\}. \tag{2.108}$$

The orbital energy at 2PN is

$$E = -\frac{\mu x c^2}{2} \left\{ 1 + \left(-\frac{3}{4} - \frac{1}{12}\eta \right) x + \left(-\frac{27}{8} + \frac{19}{8}\eta - \frac{1}{24}\eta^2 \right) x^2 + O\left(\frac{1}{c^5} \right) \right\}. \tag{2.109}$$

The correction to the gravitational wave luminosity $\mathcal{L} = -dE/dt$ is at 2PN order

$$\mathcal{L} = \frac{32c^5}{5G}\eta^2 x^5 \left\{ 1 + \left(-\frac{1247}{336} - \frac{35}{12}\eta \right) x + 4\pi x^{3/2} \right.$$
$$\left. + \left(-\frac{44711}{4072} + \frac{9271}{504}\eta + \frac{65}{18}\eta^2 \right) x^2 + O\left(\frac{1}{c^5}\right) \right\}. \quad (2.110)$$

The change in the binding energy results is a change in the orbital period and radius. From (2.109), we see that the change in E with time will also result in a change in x with time

$$\frac{dx}{dt} = \frac{dE/dt}{dE/dx} = \frac{\mathcal{L}}{dE/dx}. \quad (2.111)$$

The time variation is more conveniently parameterised by defining the dimensionless time parameter

$$\Theta = \frac{\eta c^3}{5GM}(t_c - t), \quad (2.112)$$

where t_c is the time of coalescence.

Using (2.109) and (2.110) in (2.111), we can solve for the time dependence of x due to gravitational wave radiation. To the 2PPN order, the time dependence of the orbital period parameter is

$$x(\Theta) = \frac{1}{4}\Theta^{-1/4}\left\{ 1 + \left(\frac{743}{4032} + \frac{11}{48}\eta \right)\Theta^{-1/4} + O\left(\frac{1}{c^5}\right) \right\}. \quad (2.113)$$

This increase in frequency with time is an important observable in GW signals.

The orbital parameters with up to 3.5PN corrections can be found in [8–11].

2.12   Gravitational Waveform from Sources at Cosmological Distances

In the background Minkowski metric, a gravitational wave propagating along the z-axis will be described by the metric

$$ds^2 = -dt^2 + dz^2 + \left((1 + h_+)dx^2 + (1 - h_+)dy^2 + 2h_\times dxdy \right), \quad (2.114)$$

and in radial coordinates with origin at the source the outgoing gravitational waves will be described by the metric

$$ds^2 = -dt^2 + dr^2 + r^2 \left((1 + h_+)d\theta^2 + (1 - h_+)\sin^2\theta d\phi^2 + 2h_\times \sin^2\theta d\theta d\phi \right). \quad (2.115)$$

We will next examine the metric corresponding to the gravitational waves in an expanding universe.

For sources at distance where the red-shift z is not negligible, one must account for the change in the amplitude and frequency due to cosmological expansion. The metric of the expanding homogenous and isotropic universe is taken to be of the FRLW form

$$ds^2 = -dt^2 + a^2(t) \left(\frac{dr^2}{1 - kr^2} + r^2(d\theta^2 + \sin^2\theta d\phi^2) \right). \tag{2.116}$$

Here $k = (1 - \Omega_m)a^2 H_0^2$ denotes the curvature of the spatial metric $\Omega_m = \rho_m/\rho_{cr}$ and the critical density $\rho_{cr} \equiv 3H_0^2/(8\pi G)$. Usually k is denoted in units of $\Omega_k H_0^2$ and we take $k = 0$ for Euclidean 3-space, $k = 1$ for positive curvature 3-space and $k = -1$ for the negative curvature 3-space.

In the expanding universe, the physical distance is $dl_{phy} = a(t)(\gamma_{ij}dx_i dx_j)^{1/2}$, where γ_{ij} is the metric of the three-dimensional space. The coordinate distance $dl_c = (\gamma_{ij}dx_i dx_j)^{1/2}$ is also called the comoving distance. The FRWL metric is also written as

$$ds^2 = -dt^2 + a^2(t) \left(d\chi^2 + S_k(\chi)(d\theta^2 + \sin^2\theta d\phi^2) \right),$$

$$d\chi^2 \equiv \frac{dr^2}{1 - kr^2} \tag{2.117}$$

and

$$S_k(\chi) \equiv \begin{cases} k^{-1/2}\sin(\chi\sqrt{k}), & \text{if } k > 0 \\ \chi, & k = 0 \\ |k|^{-1/2}\sinh(\chi\sqrt{|k|}), & \text{if } k, 0 \end{cases}$$

The radial comoving distance in an FRWL universe is the coming distance with $d\theta = d\phi = 0$. For light rays, the radial coming distance is

$$d_c = \int_0^{\chi'} d\chi = \int_0^t \frac{dt'}{a(t')} \tag{2.118}$$

This relation between distance and scale factor can be used to relate distance with red-shift. Due to the cosmological expansion, the frequency of emitted gravitational or light would be larger than the observed frequency by a factor which is the ratio of the scale factors which defines the red-shift z

$$\frac{f_{em}(t)}{f_{obs}(t_0)} = \frac{a(t_0)}{a(t)} \equiv 1 + z. \tag{2.119}$$

This relation can be used to relate the radial comoving distance to the red-shift z

$$d_c = \int_t^0 \frac{dt'}{a(t')} = \int_{a(t)}^{a(t_0)} \frac{1}{a} \frac{da}{aH} = \frac{1}{a(t_0)} \int_0^z \frac{1}{H(z)} dz, \qquad (2.120)$$

where we used the relation $dz = -(a(t_0)/a(t)^2)da$ from (2.119). The red-shift dependence of the Hubble parameter can be written using the Friedmann equations $H = (8\pi G/3)\rho$ and the fact that cold dark matter dilutes with the cosmological expansion as a^{-3} therefore $\rho_m(t) = \rho_m(t_0)(1 + z)^3$. Similarly for radiation that dilutes as a^{-4}, we have $\rho_r(t) = \rho_r(t_0)(1 + z)^4$ and the cosmological constant does not change and $\rho_\Lambda(t) = \rho_\Lambda)(t_0)$. Finally we define the fractional energy density of each component in the present epoch as $\Omega_i(t_0) = \rho_i(t_0)/\rho_{tot}(t_0)$. The Hubble parameter in terms of the red-shift z is therefore

$$H(z) = H_0 \left(\Omega_m(1 + z')^3 + \Omega_r(1 + z')^4 + \Omega_\Lambda \right)^{1/2}, \qquad (2.121)$$

where H_0 is the Hubble expansion rate in the present era. For application of gravitational waves from sources, in the matter era, the radiation contribution to $H(z)$ can be dropped.

In Minkowski space, the energy flux from a source falls off as r^{-2}. In the expanding universe, the energy flux $\mathcal{F}$ (energy observed per unit time per unit area) can be used to define a distance measure called the luminosity distance

$$\mathcal{F} \equiv \frac{\mathcal{L}}{4\pi d_L^2} = \frac{1}{4\pi d_L^2} \frac{dE_{em}}{dt_{em}}, \qquad (2.122)$$

where $\mathcal{L}$ is the luminosity which is the total energy radiated per unit time. Flux is the energy rate per area

$$\mathcal{F} = \frac{1}{A} \frac{dE_{obs}}{dt_{obs}}. \qquad (2.123)$$

There is a red-shift in frequency and therefore energy between the source and observer $E_{em} = (1 + z)E_{ob}$. There is also a time dilation of the interval between the passage of successive crests of a wave $\delta t_{obs} = \delta t_{em}(1 + z)$. This can be seen as follows. If the first crest of the wave is emitted at t_{em} and observed at a time t_{obs} after propagating through a coming distance χ and the next crest is emitted at $t_{em} + \delta t_{em}$ and observed at $t_{obs} + \delta t_{obs}$ by the same observer at the same comoving distance, then we have

$$\int_{t_{em}}^{t_{obs}} \frac{dt}{a(t)} = \int_{t_{em}+\delta t_{em}}^{t_{obs}+\delta t_{obs}} \frac{dt}{a(t)} = \chi. \qquad (2.124)$$

From this, we see that

$$\frac{\delta t_{em}}{\delta t_{obs}} = \frac{a(t_{obs})}{a(t_{em})}, \tag{2.125}$$

so there is a time dilation due to the cosmic expansion.

Therefore we have

$$\frac{dE_{em}}{dt_{em}} = \frac{dE_{obs}}{dt_{obs}}(1+z)^2, \tag{2.126}$$

and the energy flux red-shifts by a factor of $(1+z)^2$; one factor of $(1+z)$ is due to red-shift of frequency and the second factor of $(1+z)$ is due to time dilation.

The area of the wave-fronts in the flat expanding universe in terms of the comoving radial distance is

$$A(t_0) = a(t_0)^2 d_c^2 \int d\Omega = 4\pi a(t_0)^2 d_c^2. \tag{2.127}$$

Thus the relation between the luminosity of the source and the flux observed at a later epoch is

$$\mathcal{F} = \frac{\frac{dE_{obs}}{dt_{obs}}}{A(t_0)} = \frac{\frac{dE_{em}}{dt_{em}}\frac{1}{(1+z)^2}}{4\pi a(t_0)^2 d_c^2} = \frac{\mathcal{L}}{4\pi a(t_0)^2 d_c^2 (1+z)^2}. \tag{2.128}$$

The luminosity distance in a spatially flat universe in terms of the luminosity of the source $\mathcal{L}$ and observed flux $\mathcal{F}$ is

$$\begin{aligned} d_L(z) &= \left(\frac{\mathcal{L}}{4\pi\mathcal{F}}\right)^{1/2} = a(t_0)d_c(1+z) \\ &= \frac{1+z}{H_0}\int_0^z \frac{dz'}{\left(\Omega_m(1+z')^3 + \Omega_r(1+z')^4 + \Omega_\Lambda\right)^{1/2}}, \end{aligned}$$

and the angular distance is $d_A = d_c = d_L/(1+z)$. The extra suppression by factor $(1+z)$ of the luminosity distance is due to time dilation of the flux rate.

The amplitude of the waves falls off as $1/r$ in Minkowski space. In curved spacetime, the gravitational wave obeys the wave equation

$$\frac{1}{\sqrt{-g}}\left(\partial_\nu\sqrt{-g}\,g^{\mu\nu}\partial_\nu\right)h_{\alpha\beta} = 0. \tag{2.129}$$

For the flat FLRW universe, the equation for a gravitational wave propagating in the radial direction in the coordinates $r = 0$ at the source this equation becomes

$$\left(\partial_t^2 + 3H\partial_t - \frac{1}{a^2}\partial_r^2 - \frac{2}{a^2 r}\partial_r \right) h_{\alpha\beta} = 0. \tag{2.130}$$

The second term accounts for the dilution of graviton number in a physical volume as the universe expands. Here r is the comoving radial distance and the physical radial distance is ar. This has the solution

$$h(r, t) = \frac{A(r - t)}{ar}, \tag{2.131}$$

and therefore in the expanding universe, the amplitude of the gravitational waves falls off as $h \sim 1/d_L = 1/d_c(1 + z)$.

From (2.100), we see that by measuring the frequency f and $\dot{f}$ one can measure the chirp mass of the binary. One must take into account for the cosmological redshift in frequency $f(z) = (1+z)f$ (where f is the frequency at the source). We can go from the Minkowski space expressions for the amplitude (2.89) and the rate of change of frequency (2.100) to FLRW expressions by making the transformations [12]

$$\{f, t, \mathcal{M}_c, r\} \rightarrow \left\{ \frac{f}{(1 + z)}, t(1 + z), \mathcal{M}_c(1 + z), r(1 + z) \right\}. \tag{2.132}$$

The gravitational waves from binaries at cosmological distances will be of the form

$$h_+(t) = h \cos i \cos \Phi(t),$$

$$h_\times(t) = h \frac{\cos^2 i + 1}{2} \sin \Phi(t),$$

$$h \equiv \frac{4}{d_L(z)} (G\mathcal{M}_c(z))^{5/3} (\pi f(z))^{2/3}. \tag{2.133}$$

The amplitude h and phase $\Phi(t)$ are independent of z, which implies that measurement of the amplitude and frequency of gravitational waves cannot give us the red-shift of the source. The waveform from a binary with masses m_i at distance d_L will be the same as from binary with masses $m_i(1 + z)$ at a distance $d_L(1 + z)$. The distance to the source $d_L(z)$ needs to be determined independently.

If the inspiralling binary is from NS–NS mergers, then one may locate the associated gamma ray bursts (GRBs) from gamma ray, X-ray and radio observations. If from the optical counterpart the red-shift z of the source can be measured, then the measurement of the gravitational wave amplitude will provide the 'standard sirens' and cosmological parameters like H_0 can be determined by the multi-messenger analysis [13].

When the final state of the merger is a black hole, there is an opportunity to study strong field general relativity from the perturbations of the black hole horizon called 'ringing down', which leaves a damped oscillatory gravitational wave signal [6].

References

1. S. Weinberg, *Gravitation and Cosmology: Principles and Applications of the General Theory of Relativity* (John Wiley and Sons, New York, 1972). ISBN 978-0-471-92567-5, 978-0-471-92567-5
2. M. Maggiore, Gravitational waves, in *Theory and Experiments*, vol. 1 (Oxford University Press, Oxford, 2007)
3. S. Capozziello, M. De Laurentis, Extended theories of gravity. Phys. Rep. **509**, 167–321 (2011). [arXiv:1108.6266 [gr-qc]]
4. G.W. Horndeski, Second-order scalar-tensor field equations in a four-dimensional space. Int. J. Theor. Phys. **10**, 363–384 (1974)
5. T. Kobayashi, Horndeski theory and beyond: a review. Rep. Prog. Phys. **82**(8), 086901 (2019). [arXiv:1901.07183 [gr-qc]]
6. B.P. Abbott et al., LIGO Scientific and Virgo Collaborations, Tests of general relativity with GW150914. Phys. Rev. Lett. **116**(22), 221101 (2016). Erratum: [Phys. Rev. Lett. **121**, no. 12, 129902 (2018)]
7. I. Goldman, R.N. Mohapatra, S. Nussinov, Bounds on neutron-mirror neutron mixing from pulsar timing. Phys. Rev. D **100**(12), 123021 (2019). https://doi.org/10.1103/PhysRevD.100.123021. [arXiv:1901.07077 [hep-ph]]
8. L. Blanchet, G. Faye, B.R. Iyer, S. Sinha, The third post-Newtonian gravitational wave polarisations and associated spherical harmonic modes for inspiralling compact binaries in quasi-circular orbits. Class. Quant. Grav. **25**, 165003 (2008). [erratum: Class. Quant. Grav. **29**, 239501 (2012)]. [arXiv:0802.1249 [gr-qc]]
9. A. Le Tiec, The overlap of numerical relativity, perturbation theory and post-Newtonian theory in the binary black hole problem. Int. J. Mod. Phys. D **23**(10), 1430022 (2014). https://doi.org/10.1142/S0218271814300225. [arXiv:1408.5505 [gr-qc]]
10. S. Isoyama, R. Sturani, H. Nakano, Post-Newtonian templates for gravitational waves from compact binary inspirals [arXiv:2012.01350 [gr-qc]]
11. L. Blanchet, Gravitational radiation from post-Newtonian sources and inspiralling compact binaries. Living Rev. Rel. **17**, 2 (2014). https://doi.org/10.12942/lrr-2014-2. [arXiv:1310.1528 [gr-qc]]
12. C. Cutler, E.E. Flanagan, Gravitational waves from merging compact binaries: how accurately can one extract the binary's parameters from the inspiral wave form? Phys. Rev. D **49**, 2658 (1994)
13. B. Abbott et al., LIGO Scientific and Virgo, A gravitational-wave measurement of the Hubble constant following the second observing run of Advanced LIGO and Virgo [arXiv:1908.06060 [astro-ph.CO]]

Field Theory of Linearised Gravity

3

Abstract

In Einstein's General Relativity, the dynamics of particles are governed by the background geometry instead of an exchange of particles like the rest of the forces of nature. Here we formulate GR as a field theory of the fluctuation of the background metric which we call the graviton field. Calculations of classic GR effects like Newton's law and bending of light are done by Feynman diagram calculations. We show that the radiation of gravitational waves from massive bodies can also be treated as a field theory process. We give a field theory derivation of Einstein's quadrupole formula of gravitational radiation.

3.1 Field Theory of Linearised Gravity

The action for the graviton field $h_{\mu\nu}$ is obtained by starting with the Einstein–Hilbert action for gravity

$$S_{EH} = \int d^4x \sqrt{-g} \left[-\frac{1}{16\pi G} R \right] \tag{3.1}$$

and expanding the metric $g_{\mu\nu} = \eta_{\mu\nu} + \kappa h_{\mu\nu}$ to the linear order in $h_{\mu\nu}$, where $\kappa = \sqrt{32\pi G}$ is the gravitational coupling. Here $h_{\mu\nu}$ is treated as a quantum field with mass dimension 1 for calculations of graviton emission, absorption from exchange from classical sources. For the treatment of gravitational wave propagation and detection at interferometers, the metric is expanded as $g_{\mu\nu} = \bar{g}_{\mu\nu} + h_{\mu\nu}$, where $\bar{g}_{\mu\nu}$ is the background metric and $h_{\mu\nu}$ is a dimensionless field signifying the strain of the detector. For the field theoretic calculation of graviton radiation from binary stars, we treat gravitons as quantum fields in the background Minkowski space and study their emission from classical sources [1–5].

© The Author(s), under exclusive license to Springer Nature Switzerland AG 2023
S. Mohanty, *Gravitational Waves from a Quantum Field Theory Perspective*,
Lecture Notes in Physics 1013, https://doi.org/10.1007/978-3-031-23770-6_3

With the graviton field with mass dimension unity defined as

$$g_{\mu\nu} = \eta_{\mu\nu} + \kappa h_{\mu\nu}, \tag{3.2}$$

the inverse metric $g^{\mu\nu}$ defined as $g^{\mu\nu} g_{\mu\alpha} = \delta^\nu_\alpha$ is expanded as

$$g^{\mu\nu} = \eta^{\mu\nu} - \kappa h^{\mu\nu} + \kappa^2 h^{\mu\lambda} h^\nu_\lambda + O(\kappa^3), \tag{3.3}$$

and square root of determinant $\sqrt{-g}$ is

$$\sqrt{-g} = 1 + \frac{\kappa}{2} h + \frac{\kappa^2}{8} h^2 - \frac{\kappa^2}{4} h^{\mu\nu} h_{\mu\nu} + O(\kappa^3), \tag{3.4}$$

where $\eta_{\mu\nu} = \text{diagonal}(-1, 1, 1, 1)$ is the background Minkowski metric and $h = h^\mu_\mu$. Indices are raised and lowered by $\eta^{\mu\nu}$ and $\eta_{\mu\nu}$.

The Christoffel connection and the Ricci tensor to the linear order in $h_{\mu\nu}$ are

$$\begin{aligned}
\Gamma^\lambda_{\mu\nu} &= \frac{1}{2} g^{\lambda\rho} \left(\partial_\mu g_{\rho\nu} + \partial_\nu g_{\rho\mu} - \partial_\rho g_{\mu\nu} \right) \\
&= \frac{\kappa}{2} \eta^{\lambda\rho} \left(\partial_\mu h_{\rho\nu} + \partial_\nu h_{\rho\mu} - \partial_\rho h_{\mu\nu} \right) + O(h^2)
\end{aligned} \tag{3.5}$$

and

$$\begin{aligned}
R_{\mu\nu} &= R^\lambda{}_{\mu\lambda\nu} = \partial_\lambda \Gamma^\lambda_{\mu\nu} - \partial_\nu \Gamma^\lambda_{\lambda\mu} + O(h^2) \\
&= \frac{\kappa}{2} \left(-\partial_\lambda \partial^\lambda h_{\mu\nu} + \partial_\lambda \partial_\mu h^\lambda{}_\nu + \partial_\lambda \partial_\nu h^\lambda{}_\mu - \partial_\mu \partial_\nu h^\lambda{}_\lambda \right) + O(h^2).
\end{aligned} \tag{3.6}$$

The Ricci scalar is

$$\begin{aligned}
R = g^{\mu\nu} R_{\mu\nu} &= \frac{\kappa}{2} \left(\eta^{\mu\nu} - \kappa h^{\mu\nu} \right) \left(-\partial_\lambda \partial^\lambda h_{\mu\nu} + \partial_\lambda \partial_\mu h^\lambda{}_\nu + \partial_\lambda \partial_\nu h^\lambda{}_\mu - \partial_\mu \partial_\nu h^\lambda{}_\lambda \right) \\
&= \kappa^2 \left[-\frac{1}{2} (\partial_\mu h_{\nu\rho})^2 + \frac{1}{2} (\partial_\mu h)^2 - (\partial_\mu h)(\partial^\nu h^\mu_\nu) + (\partial_\mu h_{\nu\rho})(\partial^\nu h^{\mu\rho}) \right].
\end{aligned} \tag{3.7}$$

The action for the graviton field $h_{\mu\nu}$ obtained by linearising the Einstein–Hilbert action (3.1) is given by

$$\begin{aligned}
S_{EH} &= \int d^4x \left[-\frac{1}{2} (\partial_\mu h_{\nu\rho})^2 + \frac{1}{2} (\partial_\mu h)^2 - (\partial_\mu h)(\partial^\nu h^\mu_\nu) + (\partial_\mu h_{\nu\rho})(\partial^\nu h^{\mu\rho}) \right] \\
&\equiv \int d^4x \left[\frac{1}{2} h_{\mu\nu} \mathcal{E}^{\mu\nu\alpha\beta} h_{\alpha\beta} \right],
\end{aligned} \tag{3.8}$$

where the kinetic operator $\mathcal{E}^{\mu\nu\alpha\beta}$ has the form

$$\mathcal{E}^{\mu\nu\alpha\beta} = \left(\eta^{\mu(\alpha}\eta^{\beta)\nu} - \eta^{\mu\nu}\eta^{\alpha\beta}\right)\Box - \eta^{\mu(\alpha}\partial^{\beta)}\partial^{\nu} - \eta^{\nu(\alpha}\partial^{\beta)}\partial^{\mu}$$

$$+\eta^{\alpha\beta}\partial^{\mu}\partial^{\nu} + \eta^{\mu\nu}\partial^{\alpha}\partial^{\beta}, \tag{3.9}$$

and indices enclosed by brackets denote symmetrisation, $A^{(\mu}B^{\nu)} = \frac{1}{2}(A^{\mu}B^{\nu} + A^{\nu}B^{\mu})$. The graviton propagator $D^{(0)}_{\mu\nu\alpha\beta}$ is the inverse of the kinetic operator $\mathcal{E}^{\mu\nu\alpha\beta}$

$$\mathcal{E}^{\mu\nu\alpha\beta} D^{(0)}_{\alpha\beta\rho\sigma}(x - y) = \delta^{\mu}_{(\rho}\delta^{\mu}_{\sigma)}\delta^{4}(x - y). \tag{3.10}$$

The graviton action (3.8) has the gauge symmetry of the graviton field which follows from the invariance of general coordinate transformation in general relativity

$$x^{\mu} \rightarrow x'^{\mu} = x^{\mu} + \xi^{\mu}(x),$$

$$\Rightarrow \quad g'^{\mu\nu}(x') = \frac{\partial x'^{\mu}}{\partial x^{\rho}}\frac{\partial x'^{\nu}}{\partial x^{\sigma}} g^{\rho\sigma}(x),$$

$$\Rightarrow \quad h'_{\mu\nu}(x') = h_{\mu\nu}(x) - \left(\partial_{\mu}\xi_{\nu}(x) + \partial_{\nu}\xi_{\mu}(x)\right). \tag{3.11}$$

Due to the gauge symmetry $h_{\mu\nu} \rightarrow h_{\mu\nu} - \partial_{\mu}\xi_{\nu} - \partial_{\nu}\xi_{\mu}$, the operator $\mathcal{E}^{\mu\nu\alpha\beta}$ cannot be inverted, so the propagator cannot be determined from the relation Eq. 3.17. To invert the kinetic operator, we need to choose a gauge. The gauge choice for which the propagator has the simplest form is the de-Dhonder gauge choice in which

$$\partial^{\mu}h_{\mu\nu} - \frac{1}{2}\partial_{\nu}h = 0, \tag{3.12}$$

where $h = h^{\alpha}{}_{\alpha}$. We can impose this gauge choice by adding the following gauge fixing term to the Lagrangian equation 3.8

$$S_{gf} = -\int d^{4}x \left(\partial^{\mu}h_{\mu\nu} - \frac{1}{2}\partial_{\nu}h\right)^{2}. \tag{3.13}$$

The total action with the gauge fixing term turns out to be of the form

$$S_{EH} + S_{gf} = \int d^{4}x \left(\frac{1}{2}h_{\mu\nu}\Box h^{\mu\nu} - \frac{1}{4}h\Box h\right)$$

$$= \int d^{4}x \left(\frac{1}{2}h_{\mu\nu}\mathcal{K}^{\mu\nu\alpha\beta}h_{\alpha\beta}\right), \tag{3.14}$$

where $\mathcal{K}^{\mu\nu\alpha\beta}$ is the kinetic operator in the de-Donder gauge given by

$$\mathcal{K}^{\mu\nu\alpha\beta} = \frac{1}{2}\left(\eta^{\mu\alpha}\eta^{\nu\beta} + \eta^{\mu\beta}\eta^{\nu\alpha} - \eta^{\mu\nu}\eta^{\alpha\beta}\right)\Box. \tag{3.15}$$

The propagator in the de-Donder gauge is the inverse of the kinetic operator (3.15) and is given by

$$\mathcal{K}^{\mu\nu\alpha\beta} D^{(0)}_{\alpha\beta\rho\sigma}(x-y) = \delta^{\mu}_{(\rho}\delta^{\mu}_{\sigma)}\delta^4(x-y). \tag{3.16}$$

This relation can be used to solve for $D^{(0)}_{\alpha\beta\rho\sigma}$, which in the momentum space ($\partial_\mu = ik_\mu$) is then given

$$D^{(0)}_{\mu\nu\alpha\beta}(k) = \frac{1}{-k^2}\left(\frac{1}{2}(\eta_{\mu\alpha}\eta_{\nu\beta} + \eta_{\mu\beta}\eta_{\nu\alpha}) - \frac{1}{2}\eta_{\mu\nu}\eta_{\alpha\beta}\right). \tag{3.17}$$

We treat the graviton as a quantum field by expanding it in terms of creation and annihilation operators

$$\hat{h}_{\mu\nu}(x) = \sum_\lambda \int \frac{d^3k}{(2\pi)^3}\frac{1}{\sqrt{2\omega_k}}\left[\epsilon^\lambda_{\mu\nu}(\mathbf{k})a_\lambda(\mathbf{k})e^{-ik\cdot x} + \epsilon^{*\lambda}_{\mu\nu}(\mathbf{k})a^\dagger_\lambda(\mathbf{k})e^{ik\cdot x}\right]. \tag{3.18}$$

Here $\epsilon^\lambda_{\mu\nu}(k)$ are the polarization tensors which obey the orthogonality relation

$$\epsilon^\lambda_{\mu\nu}(\mathbf{k})\epsilon^{*\lambda'\,\mu\nu}(\mathbf{k}) = \delta_{\lambda\lambda'}, \tag{3.19}$$

while $a_\lambda(\mathbf{k})$ and $a^\dagger_\lambda(\mathbf{k})$ are graviton annihilation and creation operator which obey the canonical commutation relations

$$\left[a_\lambda(\mathbf{k}), a^\dagger_{\lambda'}(\mathbf{k}')\right] = (2\pi)^3\delta^3(\mathbf{k} - \mathbf{k}')\delta_{\lambda\lambda'}. \tag{3.20}$$

The Feynman propagator of gravitons is defined as the time ordered two point function

$$D^{(0)}_{\mu\nu\alpha\beta}(x-y) \equiv \langle 0|T(\hat{h}_{\mu\nu}(x)\hat{h}_{\alpha\beta}(y))|0\rangle, \tag{3.21}$$

which may be evaluated using Eq. 3.18 to give

$$D^{(0)}_{\mu\nu\alpha\beta}(x-y) = \int \frac{d^4k}{(2\pi)^4}\frac{1}{-k^2 + i\epsilon}e^{ik(x-y)}\sum_\lambda \epsilon^\lambda_{\mu\nu}(k)\epsilon^{*\lambda}_{\alpha\beta}(k). \tag{3.22}$$

Comparing Eqs. 3.17 and 3.22, we have the expression for the polarization sum of massless spin-2 gravitons

$$\sum_{\lambda=1}^{2} \epsilon_{\mu\nu}^{\lambda}(k)\epsilon_{\alpha\beta}^{*\lambda}(k) = \frac{1}{2}(\eta_{\mu\alpha}\eta_{\nu\beta} + \eta_{\mu\beta}\eta_{\nu\alpha}) - \frac{1}{2}\eta_{\mu\nu}\eta_{\alpha\beta}. \tag{3.23}$$

This relation will be used in the computation of processes with a graviton in the external leg.

3.2 Graviton–Matter Coupling

The graviton couples to the stress tensor $T_{\mu\nu}$ with the universal coupling $\kappa/2$ for all forms of matter. This can be seen by starting adding to the Einstein–Hilbert Lagrangian (3.1) the covariant Lagrangian of the matter field

$$S = S_{EH}(g_{\mu\nu}) + S_m(\phi_i, g_{\mu\nu}) \tag{3.24}$$

and expanding the matter Lagrangian to the linear order in the graviton field, $g_{\mu\nu} = \bar{g}_{\mu\nu} + \kappa h_{\mu\nu}$ around the background metric $\bar{g}_{\mu\nu}$. Taylor expanding the matter action around the background metric, we have

$$\begin{aligned}
\delta S_m &= \int d^4x \, \delta \left[\sqrt{-g} \mathcal{L}_m(\phi_i, g_{\mu\nu}) \right] \\
&= \int d^4x \sqrt{-g} \left[\frac{\delta \mathcal{L}_m}{\delta g_{\mu\nu}} - \frac{1}{2} \mathcal{L}_m g_{\mu\nu} \right]_{g_{\mu\nu}=\bar{g}_{\mu\nu}} \delta g_{\mu\nu} \\
&= \int d^4x \sqrt{-\bar{g}} \left(\frac{-1}{2} \right) T^{\mu\nu} \left(\kappa h_{\mu\nu} \right)
\end{aligned} \tag{3.25}$$

where we used the relation $\delta(\sqrt{-g}) = (-1/2)\sqrt{-g}g_{\mu\nu}\delta g_{\mu\nu}$, and the stress tensor of the matter fields is defined as

$$T^{\mu\nu} = -2 \left[\frac{\delta \mathcal{L}_m}{\delta g_{\mu\nu}} - \frac{1}{2} \mathcal{L}_m g_{\mu\nu} \right]_{g_{\mu\nu}=\bar{g}_{\mu\nu}}. \tag{3.26}$$

This gives us the graviton–matter interaction term in the Lagrangian

$$\mathcal{L}_{int}(\phi_i, h_{\mu\nu}) = -\frac{\kappa}{2} T^{\mu\nu} h_{\mu\nu}, \quad \kappa = \sqrt{32\pi G}. \tag{3.27}$$

To calculate the gravitational wave radiation from compact binary stars, we treat them as point masses. For a collection of point masses, action is given by

$$S = -\sum_n m_n \int d\tau_n = -\sum_n m_n \int \sqrt{g_{\mu\nu} \dot{x}_n^\mu \dot{x}_n^\nu}, \qquad (3.28)$$

and the stress tensor is given by

$$
\begin{aligned}
T^{\alpha\beta}(x^\mu) &= \sum_n \int m_n U_n^\alpha U_n^\beta \, \delta^4(x^\mu - x_n^\mu(\tau_n)) d\tau_n \\
&= \sum_n m_n \gamma_n U_n^\alpha U_n^\beta \, \delta^3(\mathbf{x} - \mathbf{x}_n(x^0)) \\
&= \sum_n \frac{p_n^\alpha p_n^\beta}{p_n^0} \delta^3(\mathbf{x} - \mathbf{x}_n(x^0)),
\end{aligned}
\qquad (3.29)
$$

where $U_n = \gamma_n(1, \mathbf{v}_n)$ is the four-velocity and $p_n^\alpha = m_n U_n^\alpha$ the four-momentum of the n-th point mass.

In the early universe, gravitational waves are generated during phase transitions or during inflation from scalar fields. The Lagrangian for a scalar field in curved spacetime is

$$\mathcal{L}(\phi) = -\frac{1}{2} g^{\mu\nu} \partial_\mu \phi \partial_\nu \phi - V(\phi). \qquad (3.30)$$

The corresponding stress tensor of a scalar field in the background metric $\bar{g}_{\mu\nu}$ is given by

$$T_{\mu\nu} = \partial_\mu \phi \partial_\nu \phi - \frac{1}{2} \bar{g}_{\mu\nu} \partial^\alpha \phi \partial_\alpha \phi - \bar{g}_{\mu\nu} V(\phi). \qquad (3.31)$$

There is an effect of graviton–matter interaction during propagation of gravitational waves through matter. When the wavelength of gravitational waves is larger than mean free path of the microscopic particles, the medium can be treated as a thermodynamic fluid and the general describing a fluid element is given by

$$T_{\mu\nu} = (\rho + p) u_\mu u_\nu + p g_{\mu\nu} - 2\eta \sigma_{\mu\nu} - \xi \theta h_{\mu\nu}, \qquad (3.32)$$

where ρ is the energy density, p the pressure, u_μ the four-velocity of the fluid element, η the shear viscosity, ξ the bulk viscosity, $\sigma_{\mu\nu}$ the shear, θ the bulk expansion and $h_{\mu\nu} = g_{\mu\nu} + u_\mu u_\nu$ the projection operator. We will study the propagation of gravitational waves through a general fluid medium and show that the presence of shear viscosity η causes a damping of the gravitational waves.

3.3 Gauge Invariance and Conservation of Stress Tensor

In the full non-linear theory of gravity, Einstein's equations

$$G_{\mu\nu} \equiv R_{\mu\nu} - \frac{1}{2}g_{\mu\nu}R = 8\pi G T_{\mu\nu} \tag{3.33}$$

relate Einstein's tensor $G_{\mu\nu}$ with the stress tensor of matter $T_{\mu\nu}$. Einstein's tensor $G_{\mu\nu}$ obeys the Bianchi identity $\nabla^\mu G_{\mu\nu} = 0$. From this, it follows that the stress tensor obeys the conservation equation

$$\nabla^\mu T_{\mu\nu} = 0, \tag{3.34}$$

where ∇^μ is the covariant derivative. In the linearised gravity theory, the graviton field has a universal coupling with the matter with the interaction Lagrangian given by

$$\mathcal{L}_{int} = -\frac{\kappa}{2}T^{\mu\nu}h_{\mu\nu}. \tag{3.35}$$

Einstein's gravity has the gauge symmetry

$$x^\mu \rightarrow x^\mu + \xi^\mu(x),$$

$$h_{\mu\nu}(x) \rightarrow h_{\mu\nu}(x) - \left(\partial_\mu \xi_\nu(x) + \partial_\nu \xi_\mu(x)\right). \tag{3.36}$$

Under this gauge transformation, the interaction Lagrangian changes as

$$-\frac{\kappa}{2}T^{\mu\nu}h_{\mu\nu} \rightarrow -\frac{\kappa}{2}T^{\mu\nu}\left(h_{\mu\nu} - \partial_\mu \xi_\nu - \partial_\nu \xi_\mu\right)$$

$$= -\frac{\kappa}{2}\left(T^{\mu\nu}h_{\mu\nu} + \partial_\mu T^{\mu\nu}\xi_\nu + \partial_\nu T^{\mu\nu}\xi_\mu\right) \tag{3.37}$$

where we have integrated by parts in the last step. The interaction Lagrangian should be invariant under the gauge transformation (3.36), which implies that the stress tensor obeys the conservation equations

$$\partial_\mu T^{\mu\nu} = \partial_\nu T^{\mu\nu} = 0. \tag{3.38}$$

3.4 Newtonian Gravity

Linearised Einstein's gravity for non-relativistic bodies gives the same interaction potential as Newtonian gravity. Consider the Feynman diagram shown in Fig. 3.1 of a graviton exchange between two massive bodies. The amplitude of this process is the effective action in the second order in coupling $(\kappa/2)^2$ and is given by

$$iS^{(2)} = \frac{\kappa^2}{4}\int d^4x d^4x' T_1^{\mu\nu}(x)G_{\mu\nu\alpha\beta}(x-x')T_2^{\alpha\beta}(x'). \tag{3.39}$$

Fig. 3.1 Graviton exchange
between two massive bodies

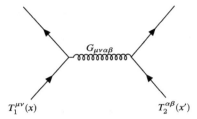

$G_{\mu\nu\alpha\beta}(x - x')$ is the graviton propagator in position space. This is the Fourier transform of the momentum space propagator (3.17)

$$G_{\mu\nu\alpha\beta}(x, x') = \int \frac{dk_0}{(2\pi)} \frac{d^3k}{(2\pi)^3} \frac{-i\, P_{\mu\nu\alpha\beta}}{k_0^2 - |\mathbf{k}|^2 + i\epsilon} e^{-ik_0(t-t')} e^{i\mathbf{k}\cdot(\mathbf{x}-\mathbf{x}')}, \qquad (3.40)$$

where

$$P_{\mu\nu\alpha\beta} = \frac{1}{2} \left(\eta_{\mu\alpha}\eta_{\nu\beta} + \eta_{\mu\beta}\eta_{\nu\alpha} - \eta_{\mu\nu}\eta_{\alpha\beta} \right). \qquad (3.41)$$

The stress tensor for each body is

$$T_1^{\mu\nu}(x) = M_1\, \dot{q}_1^{\mu}\, \dot{q}_1^{\nu}\, \delta^3(\mathbf{x} - \mathbf{q}_1(t)), \quad T_2^{\alpha\beta}(x') = M_2\, \dot{q}_2^{\alpha}\, \dot{q}_2^{\beta}\, \delta^3(\mathbf{x}' - \mathbf{q}_2(t)), \qquad (3.42)$$

where $\dot{q}_1 = u_1^{\mu}$, and the four-velocity of the body "1" and body "2" has mass M_2 and four-velocity $\dot{q}_2^{\mu}$. Substituting these expressions in (3.39), we see that we can trivially do the d^3x and d^3x' integrals using the delta functions in (3.42) to obtain

$$S^{(2)} = \frac{\kappa^2}{4} M_1 M_2 \int dt dt'\, \dot{q}_1^{\mu}\, \dot{q}_1^{\nu}\, \dot{q}_2^{\alpha}\, \dot{q}_2^{\beta}\, P_{\mu\nu\alpha\beta}$$

$$\times \int \frac{dk_0}{(2\pi)} \frac{d^3k}{(2\pi)^3} \frac{-1}{k_0^2 - |\mathbf{k}|^2 + i\epsilon} e^{-ik_0(t-t')} e^{i\mathbf{k}\cdot(\mathbf{q}_1(t)-\mathbf{q}_2(t'))}. \qquad (3.43)$$

For non-relativistic bodies, we can do an expansion of

$$q_a^{\mu} = (1 - v_a^2)^{-1/2}(1, \mathbf{v}_a), \qquad a = 1, 2, \qquad (3.44)$$

as series in the three-velocity v_a^i and evaluate $S^{(2)}$ order by order in v_a^i. If the two bodies are in a bound state, the orbital frequency is $\sim v/r$ for a separation r. The four momenta of the exchanged graviton in this case would be $k_0 \sim v/r$ and $|\mathbf{k}| \sim r$. Therefore, in the non-relativistic expansion, we have $k_0 \ll |\mathbf{k}|$, and in the leading order in v the k_0^2 can be dropped from the propagator.

The action to the zeroeth order in v_a^i is

$$S_{v0}^{(2)} = \frac{\kappa^2}{8} M_1 M_2 \int dt\, dt' \int \frac{dk_0}{(2\pi)} \frac{d^3k}{(2\pi)^3} \frac{1}{|\mathbf{k}|^2 + i\epsilon} e^{-ik_0(t-t')} e^{i\mathbf{k}\cdot(\mathbf{q}_1(t) - \mathbf{q}_2(t'))} \tag{3.45}$$

where we used the result $P_{0000} = 1/2$. The dk_0 integral of the exponential function gives $2\pi\delta(t - t')$, which can be used to perform dt' integral, and we are finally left with

$$S_{v0}^{(2)} = \frac{\kappa^2}{8} M_1 M_2 \int dt \int \frac{d^3k}{(2\pi)^3} \frac{1}{|\mathbf{k}|^2 + i\epsilon} e^{i\mathbf{k}\cdot(\mathbf{q}_1(t) - \mathbf{q}_2(t))}$$

$$= G M_1 M_2 \int dt\, \frac{1}{|\mathbf{q}_1(t) - \mathbf{q}_2(t)|} \tag{3.46}$$

where we used the Fourier transform

$$\int \frac{d^3k}{(2\pi)^3} \frac{1}{|\mathbf{k}|^2} e^{i\mathbf{k}\cdot\mathbf{x}} = \frac{1}{4\pi} \frac{1}{|\mathbf{x}|}. \tag{3.47}$$

The effective potential from the two-body point particle action (3.46) is therefore[1]

$$V(\mathbf{q}_1, \mathbf{q}_2) = -G M_1 M_2 \frac{1}{|\mathbf{q}_1 - \mathbf{q}_2|}, \tag{3.48}$$

which is the Newtonian potential.

3.5 Non-relativistic Effective Theory of Gravity: NRGR

We expand the metric $g_{\mu\nu} = \eta_{\mu\nu} + \kappa h_{\mu\nu}$ and do an expansion in velocity $v = |\mathbf{v}|$ of the point particle action (3.5) to obtain the effective non-relativistic Lagrangian for to describe interaction of binary stars with the gravitational waves [6–8].

We can do the binary star radiation calculation in the non-relativistic effective theory which is an expansion in velocity of the point particle action

$$S = -m \int \sqrt{dx^\alpha dx^\beta g_{\alpha\beta}} \tag{3.49}$$

and the corresponding point particle stress tensor is

$$T^{\alpha\beta}(\mathbf{x}', \tau) = m \frac{u^\alpha u^\beta}{\sqrt{-g u^0}} \delta^3(\mathbf{x}' - \mathbf{x}(\tau)), \tag{3.50}$$

where $u^\alpha \equiv dx^\alpha/d\tau = (u^0, \mathbf{v})$ is the four-velocity of the particles.

---

[1] The point particle action is $S_{pp} = \int dt\, (T - V)$.

We expand (3.5) in the graviton field in the Minkowski background $g_{\mu\nu} = \eta_{\mu\nu} + \kappa h_{\mu\nu}$

$$
\begin{aligned}
S &= -m \int \sqrt{dx^\alpha dx^\beta \eta_{\alpha\beta} + \kappa dx^\alpha dx^\beta h_{\alpha\beta}} \\
&= -m \int d\tau \left(1 + \kappa \frac{dx^\alpha}{d\tau} \frac{dx^\beta}{d\tau} h_{\alpha\beta} \right)^{1/2} \\
&= -m \int d\tau \left(1 + \frac{1}{2}\kappa \frac{dx^\alpha}{d\tau} \frac{dx^\beta}{d\tau} h_{\alpha\beta} - \frac{\kappa^2}{8} \left(\frac{dx^\alpha}{d\tau} \frac{dx^\beta}{d\tau} h_{\alpha\beta} \right)^2 + O((\kappa h_{\alpha\beta})^3) \right),
\end{aligned}
\tag{3.51}
$$

where $d\tau = (dx^\alpha dx^\beta \eta_{\alpha\beta})^{1/2}$ is the proper time. We can expand (3.51) in terms of the three-velocity v as follows:

$$
d\tau = dt(1 - v^2)^{1/2} = dt \left(1 - \frac{v^2}{2} - \frac{v^4}{8} + O(v^6) \right)
$$

$$
\frac{dx^\mu}{d\tau} = \frac{1}{(1-v^2)^{1/2}} \frac{dx^\mu}{dt} = \left(1 + \frac{v^2}{2} + \frac{3}{8}v^4 + O(v^6) \right).
\tag{3.52}
$$

Using (3.52) in (3.51), we obtain the coupling of gravitons with non-relativistic point particles as a series in the velocity v

$$
\begin{aligned}
L_{pp} &= -m \int dt \left[\left(1 - \frac{1}{2}v^2 - \frac{1}{8}v^4 \right) + \frac{\kappa}{2} \left(h_{00} + 2h_{0i}v_i + \frac{1}{2}h_{00}v^2 + h_{ij}v_i v_j \right. \right. \\
&\quad \left. \left. + h_{00}\frac{v^2}{2} + h_{0i}v^i v^2 + \frac{h_{ij}}{2}v^i v^j v^2 + \frac{3}{8}h_{00}v^4 \right) + O(\kappa^2) \right].
\end{aligned}
\tag{3.53}
$$

These are the leading order relativistic corrections to the point particles in an expansion in v.

3.6   First Post-newtonian Correction: Einstein–Infeld–Hoffmann Action

Next we compute the order v^2 corrections to the Newtonian potential. Each v^2 correction counts as 1PN (post-Newtonian) correction and the 1PN correction was calculated by Einstein, Infeld and Hoffmann in 1938 [9].

To derive the 1PN correction from a Feynman diagram calculation, we consider the effective action (3.43) which arises from diagram shown in Fig. 3.1

$$S^{(2)} = \frac{\kappa^2}{4} M_1 M_2 \int dt dt' \, \dot{q}_1^\mu \, \dot{q}_1^\nu \, \dot{q}_2^\alpha \, \dot{q}_2^\beta \, P_{\mu\nu\alpha\beta}$$

$$\times \int \frac{dk_0}{(2\pi)} \frac{d^3 k}{(2\pi)^3} \frac{-1}{k_0^2 - |\mathbf{k}|^2 + i\epsilon} e^{-ik_0(t-t')} e^{i\mathbf{k}\cdot(\mathbf{q}_1(t) - \mathbf{q}_2(t'))}. \quad (3.54)$$

As discussed in the last section, the graviton frequency $k_0 \sim v/r$, and to compute the v^2 corrections we expand the propagator in a series in k_0 and keep the leading order term in k_0^2

$$\frac{1}{k_0^2 - |\mathbf{k}|^2} = -\frac{1}{|\mathbf{k}|^2} \left(1 + \frac{k_0^2}{|\mathbf{k}|^2} + \cdots \right). \quad (3.55)$$

We can now perform the k_0 integral in (3.54)

$$\int \frac{dk_0}{2\pi} \left(1 + \frac{k_0^2}{|\mathbf{k}|^2} \right) e^{-ik_0(t-t')} = \int \frac{dk_0}{2\pi} \left(1 - \frac{1}{|\mathbf{k}|^2} \frac{\partial}{\partial t} \frac{\partial}{\partial t'} \right) e^{-ik_0(t-t')}$$

$$= \left(1 - \frac{1}{|\mathbf{k}|^2} \frac{\partial}{\partial t} \frac{\partial}{\partial t'} \right) \delta(t - t') \quad (3.56)$$

On integrating (3.54) by parts and dropping the total time derivative in the action, the time derivatives act on $e^{i\mathbf{k}\cdot(\mathbf{q}_1(t) - \mathbf{q}_2(t'))}$ to give

$$-\frac{1}{|\mathbf{k}|^2} \frac{\partial}{\partial t} \frac{\partial}{\partial t'} \delta(t - t') e^{i\mathbf{k}\cdot(\mathbf{q}_1(t) - \mathbf{q}_2(t'))} = \frac{1}{|\mathbf{k}|^2} (k_i v_1^i k_j v_2^j) e^{i\mathbf{k}\cdot(\mathbf{q}_1(t) - \mathbf{q}_2(t'))} \delta(t - t').$$

$$(3.57)$$

The integral over t' can now be performed using the delta function. The momentum integral of the $k_i k_j$ terms can be performed using

$$\int \frac{d^3 k}{(2\pi)^3} k_i k_j e^{i\mathbf{k}\cdot(\mathbf{q}_1(t) - \mathbf{q}_2(t))} = \frac{1}{8\pi r^3} \left(r^2 \delta_{ij} - r^i r^j \right), \quad (3.58)$$

where $\mathbf{r}(t) \equiv (\mathbf{q}_1(t) - \mathbf{q}_2(t))$.

Therefore, in (3.45), if one keeps terms of order k_0^2 in the propagator while taking the zeroeth order terms in v from $T_1^{\mu\nu} T_2^{\alpha\beta}$, we have the extra contribution $O(v^2)$ to (3.46) of the form

$$S_{v^2}^2(1) = \int dt \frac{1}{2} \frac{G M_1 M_2}{r} \left(\mathbf{v}_1 \cdot \mathbf{v}_2 - \frac{(\mathbf{v}_1 \cdot \mathbf{r})(\mathbf{v}_2 \cdot \mathbf{r})}{r^2} \right). \quad (3.59)$$

The other $O(v^2)$ contributions are obtained by keeping the v^2 dependence in the stress tensors $T_1^{\mu\nu} T_2^{\alpha\beta}$

$$\dot{q}_1^\mu \dot{q}_1^\nu \dot{q}_2^\alpha \dot{q}_2^\beta P_{\mu\nu\alpha\beta} = \frac{1}{2} + \frac{3}{2}(v_1^2 + v_2^2) - 4v_1 \cdot v_2 \tag{3.60}$$

The $O(v^2)$ terms in (3.60) contribute to the action

$$S_{v^2}^2(2) = \int dt \frac{GM_1M_2}{2r} \left(3(v_1^2 + v_2^2) - 8v_1 \cdot v_2\right) \tag{3.61}$$

The total $O(v^2)$ to the second action is the sum of (3.59) and (3.61) and is given by

$$S_{v^2}^2 = \int dt \frac{GM_1M_2}{2r} \left(3(v_1^2 + v_2^2) - 7v_1 \cdot v_2 - \frac{(v_1 \cdot r)(v_2 \cdot r)}{r^2}\right). \tag{3.62}$$

Therefore the two-body potential at 1PN is

$$V(r) = -\frac{GM_1M_2}{r}\left[1 + \frac{1}{2}\left(3(v_1^2 + v_2^2) - 7v_1 \cdot v_2 - \frac{(v_1 \cdot r)(v_2 \cdot r)}{r^2}\right)\right], \tag{3.63}$$

which is the Einstein–Born–Infeld potential [9].

3.7   Potentials and Waves

Consider the Feynman propagator for the graviton

$$G_{\mu\nu\alpha\beta}(x - x')_F = \int \frac{d^4k}{(2\pi)^4} e^{ik\cdot(x-x')} \frac{1}{\omega^2 - |\mathbf{k}|^2 + i\epsilon} P_{\mu\nu\alpha\beta}, \tag{3.64}$$

where $k = (\omega, |\mathbf{k}|)$. For the gravitational waves propagating in vacuum $k_0 = |\mathbf{k}|$ and the propagator for different choice of poles are given by

$$G_{\mu\nu\alpha\beta}(x - x')_{R,A} = \int \frac{d^4k}{(2\pi)^4} \frac{e^{ik\cdot(x-x')}}{(\omega \pm i\epsilon)^2 - |\mathbf{k}|^2} P_{\mu\nu\alpha\beta}$$

$$= -\frac{1}{4\pi}\left[\delta((t - t') \mp |\mathbf{x} - \mathbf{x'}|)\right] P_{\mu\nu\alpha\beta}, \tag{3.65}$$

where the $-(+)$ signs in the delta function are for the retarded (advanced) propagator.

When gravitons are not on-shell, i.e. $\omega \neq |\mathbf{k}|$, Green's function can be expanded as series as

$$\frac{1}{(|\mathbf{k}|^2 - \omega^2)} = \frac{1}{|\mathbf{k}|^2}\left(1 + \frac{\omega^2}{|\mathbf{k}|^2} + \cdots\right). \tag{3.66}$$

This is an expansion in (Gmv/r) of the source. For a static source, this represents the Newtonian potential. The graviton field of the potentials is denoted as $H_{\mu\nu}(x)$ and these dominated in the near field regions where $r \ll \lambda$.

In momentum space, the graviton propagator for slow moving sources $k_0 \ll |\mathbf{k}|$ is

$$\langle H_{\mu\nu}(x^0, \mathbf{p})H_{\alpha\beta}(y^0, \mathbf{q})\rangle = -i\delta(x_0 - y_0)\frac{1}{|\mathbf{q}|^2}(2\pi)^3 P_{\mu\nu\alpha\beta}, \tag{3.67}$$

and in position representation this is

$$\langle H_{\mu\nu}(x^0, \mathbf{x})H_{\alpha\beta}(y^0, \mathbf{y})\rangle = -i\delta(x_0 - y_0)\frac{1}{|\mathbf{x} - \mathbf{y}|}P_{\mu\nu\alpha\beta} \tag{3.68}$$

This is the instantaneous Newtonian potential in the near field region.

In the generation of gravitational waves, the source potentials can scatter the gravitational waves generated by the bodies. This coupling of the potential field to the radiation field is called the tail effect.

3.8 Quantum Gravity Corrections to Newtonian Potential

Adding loop diagrams to the process of non-relativistic particles scattering by a graviton exchange, we have the quantum corrections to the Newtonian potential [10]

$$V(r) = -\frac{GM_1M_2}{r}\left[1 + 3\frac{G(M_1 + M_2)}{c^2r} + \frac{41}{10\pi}\frac{G\hbar}{c^3r^2}\right]. \tag{3.69}$$

At the innermost stable circular orbits (ISCOs) of binary black holes, $r = 6GM$. For two solar mass black holes the correction from the second term, the post-Newtonian GR correction at the ISCO is

$$3\frac{G(M_1 + M_2)}{c^2r} \simeq 1. \tag{3.70}$$

The quantum correction term proportional to $\hbar$ is

$$\frac{41}{10\pi}\frac{G\hbar}{c^3r^2} \simeq 5 \times 10^{-78}. \tag{3.71}$$

Quantum corrections to the dynamics of massive bodies are out of current experimental reach.

3.9   Massive Gravity Theories

Adding a graviton mass term to the Einstein–Hilbert–Lagrangian breaks the gauge invariance of gravitons. In addition quadratic terms in $h_{\mu\nu}$ in the action give rise to negative kinetic energy terms (ghost fields) in the scalar mode of graviton. One theory that adds the quadratic terms in a specific combination is ghost free at the tree level and in flat background. This is the Fierz–Pauli theory [11] described by the action

$$S = \int d^4x \left[-\frac{1}{2}(\partial_\mu h_{\nu\rho})^2 + \frac{1}{2}(\partial_\mu h)^2 - (\partial_\mu h)(\partial^\nu h^\mu_\nu) + (\partial_\mu h_{\nu\rho})(\partial^\nu h^{\mu\rho}) \right.$$
$$\left. + \frac{1}{2}m_g^2 \left(h_{\mu\nu}h^{\mu\nu} - h^2 \right) + \frac{\kappa}{2} h_{\mu\nu} T^{\mu\nu} \right].$$

(3.72)

The mass term breaks the gauge symmetry $h_{\mu\nu} \rightarrow h_{\mu\nu} - \partial_\mu \xi_\nu - \partial_\nu \xi_\mu$. We will make the separate assumption that the energy–momentum of the matter is conserved, $\partial_\mu T^{\mu\nu} = 0$.

The e.o.m. for $h_{\mu\nu}$ from (3.72) is given by

$$\left(\Box + m_g^2 \right) h_{\mu\nu} - \eta_{\mu\nu} \left(\Box + m_g^2 \right) h$$
$$- \partial_\mu \partial^\alpha h_{\alpha\nu} - \partial_\nu \partial^\alpha h_{\alpha\mu} + \eta_{\mu\nu} \partial^\alpha \partial^\beta h_{\alpha\beta} + \partial_\mu \partial_\nu h = -\kappa T_{\mu\nu}.$$

(3.73)

Taking the divergence of (3.73), we have

$$m_g^2 \left(\partial^\mu h_{\mu\nu} - \partial_\nu h \right) = 0.$$

(3.74)

These are 4 constraint equations which reduce the independent degrees of freedom of the graviton from 10 to 6.

Using (3.74) in (3.73), we obtain

$$\Box h_{\mu\nu} - \partial_\mu \partial_\nu h + m_g^2 \left(h_{\mu\nu} - \eta_{\mu\nu} h \right) = -\kappa T_{\mu\nu}.$$

(3.75)

Taking the trace of this equation, we obtain the relation

$$h = \frac{\kappa}{3m_g^2} T.$$

(3.76)

Therefore trace h is not a propagating mode but is determined algebraically from the trace of the stress tensor. This is the ghost mode as the kinetic term for h in e.o.m. (3.73) appears with the wrong sign. Thus, in the Fierz–Pauli theory, the ghost mode is not a dynamical degree of freedom. The number of independent propagating degrees of freedom of the Fierz Pauli theory is therefore 5. These are the 2 tensor modes, 2 three-vector d.o.f. which do not couple to the energy–momentum tensor and 1 scalar which is still a dynamical mode and which couples to the trace of the energy–momentum tensor.

Going to momentum space ($\partial_\mu \to ik_\mu$), we can solve for the propagator which to be

$$D^{(m)}_{\alpha\beta\rho\sigma}(k) = \frac{1}{-k^2 + m_g^2}\left(\frac{1}{2}(P_{\alpha\rho}P_{\beta\sigma} + P_{\alpha\sigma}P_{\beta\rho}) - \frac{1}{3}P_{\alpha\beta}P_{\rho\sigma}\right), \qquad (3.77)$$

where

$$P_{\alpha\beta} \equiv \eta_{\alpha\beta} - \frac{k_\alpha k_\beta}{m_g^2}. \qquad (3.78)$$

In tree level processes where there is a graviton exchange between conserved currents, the amplitude is of the form

$$\mathscr{A}_{FP} = \frac{\kappa^2}{4}T^{\alpha\beta}D^{(m)}_{\alpha\beta\mu\nu}T'^{\mu\nu}. \qquad (3.79)$$

The momentum dependent terms will vanish due to conservation of the stress tensor $k_\mu T^{\mu\nu} = k_\nu T^{\mu\nu} = 0$. Hence, for tree level calculations, one may drop the momentum dependent terms in (3.77), and the propagator for the FP theory may be written as

$$D^{(m)}_{\mu\nu\alpha\beta}(p) = \frac{1}{-k^2 + m_g^2}\left(\frac{1}{2}(\eta_{\alpha\mu}\eta_{\beta\nu} + \eta_{\alpha\nu}\eta_{\beta\mu}) - \frac{1}{3}\eta_{\alpha\beta}\eta_{\mu\nu}\right). \qquad (3.80)$$

The last term in the brackets is (1/3) instead of (1/2) as in the case of the massless graviton.

When the graviton is treated as a quantum field, the Feynman propagator is defined as in the massless theory (3.22)

$$D^{(m)}_{\mu\nu\alpha\beta}(x - y) = \langle 0|T(\hat{h}_{\mu\nu}(x)\hat{h}_{\alpha\beta}(y))|0\rangle$$

$$= \int \frac{d^4k}{(2\pi)^4}\frac{1}{-k^2 + m_g^2 + i\epsilon}e^{ik(x-y)}\sum_\lambda \epsilon^\lambda_{\mu\nu}(k)\epsilon^{*\lambda}_{\alpha\beta}(k). \qquad (3.81)$$

Comparing (3.80) and (3.81), we see that the polarisation sum for the FP massive gravity theory can be written as

$$\sum_\lambda \epsilon_{\mu\nu}^\lambda(k)\epsilon_{\alpha\beta}^{*\lambda}(k) = \frac{1}{2}(\eta_{\mu\alpha}\eta_{\nu\beta} + \eta_{\nu\alpha}\eta_{\mu\beta}) - \frac{1}{3}\eta_{\alpha\beta}\eta_{\mu\nu}) \tag{3.82}$$

We see that when the propagator (3.17) and polarisation sum (3.23) of the massless graviton theory is different compared with the corresponding expressions (3.80) and (3.82) in the Fierz–Pauli massive gravity theory even in the $m_g \to 0$ limit. The FP action (3.72) however goes to the Einstein–Hilbert action (3.8). This discrepancy was first pointed out by van Dam and Veltman [12] and Zakharov [13] and is called the van Dam–Veltman–Zakharov (vDVZ) discontinuity. This discontinuity is due to the contribution of the scalar degree of freedom of $g_{\mu\nu}$ which does not decouple from the stress tensor in the $m_g \to 0$ limit.

Consider the Newtonian potential between two massive bodies in the Fierz–Pauli theory. The one graviton exchange amplitude (3.79) is

$$\mathcal{A}_{FP} = \frac{\kappa^2}{4}\frac{1}{-k^2 + m_g^2}\left(T_{\mu\nu} - \frac{1}{3}\eta_{\mu\nu}T_\alpha^\alpha\right)T'^{\mu\nu}, \tag{3.83}$$

and the gravitational potential between two massive bodies in the FP theory is

$$\begin{aligned} V_{FP} &= = \frac{\kappa^2}{4}\int\frac{d^3k}{(2\pi)^3}e^{ik\cdot r}\frac{1}{-k^2 + m_g^2}\left(T_{\mu\nu} - \frac{1}{3}\eta_{\mu\nu}T_\alpha^\alpha\right)T'^{\mu\nu} \\ &= \left(\frac{4}{3}\right)\frac{GM_1M_2}{r}e^{-m_g r}. \end{aligned} \tag{3.84}$$

The FP theory gives a Yukawa potential as expected, and however in the $m_g \to 0$ limit the gravitational potential between massive bodies in the FP theory is a factor $4/3$ larger than the Newtonian potential arising from GR. This is ruled out from solar system tests of gravity [14] even in the $m_g \to 0$ limit. We note here that the bending of light by massive bodies is unaffected (in $m_g \to 0$ limit) as the stress tensor for photons $T_\nu^\mu = (\omega, 0, 0, -\omega)$ is traceless and the scattering amplitudes $\mathcal{A}_{FP}(m_g \to 0) = \mathcal{A}_{GR}$. Experimental observations [15] of the bending of radio waves by the Sun match GR to 1%. The two observations together imply that the extra factor of $(4/3)$ in the Newtonian potential of FP theory cannot be absorbed by redefining G.

It has been pointed out by Vainshtein [16, 17] that the linear FP theory breaks down at distances much larger than the Schwarzschild radius $R_s = 2GM$ below which the linearised GR is no longer valid as ($\kappa h_{\mu\nu} \sim$ below this distance). The scalar mode in FP theory becomes strongly coupled with decreasing m_g and the minimum radius from a massive body at which the linearised FP theory is valid is called the Vainshtein radius and is given by $r_V = (R_s/m_g^4)^{1/5}$. The radius within which the FP theory is not sufficient increases as $m_g \to 0$.

3.10 Bending of Light

For relativistic particles, the predictions from Einstein's theory are very different from Newton's. The best example of this is bending of light by a massive body. Here Einstein's gravity predicts the deflection angle to be $\theta_{gr} = 4GM/b$, where b is the impact parameter. The bending angle depends upon the mass of the heavy body but is independent of the energy of the photon. The prediction from Newton's theory is half of this and measurements of the shift in the position of stars when the Sun is at a grazing distance from the path of the light rays compared to when the Sun is away confirms Einstein's theory. In this section we will derive the bending angle using linearised gravity [18].

The Feynman diagram describing the scattering of light by the gravitational field of the sun is shown in Fig. 3.2. The heavy body is non-relativistic and is described by the classical stress tensor $T_1^{\alpha\beta} = Mu^\alpha u^\beta$ and $T_1^{00} = M$ and all other components are zero. The classical stress tensor of the photon is given by

$$T_\gamma^{\mu\nu} = -F_{\mu\alpha}F_\nu{}^\alpha + \frac{1}{4}\eta_{\mu\nu}F_{\alpha\beta}F^{\alpha\beta} \tag{3.85}$$

The photon–graviton vertex in Fig. 3.2 is given by

$$\frac{\kappa}{2}\langle\epsilon_2^{*\mu}p_2|\hat{T}^{\mu\nu}|\epsilon_1^\nu p_1\rangle = \frac{i\kappa}{2}\left[\epsilon_2^* \cdot \epsilon_1(p_1^\mu p_2^\nu + p_1^\nu p_2^\mu)\right] + \cdots \tag{3.86}$$

where we have dropped the terms which are small in the forward scattering limit. The amplitude of the scattering process is

$$\mathcal{M} = \frac{i\kappa}{2}M\frac{-iP_{00\mu\nu}}{-|\mathbf{k}|^2}\epsilon_2^* \cdot \epsilon_1(p_1^\mu p_2^\nu + p_1^\nu p_2^\mu), \tag{3.87}$$

where we have taken $k_0 \simeq \ll |\mathbf{k}|$ for static source. From momentum conservation at the photon vertex we have $k = p_2 - p_1$, as $k_0 = 0$ we have $p_1^0 = p_2^0 = \omega$ the photon energy and $\mathbf{k} = \mathbf{p}_2 - \mathbf{p}_1$. If the deflection angle between $\mathbf{p}_1$ and $\mathbf{p}_2$ is θ then we have $\mathbf{k}^2 = (\mathbf{p}_2 - \mathbf{p}_1)^2 = \omega^2(1 - \cos\theta) = \frac{\omega^2}{2}\sin^2(\frac{\theta}{2})$.

Fig. 3.2 Bending of light by a massive body

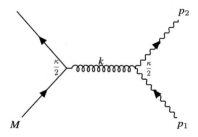

The spin averaged amplitude squares are therefore

$$\langle |\mathcal{M}|^2 \rangle = \frac{\kappa^2}{2} M^2 \frac{1}{\sin^4\left(\frac{\theta}{2}\right)}, \tag{3.88}$$

and we see that the photon energy ω cancels between the numerator and the denominator.

The differential cross section in the cm frame is related to the amplitude squared as

$$\frac{d\sigma}{d\Omega} = \frac{\langle |\mathcal{M}|^2 \rangle}{8\pi M}, \tag{3.89}$$

so for small angles we have the differential cross section of light scattering by a massive body given by

$$\frac{d\sigma}{d\Omega} = \frac{16 G^2 M^2}{\theta^4}. \tag{3.90}$$

We can relate the cross section to the impact parameter by the relation for the geometrical cross section $\sigma = \pi b^2$ writing this in differential form $b\,db = -(d\sigma/d\omega)\sin\theta\,d\theta$. Substituting from (3.90) and integrating, we have

$$\int_0^b b'\,db' = -\int \frac{d\sigma}{d\Omega'}\sin\theta'\,d\theta' = -16 G^2 M^2 \int_0^\theta \frac{1}{\theta'^3}\,d\theta'. \tag{3.91}$$

Integrating both sides, we have the relation for the bending angle

$$\theta = \frac{4 G M}{b}. \tag{3.92}$$

The bending angle in the linearised theory matches the result of the full GR.

We see from these examples that for weak field processes the tree level quantum field theory calculations give the same result as classical GR.

3.11 Eikonal Method

In this section we employ a different method for computing the bending and time delay of light in the vicinity of a star or a black hole. The propagation of light in the gravitational field will be treated as a quantum mechanical problem of a series of scatterings where in each scattering there is a transfer of momentum. The product of these scatterings amplitudes gives rise to the exponential phase or eikonal. The bending angle or time delay is then the determined by extremising the eikonal [19–23].

Fig. 3.3 Bending of light eikonal method

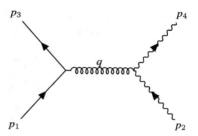

Consider the process $S(p_1) + \gamma(p_2) \to S(p_3) + \gamma(p_4)$ of the scattering of photons by a star of mass M through graviton exchange (Fig. 3.3). The momenta p_1 and p_2 are incoming and p_3 and p_4 are outgoing. The Mandelstam variables are $s = (p_1 + p_2)^2 = (p_3 + p_4)^2$, $t = (p_1 - p_3)^2 = (p_2 - p_4)^2$ and $u = (p_1 - p_4)^2 = (p_2 - p_3)^2$. From energy momentum conservation and the on-shell condition for each external particle $(p_a^\mu)^2 = m_a^2$, we have the relation $s + t + u = \sum_a m_a^2 = 2M^2$.

In this process there is only the t channel diagram and no s and u channel diagrams. We choose the in and out four-momenta of the star and photon in the center of mass frame as

$$p_1 = (E_1, 0, 0, p_1'), \qquad\qquad p_2 = (E_2, 0, 0, -p_1'),$$
$$p_3 = (E_3, 0, p_2' \sin\theta, p_2' \cos\theta), \qquad p_4 = (E_4, 0, -p_2' \sin\theta, -p_2' \cos\theta).$$

$$(3.93)$$

Since $p_2^2 = p_4^2 = 0$, we have $p_1' = E_2$ and $p_2' = E_4$. The graviton momentum is $q = (p_2 - p_4)$ and therefore

$$q^2 = t = -2p_2 \cdot p_4 = -2E_2 E_4 (1 - \cos\theta) = -4E_2 E_4 \sin^2 \frac{\theta}{2}. \qquad (3.94)$$

We have $q_0 = 0$ for a static body and we have $|\mathbf{q}| \ll E_2 \ll M$, so we can take $E_1 \simeq E_3 \simeq M$ and $E_2 \simeq E_4 \equiv E$. The momentum transferred is $\mathbf{q}^2 = -4E^2 \sin^2 \frac{\theta}{2}$.

The amplitude of a single graviton exchange process as calculated in the previous section is

$$\mathcal{M}_1 = -\kappa^2 \frac{EM}{q^2}. \qquad (3.95)$$

From the hierarchy of energies, we see that the scattering angle $\theta \ll 1$ and $\mathbf{q} \simeq \mathbf{q}_\perp$ the component perpendicular to $\mathbf{p}_2$. The amplitude of the graviton exchange diagram is

$$\mathcal{M}_1(\mathbf{q}_\perp) = -\kappa^2 \frac{EM}{\mathbf{q}_\perp^2}. \qquad (3.96)$$

It will be useful to define the Fourier transform of $\mathcal{M}_1(\mathbf{q}_\perp)$ w.r.t. the two-dimensional parameter $\mathbf{b}$ (which is transverse to the initial photon momentum)

$$
\begin{aligned}
\chi_0(b) &= \int \frac{d^2 q_\perp}{(2\pi)^2} e^{i\mathbf{q}_\perp \cdot b} \mathcal{M}_1(\mathbf{q}_\perp) , \\
&= -\frac{\kappa^2 M E}{4} \int \frac{d^2 q_\perp}{(2\pi)^2} e^{-i\mathbf{q}_\perp \cdot b} \, \mathbf{q}_\perp^2 \\
&= -4\kappa^2 M E \left[\frac{1}{d-4} - \log b + \gamma_E \right] .
\end{aligned}
\tag{3.97}
$$

We can sum over n graviton exchanges and denote the amplitude of the total as

$$
\mathcal{M}_{tot}(\mathbf{q}_\perp) = (4\pi)^2 M E \sum_{n=1}^{\infty} \frac{(i\kappa^2 M E)^n}{n!} \prod_{i=1}^{n} \int \frac{d^{d-2} k_i}{(2\pi)^2} \frac{1}{\mathbf{k}_i^2} \, \delta^2\left(\sum_{j=1}^{n} \mathbf{k}_j - \mathbf{q}_\perp\right).
\tag{3.98}
$$

We define the Fourier transformation of $\mathcal{M}_{tot}(\mathbf{q}_\perp)$

$$
\mathcal{M}(b) = \int \frac{d^2 q_\perp}{(2\pi)^2} e^{i\mathbf{q}_\perp \cdot b} \mathcal{M}_{tot}(\mathbf{q}_\perp),
\tag{3.99}
$$

which turns out to be

$$
\begin{aligned}
\mathcal{M}(b) &= 4\pi M E \sum_{n=1}^{\infty} \frac{1}{n!} \left(i \chi_0(b) \right)^n \\
&= 4\pi M E \left(e^{i\chi_0(b)} - 1 \right),
\end{aligned}
\tag{3.100}
$$

where χ_0 is the FT of the one particle amplitude given in (3.97). Taking the inverse transform,

$$
\begin{aligned}
\mathcal{M}_{tot}(\mathbf{q}_\perp) &= \int \frac{d^2 b}{(2\pi)^2} e^{-i\mathbf{q}_\perp \cdot b} \mathcal{M}(b) \\
&= 2(s - M^2) \int \frac{d^2 b}{(2\pi)^2} e^{-i\mathbf{q}_\perp \cdot b} \left(e^{i\chi_0(b)} - 1 \right).
\end{aligned}
\tag{3.101}
$$

The integral can be computed by the saddle point method. The maximum contribution to the integral comes from the stationary pint of the phase given by

$$
\frac{\partial}{\partial b} \left(-\mathbf{q}_\perp \cdot b + \chi_0(b) \right) = 0,
\tag{3.102}
$$

which using (3.97) for $\chi_0(b)$ gives us

$$2E \sin \frac{\theta}{2} = -(-4\kappa^2 M E)\frac{\partial}{\partial b}\log b \qquad (3.103)$$

and this gives us for small angles the relation

$$\theta = \frac{4GM}{b}. \qquad (3.104)$$

So we see that the multi-graviton scattering calculated with the eikonal method gives the same result as GR.

Taking into account the spin of the massive body in the scattering amplitude, the angle deflection of light is given by Bastianelli et al. [24]

$$\theta = 4GM \left(\frac{1}{b} - \frac{s}{b^2} + \frac{s^3}{b^3}\right). \qquad (3.105)$$

3.12 Gravitational Waveform from Scattering Amplitude

In the field theory treatment of gravity, we compute the amplitude of processes like graviton emission from binary stars. We then calculate gravitational wave signal at the detector by using the relation

$$h_{\mu\nu}(r, t) = -\frac{1}{4\pi r} \sum_{\lambda=\pm 2} \int \frac{dk^0}{2\pi} e^{-ik^0(t-r)} \times \epsilon^\lambda_{\mu\nu}(k)\mathcal{A}_\lambda(k)\bigg|_{k^\mu=(k^0,\mathbf{n}k^0)}. \qquad (3.106)$$

Here $\mathcal{A}_\lambda(k)$ is the amplitude for the graviton emission by a source. $\epsilon^{\mu\nu}_\lambda$ is the polarisation tensor of graviton, and $\mathbf{n}$ and r are the direction and distance of the observer w.r.t. the source.

This relation can be derived as follows. We start with the wave equation of gravitons $h_{\mu\nu} = g_{\mu\nu} - \eta_{\mu\nu}$ with a source term

$$\Box h_{\mu\nu} = -\frac{\kappa}{2}T_{\mu\nu} \qquad (3.107)$$

from which we have the Green function relation between gravitational wave signal measured by the detector at spacetime point x which is radiated by a source at point x' by the given by

$$h_{\alpha\beta}(x) = -\frac{\kappa}{2}\int d^4x' \, G_{\alpha\beta,\mu\nu}(x, x')\, T^{\mu\nu}(x'). \qquad (3.108)$$

The Green's function is the Fourier transform of graviton propagator (3.22) and is given by

$$
G_{\alpha\beta,\mu\nu}(x,x') = \sum_{\lambda=1}^{2} \int \frac{d^4q}{(2\pi)^4} \frac{\epsilon_{\alpha\beta}^{\lambda}(\mathbf{n})\epsilon_{\mu\nu}^{*\lambda}(\mathbf{n})}{(q_0+i\epsilon)^2 - \mathbf{q}^2} e^{-i(q_0(t-t')+i\mathbf{q}\cdot(\mathbf{x}-\mathbf{x}'))}. \quad (3.109)
$$

We have only the emission of gravitational waves from the source and not absorption, so consider the retarded Green function which is non-zero for $t > t'$ which is the for the choice of the $i\epsilon$ sign in the denominator.

Performing the momentum space integrals, we obtain

$$
G_{\alpha\beta,\mu\nu}(x,x') = -\frac{1}{4\pi} \frac{\delta\left(t' - (t - |\mathbf{x}-\mathbf{x}'|)\right)}{|\mathbf{x}-\mathbf{x}'|} \sum_{\lambda=1}^{2} \epsilon_{\alpha\beta}^{\lambda}(\mathbf{n})\epsilon_{\mu\nu}^{*\lambda}(\mathbf{n}), \quad (3.110)
$$

where $\mathbf{n}$ is the direction of the graviton propagation (in the coordinates with the origin at the location of the s source). For distant observers, we have $|\mathbf{x}| \gg |\mathbf{x}'|$ and we can make a series expansion

$$
|\mathbf{x}| \gg |\mathbf{x}'| = r - \mathbf{x}' \cdot \mathbf{n} + \cdots, \quad (3.111)
$$

and we can write

$$
\frac{1}{|\mathbf{x}-\mathbf{x}'|} \delta\left(t' - (t - |\mathbf{x}-\mathbf{x}'|)\right) = \frac{1}{r} \delta\left(t' - (t - r + \mathbf{x}' \cdot \mathbf{n})\right). \quad (3.112)
$$

Next we transform the stress tensor of the source in (3.108) to momentum space

$$
T^{\mu\nu}(x') = \int \frac{d^4k}{(2\pi)^4} \tilde{T}^{\mu\nu}(k_0, \mathbf{k}) e^{-ik_0t'+i\mathbf{k}\cdot\mathbf{x}'} \quad (3.113)
$$

Using the expression in (3.108), we obtain

$$
h_{\alpha\beta}(\mathbf{x}, t) = \frac{\kappa}{8\pi} \frac{1}{r} \int d^4x' \frac{d^4k}{(2\pi)^4} \sum_{\lambda=1}^{2} \epsilon_{\alpha\beta}^{\lambda}(\mathbf{n})\epsilon_{\mu\nu}^{*\lambda}(\mathbf{n})
$$

$$
\times \delta\left(t' - (t - r + \mathbf{x}' \cdot \mathbf{n})\right) \tilde{T}^{\mu\nu}(k_0, \mathbf{k}) e^{-ik_0t'+i\mathbf{k}\cdot\mathbf{x}'}.
$$

$$
(3.114)
$$

Doing the dt' integration, we obtain

$$
h_{\alpha\beta}(\mathbf{x}, t) = \frac{\kappa}{8\pi} \frac{1}{r} \int d^3x' \frac{d^4k}{(2\pi)^4} \sum_{\lambda=1}^{2} \epsilon_{\alpha\beta}^{\lambda}(\mathbf{n})\epsilon_{\mu\nu}^{*\lambda}(\mathbf{n})\tilde{T}^{\mu\nu}(k_0, \mathbf{k}) e^{-ik_0'(t-r)+i(\mathbf{k}-k_0\mathbf{n})\cdot\mathbf{x}'}.
$$

$$
(3.115)
$$

The d^3x' integral gives a delta function

$$\int d^3x' e^{i(\mathbf{k}-k_0\mathbf{n})\cdot\mathbf{x}'} = (2\pi)^3 \delta^3(\mathbf{k} - k_0\mathbf{n}), \tag{3.116}$$

which can be used to perform the d^3k in (3.117) integral and we obtain

$$h_{\alpha\beta}(\mathbf{x}, t) = \frac{\kappa}{8\pi} \frac{1}{r} \int \frac{dk_0}{(2\pi)} \sum_{\lambda=1}^{2} \epsilon_{\alpha\beta}^{\lambda}(\mathbf{n}) \epsilon_{\mu\nu}^{*\lambda}(\mathbf{n}) \tilde{T}^{\mu\nu}(k_0, k_0\mathbf{n}) e^{-ik_0(t-r)}. \tag{3.117}$$

The probability amplitude of emitting a graviton of polarisation $\epsilon_{\mu\nu}^{\lambda}(\mathbf{n})$ from the source with stress tensor (in the momentum space) $\tilde{T}^{\mu\nu}(k)$ is given by

$$\mathcal{A}_{\lambda}(k_0, \mathbf{n}k_0) = -i\frac{\kappa}{2}\epsilon_{\mu\nu}^{*\lambda}(\mathbf{n})\tilde{T}^{\mu\nu}(k_0, \mathbf{n}k_0). \tag{3.118}$$

Using (3.118) in (3.117), we can express the gravitational wave metric observed at the detector in terms of the probability amplitude of a graviton emission by a source (like a binary pulsar) as

$$h_{\alpha\beta}(\mathbf{x}, t) = \frac{1}{4\pi r} \int \frac{dk_0}{(2\pi)} \sum_{\lambda=1}^{2} \epsilon_{\alpha\beta}^{\lambda}(\mathbf{n}) \mathcal{A}_{\lambda}(k_0, \mathbf{n}k_0) e^{-ik_0(t-r)}. \tag{3.119}$$

The graviton field in (3.119) is a canonical spin-2 field with mass dimension 1 as it is defined as an expansion of the metric $g_{\mu\nu} = \eta_{\mu\nu} + \kappa h_{\mu\nu}$, where $\kappa = \sqrt{32\pi G}$. The metric perturbation identified as gravitational wave is the dimensionless quantity $\tilde{h}_{\mu\nu} \equiv g_{\mu\nu} - \eta_{\mu\nu} = \kappa h_{\mu\nu}$. The expression for the dimensionless gravitational wave as a function of the amplitude is therefore from (3.119) given by

$$\tilde{h}_{\alpha\beta}(\mathbf{x}, t) = \frac{\kappa}{4\pi r} \int \frac{dk_0}{(2\pi)} \sum_{\lambda=1}^{2} \epsilon_{\alpha\beta}^{\lambda}(\mathbf{n}) \mathcal{A}_{\lambda}(k_0, \mathbf{n}k_0) e^{-ik_0(t-r)}. \tag{3.120}$$

This relates the waveform at the detector to the probability amplitude of graviton emission by the source.

To relate the waveform at the detector to the source stress tensor, we substitute the expression for the amplitude (3.118) in (3.120) to obtain

$$\tilde{h}_{\alpha\beta}(\mathbf{x}, t) = -\frac{\kappa^2}{8\pi r} \int \frac{dk_0}{2\pi} \sum_{\lambda=1}^{2} \epsilon_{\alpha\beta}^{\lambda}(\mathbf{n}) \epsilon_{\mu\nu}^{*\lambda}(\mathbf{n}) \tilde{T}^{\mu\nu}(k_0, \mathbf{n}k_0) e^{-ik_0(t-r)}$$

$$= -\frac{4G}{r} \int \frac{dk_0}{2\pi} \left(\tilde{T}_{\alpha\beta}(k_0, \mathbf{n}k_0) - \frac{1}{2}\eta_{\alpha\beta}\tilde{T}^{\mu}_{\mu}(k_0, \mathbf{n}k_0) \right) e^{-ik_0(t-r)}, \tag{3.121}$$

where we made use of the completeness relation (3.23). In (3.121), we need to project the TT components h_{ij}^{TT} of $\tilde{h}_{\alpha\beta}$ to obtain the propagating degrees of freedom.

$$\tilde{h}_{ij}^{TT}(\mathbf{x}, t) = -\frac{4G}{r}\Lambda_{ij,kl}(\mathbf{n}) \int \frac{dk_0}{2\pi} \left(\tilde{T}_{kl}(k_0, \mathbf{n}k_0) - \frac{1}{2}\eta_{kl}\, \tilde{T}^{\mu}{}_{\mu}(k_0, \mathbf{n}k_0) \right) e^{-ik_0(t-r)}.$$
(3.122)

The $\tilde{T}^{\mu}{}_{\mu}$ part drops out as $\Lambda_{ij,kl}\,\eta_{kl} = 0$, and we obtain

$$\tilde{h}_{ij}^{TT}(\mathbf{x}, t) = -\frac{4G}{r}\Lambda_{ij,kl}(\mathbf{n}) \int \frac{dk_0}{2\pi} T_{kl}(k_0, \mathbf{n}k_0)e^{-ik_0(t-r)}.$$
(3.123)

We shall use the relation (3.121) to compute the gravitational waveform from various sources, like compact binaries in bound and unbound orbits.

3.13 Gravitational Radiation: Fermi's Golden Rule

We treat the graviton $\hat{h}^{\mu\nu}(k)$ as a quantum field which is expanded in terms of creation and annihilation operators as

$$\hat{h}_{\mu\nu}(x) = \sum_{\lambda'} \int \frac{d^3\mathbf{q}}{(2\pi)^3} \frac{1}{\sqrt{2\omega_q}} \left[\epsilon_{\mu\nu}^{\lambda'}(\mathbf{q})a_{\lambda'}(\mathbf{q})e^{-iq\cdot x} + \epsilon_{\mu\nu}^{*\lambda'}(\mathbf{q})a_{\lambda'}^{\dagger}(\mathbf{q})e^{iq\cdot x} \right],$$
(3.124)

where $\omega_q = |\mathbf{q}|$, $\epsilon_{\mu\nu}^{\lambda}(\mathbf{q})$ are the polarization tensors which obey the orthogonality relation

$$\epsilon_{\mu\nu}^{\lambda}(\mathbf{k})\epsilon^{*\lambda'\mu\nu}(\mathbf{k}) = \delta_{\lambda\lambda'},$$
(3.125)

and $a_{\lambda}(\mathbf{q})$ and $a_{\lambda}^{\dagger}(\mathbf{q})$ are graviton annihilation and creation operator which obey the canonical commutation relations

$$\left[a_{\lambda}(\mathbf{k}), a_{\lambda'}^{\dagger}(\mathbf{k}') \right] = (2\pi)^3\delta^3(\mathbf{k} - \mathbf{k}')\delta_{\lambda\lambda'}.$$
(3.126)

The one graviton state is defined as

$$|\epsilon_{\lambda}^{\mu\nu}(\mathbf{k})\rangle = \sqrt{2\omega_k}a_{\lambda}^{\dagger}(\mathbf{k})|0\rangle.$$
(3.127)

The states are normalised as

$$\langle\epsilon_{\lambda}^{\mu\nu}(\mathbf{k})|\epsilon_{\lambda}^{\mu\nu}(\mathbf{k}')\rangle = 2\omega_k(2\pi)^3\delta^3(\mathbf{k} - \mathbf{k}'),$$
(3.128)

and the completeness relation is

$$1 = \int \frac{d^3\mathbf{k}}{(2\pi)^3} \frac{1}{2\omega_k} |\epsilon_\lambda^{\mu\nu}(\mathbf{k})\rangle\langle\epsilon_\lambda^{\mu\nu}(\mathbf{k})|. \tag{3.129}$$

Gravitational wave radiation can be treated as a one-vertex emission process of a quanta of graviton from a classical source. The transition matrix element between the vacuum state and the one graviton state is

$$S_{if} = -i\langle\epsilon_\lambda^{\mu\nu}(\mathbf{k})| \int d^4x \, \mathcal{L}_{int}(x)|0\rangle$$

$$= -i\frac{\kappa}{2} \int d^4x \, \langle 0|\sqrt{2\omega_k}a_\lambda(\mathbf{k})T_{\mu\nu}(x)\hat{h}^{\mu\nu}(x)|0\rangle. \tag{3.130}$$

The graviton–matter vertex is

$$= -i\frac{\kappa}{2}T_{\mu\nu}(\mathbf{k},\omega_n')\epsilon_\lambda^{*\mu\nu}(\mathbf{k}), \tag{3.131}$$

where the double line indicates that the source is a macroscopic object. There is no propagator associated with the macroscopic sources. In the diagrams they count only as sources but not as propagators.

The sources represented by $T_{\mu\nu}(x)$ are not plane waves as in particle decays but have a periodicity in time. For periodic sources, the frequencies are discrete and we can express the stress tensor as a sum over the discrete harmonics ω_n' in frequency space

$$T^{\mu\nu}(\mathbf{x},t) = \sum_n \int \frac{d^3\mathbf{k}'}{(2\pi)^3} T^{\mu\nu}(\mathbf{k}',\omega_n')e^{-i\mathbf{k}'\cdot\mathbf{x}}e^{i\omega_n't}. \tag{3.132}$$

We substitute (3.124) and (3.132) in (3.130) and use the orthogonality and commutation relations (3.125) and (3.126). The commutation relations give

$$\langle 0|a_\lambda(\mathbf{k})a_\lambda^\dagger(\mathbf{q})|0\rangle = \langle 0|[a_\lambda(\mathbf{k}),(a_\lambda^\dagger(\mathbf{q})] + a_\lambda^\dagger(\mathbf{q})a_\lambda(\mathbf{k})|0\rangle = (2\pi)^3\delta^3(\mathbf{k}-\mathbf{q}), \tag{3.133}$$

and we obtain the expression for the transition matrix (3.130)

$$S_{if} = -i\frac{\kappa}{2}\sum_n \int d^4x \int \frac{d^3\mathbf{q}}{(2\pi)^3} \int \frac{d^3\mathbf{k}'}{(2\pi)^3} \frac{1}{\sqrt{2\omega_q}}\sqrt{2\omega_k}(2\pi)^3\delta^3(\mathbf{k}-\mathbf{q})$$

$$\times \epsilon_\lambda^{*\mu\nu}(\mathbf{k})T^{\mu\nu}(\mathbf{k}',\omega_n')e^{i(\mathbf{q}-\mathbf{k}')\cdot\mathbf{x}}e^{-i(\omega-\omega_n')t}. \tag{3.134}$$

The $d^3\mathbf{q}$ integral can be performed using the delta function. The volume integral d^4x gives a delta function $(2\pi)^4\delta^4(\mathbf{k} - \mathbf{k}')\delta(\omega - \omega_n')$. We then perform the $d^3\mathbf{k}'$ integral using the delta function. After all these simplifications, we obtain

$$S_{if} = -i\frac{\kappa}{2}\sum_n T_{\mu\nu}(\mathbf{k}, \omega_n')\epsilon_\lambda^{*\mu\nu}(\mathbf{k})\,(2\pi)\delta(\omega - \omega_n')\,. \tag{3.135}$$

The coefficient of the delta function is the amplitude of emission from an initial state of the stress tensor of a given frequency ω_n' of a graviton of polarisation $\epsilon_\lambda^{*\mu\nu}(\mathbf{k})$ and frequency ω

$$\mathcal{M}_\lambda(\omega, \omega_n') = -i\frac{\kappa}{2}T_{\mu\nu}(\mathbf{k}, \omega_n')\epsilon_\lambda^{*\mu\nu}(\mathbf{k})\,. \tag{3.136}$$

The probability of graviton emission per unit time is the incoherent sum of the rate of emission from each initial state labeled by ω_n and is given by the Fermi's Golden Rule—the probability is the amplitude mod-squared times the phase space volume of the outgoing states

$$\dot{P} = \frac{|S_{fi}|^2}{T} = \frac{\kappa^2}{4}\sum_n \frac{1}{T}\left|T_{\mu\nu}(\mathbf{k}, \omega_n')\epsilon_\lambda^{*\mu\nu}(\mathbf{k})\,(2\pi)\delta(\omega - \omega_n')\right|^2\,. \tag{3.137}$$

There are no interference terms between the different initial states $|T_{\mu\nu}(\mathbf{k}, \omega_n')\rangle$ as the initial states consisting of orbiting binary stars are classical. The time interval T cancels with one of the delta functions as

$$T\delta(\omega - \omega_n') = \int_{-T/2}^{T/2} dt\, e^{i(\omega - \omega_n')t}\delta(\omega - \omega_n') = 2\pi[\delta(\omega - \omega_n')]^2$$

$$\Rightarrow T = 2\pi\delta(\omega - \omega_n')\,, \tag{3.138}$$

and we get from (3.137) and (3.138)

$$\dot{P} = \frac{|S_{fi}|^2}{T} = \frac{\kappa^2}{4}\sum_n \left|T_{\mu\nu}(\mathbf{k}, \omega_n')\epsilon_\lambda^{*\mu\nu}(\mathbf{k})\right|^2 (2\pi)\delta(\omega - \omega_n')\,. \tag{3.139}$$

The rate of graviton emission summed over the final state graviton polarisation and phase space volume is the decay width given by

$$\Gamma = \sum_\lambda \int \frac{|S_{fi}|^2}{T}\frac{d^3\mathbf{k}}{(2\pi)^3 2\omega} = \sum_n\sum_\lambda \int |\mathcal{M}_\lambda(\omega, \omega_n')|^2\,(2\pi)\delta(\omega - \omega_n')\frac{d^3\mathbf{k}}{(2\pi)^3 2\omega}$$

$$= \frac{\kappa^2}{4}\sum_n\sum_\lambda \int \left|T_{\mu\nu}(\mathbf{k}, \omega_n')\epsilon_\lambda^{*\mu\nu}(\mathbf{k})\right|^2 (2\pi)\delta(\omega - \omega_n')\frac{d^3\mathbf{k}}{(2\pi)^3 2\omega_k} \tag{3.140}$$

The rate of energy radiated is obtained from the probability of radiation (3.140) by including an extra factor of $\omega_k = |\mathbf{k}|$ in the integral

$$\frac{dE_{gw}}{dt} = \frac{\kappa^2}{4} \sum_n \sum_\lambda \int \left| T_{\mu\nu}(\mathbf{k}, \omega'_n) \epsilon_\lambda^{*\mu\nu}(\mathbf{k}) \right|^2 (2\pi)\delta(\omega - \omega'_n) \ \omega_k \ \frac{d^3\mathbf{k}}{(2\pi)^3 2\omega}$$

(3.141)

We can open the modulus squared (3.141) and simplify using the polarisation sum relation (3.23)

$$\sum_\lambda \left| T_{\mu\nu}(\mathbf{k}, \omega'_n) \epsilon_\lambda^{*\mu\nu}(\mathbf{k}) \right|^2 = \sum_\lambda \left(T_{\mu\nu}(\mathbf{k}, \omega'_n) T_{\alpha\beta}^*(\mathbf{k}, \omega'_n) \right) \left(\epsilon_\lambda^{*\mu\nu}(\mathbf{k}) \epsilon_\lambda^{\alpha\beta}(\mathbf{k}) \right)$$

$$= \left(T_{\mu\nu}(\mathbf{k}, \omega'_n) T_{\alpha\beta}^*(\mathbf{k}, \omega'_n) \right)$$

$$\times \left(\frac{1}{2}(\eta^{\mu\alpha}\eta^{\nu\beta} + \eta^{\mu\beta}\eta^{\nu\alpha} - \eta^{\mu\nu}\eta^{\alpha\beta}) \right)$$

$$= T_{\mu\nu}(\mathbf{k}, \omega'_n) T^{*\nu\mu}(\mathbf{k}, \omega'_n) - \frac{1}{2}\left| T^\mu{}_\mu(\mathbf{k}, \omega'_n) \right|^2 \quad (3.142)$$

where we made use of the completeness relation

$$\sum_\lambda \epsilon_\lambda^{*\mu\nu}(\mathbf{k}) \epsilon_\lambda^{\alpha\beta}(\mathbf{k}) = \frac{1}{2}\left(\eta^{\mu\alpha}\eta^{\nu\beta} + \eta^{\mu\beta}\eta^{\nu\alpha} - \eta^{\mu\nu}\eta^{\alpha\beta}\right) \equiv P_{\mu\nu;\alpha\beta} . \quad (3.143)$$

We can use the conserved current relation $k_\mu T^{\mu\nu} = 0$, to express the T^{00} and T^{i0} components of the stress tensor in terms of the T^{ij} components

$$T_{0j} = -\hat{k}^i T_{ij}, \quad T_{00} = \hat{k}^i \hat{k}^j T_{ij}. \quad (3.144)$$

Using these relations, we write (3.142) as

$$\left| T_{\mu\nu}(\mathbf{k}, \omega'_n) \right|^2 - \frac{1}{2}\left| T^\mu{}_\mu(\mathbf{k}, \omega'_n) \right|^2 = T_{ij}T^{*ji} + T_{00}T^{*00} + T_{0i}T^{*i0} + T_{i0}T^{*0i}$$

$$- \frac{1}{2}\left(T^0{}_0 + T^i{}_i \right)\left(T^{*0}{}_0 + T^{*j}{}_j \right)$$

$$= \left(T_{ij}T^{*ji} - \frac{1}{2}T^i{}_i T^{*j}{}_j \right) + \frac{1}{2}\hat{k}^i \hat{k}^j \hat{k}^l \hat{k}^m T_{ij}T_{lm}^*$$

$$- \left(\hat{k}^l \hat{k}^m T_{il}T_{mi}^* + \hat{k}^l \hat{k}^m T_{il}T_{mi}^* \right) + \frac{1}{2}\left(\hat{k}^l \hat{k}^m T_{lm}^* T^i{}_i + \hat{k}^l \hat{k}^m T_{lm}T^{*j}{}_j \right).$$

(3.145)

In the quadrupole approximation of the source, for sources with size small than the wavelength of the gravitational waves, $\mathbf{k} \cdot \mathbf{x} \ll 1$, the stress tensor in momentum space $T_{\mu\nu}(\mathbf{k}, \omega'_n)$ has no explicit $\hat{\mathbf{k}}$ dependence. Therefore on substituting (3.145) in (3.141), one can perform the angular integrations using the relations

$$\int d\Omega_k = 4\pi, \quad \int d\Omega_k \, \hat{k}^i \hat{k}^j = \frac{4\pi}{3} \delta_{ij},$$

$$\int d\Omega_k \, \hat{k}^i \hat{k}^j \hat{k}^l \hat{k}^m = \frac{4\pi}{15} \left(\delta_{ij}\delta_{lm} + \delta_{il}\delta_{jm} + \delta_{im}\delta_{jl} \right) \tag{3.146}$$

to obtain

$$\int d\Omega_k \left[|T_{\mu\nu}(\mathbf{k}, \omega'_n)|^2 - \frac{1}{2}|T^\mu{}_\mu(\mathbf{k}, \omega'_n)|^2 \right] = \frac{8\pi}{5} \left(T_{ij}(\omega'_n) T^*_{ji}(\omega'_n) - \frac{1}{3}|T^i{}_i(\omega'_n)|^2 \right). \tag{3.147}$$

Substituting (3.147) in (3.141), we have the expression for the energy radiated by a source in terms of the source stress tensor,

$$\frac{dE_{gw}}{dt} = \frac{\kappa^2}{4} \sum_n \int \frac{8\pi}{5} \left(T_{ij}(\omega'_n) T^*_{ji}(\omega'_n) - \frac{1}{3}|T^i{}_i(\omega'_n)|^2 \right) \omega^2 \, 2\pi\delta(\omega'_n - \omega) \frac{d\omega}{(2\pi)^3 2\omega}. \tag{3.148}$$

We shall now calculate the energy loss from binary star orbits. Binary neutron stars or binary black holes orbiting each other at non-relativistic velocities can be treated as point masses for the calculation of gravitational radiation [6–8]. The size of the compact stars is given by the Schwarzschild radius $r_s = 2GM$. If the separation of the objects is r, then for the bound system we have the relation $r \simeq r_s/v^2$. The frequency and wavenumber of the gravitons emitted will be $(k^0, |\mathbf{k}|) \simeq (v/r, v/r)$. For the point mass approximation to hold, we must have the following hierarchy of scales

$$r_s \ll r \ll \lambda_{gr} \Rightarrow r_s \ll \frac{r_s}{v^2} \ll \frac{r_s}{v^3}, \tag{3.149}$$

which clearly holds for $v \ll 1$.

3.14 Gravitational Radiation: Imaginary Part of Second Order Action

In Schwinger's source theory formulation [25] of quantum theory of fields, Schwinger dispenses with the canonical quantisation and creation and annihilation of fields and derives interactions as Green's function between sources-which we measure. In this method the problem of radiation from a classical source is then derived from the imaginary part of the action. This method gives a more elegant derivation of radiation from classical sources processes like synchrotron and Cerenkov radiation [26]. In this section we will derive the gravitational energy loss formula using Scwinger's Greens function method [27].

We start with the transition amplitude from the past vacuum to future vacuum in the presence of sources $T_{\mu\nu}$. This is given by Schwinger [25]

$$\langle 0_+|0_-\rangle = \exp\left\{-\frac{i}{\hbar}W\right\} \tag{3.150}$$

where the action for the process of emission and absorption of gravitons by sources $T^{\mu\nu}(x)$ and $T^{\alpha\beta}(y)$ is

$$W = 8\pi G c^2 \frac{1}{2}\int d^4x\, d^4y\, T_{\mu\nu}(x)G_{\mu\nu\alpha\beta}(x-y)T_{\alpha\beta}(y) \tag{3.151}$$

where the constant $8\pi G/c^2$ is chosen to give the correct Newtonian potential between static sources.

This corresponds to the Feynman diagram

$$W = \quad\underbrace{}_{k} \tag{3.152}$$

Green's function $G_{\mu\nu\alpha\beta}$ is the Feynman propagator

$$G_{\mu\nu\alpha\beta}(x-y) = \int \frac{d^4k}{(2\pi)^4}e^{ik\cdot(x-y)}\frac{P_{\mu\nu\alpha\beta}}{|\mathbf{k}|-\omega^2-i\epsilon} \qquad \epsilon \to +0, \tag{3.153}$$

where $P_{\mu\nu\alpha\beta} = \sum_\lambda \epsilon_\lambda^{*\mu\nu}(\mathbf{n})\epsilon_\lambda^{\alpha\beta}(\mathbf{n})$ is the tensor given in (3.143).
Using (3.153) in (3.151), we have

$$W = \frac{8\pi G}{c^2}\frac{1}{2}\int \frac{d^4k}{(2\pi)^4}d^4x\, d^4y\, T_{\mu\nu}(x)P_{\mu\nu\alpha\beta}T_{\alpha\beta}(y)\frac{e^{ik\cdot(x-y)}}{|\mathbf{k}|-\omega^2-i\epsilon}. \tag{3.154}$$

The spacetime integrals with the exponential functions give the Fourier transforms of $T^{\mu\nu}$

$$T_{\mu\nu}(k) = \int d^4x\, e^{ikx}\, T^{\mu\nu}(x), \tag{3.155}$$

and therefore we have

$$
\begin{aligned}
W &= \frac{8\pi G}{c^2} \frac{1}{2} \int \frac{d^4 k}{(2\pi)^4} \, T_{\mu\nu}(k) \, P_{\mu\nu\alpha\beta} \, T^*_{\alpha\beta}(k) \, \frac{1}{|\mathbf{k}| - \omega^2 - i\epsilon} \\
&= \frac{8\pi G}{c^2} \frac{1}{2} \int \frac{d^4 k}{(2\pi)^4} \sum_\lambda \left| \epsilon^{*\mu\nu}_\lambda(\mathbf{n}) T_{\mu\nu}(k) \right|^2 \frac{1}{|\mathbf{k}| - \omega^2 - i\epsilon},
\end{aligned}
\tag{3.156}
$$

where $\mathbf{n} = \mathbf{k}/|\mathbf{k}|$.

The probability for the vacuum state to remain as vacuum state (or the persistence of vacuum [25]) is

$$
\left| \langle 0_+ | 0_- \rangle \right|^2 = \exp\left\{ -\frac{2}{\hbar} \mathrm{Im}\, W \right\}.
\tag{3.157}
$$

We interpret the exponent as the expectation value of number of gravitons produced

$$
\langle N \rangle = \frac{2}{\hbar} \mathrm{Im}\, W,
\tag{3.158}
$$

and the probability of vacuum state to remain vacuum is unity when there are no gravitons produced. The graviton emission probability is therefore a Poisson distribution.[2]

From the expression (3.156) for W, we see that the imaginary contribution will come from the propagator

$$
\frac{1}{|\mathbf{k}|^2 - \omega^2 - i\epsilon} = P\left(\frac{1}{|\mathbf{k}|^2 - \omega^2 - i\epsilon} \right) + i\pi\delta(|\mathbf{k}|^2 - \omega^2).
\tag{3.160}
$$

Therefore we have

$$
\begin{aligned}
\langle N \rangle &= \frac{2}{\hbar} \mathrm{Im}\, W = \frac{8\pi G}{c^2} \int \frac{d^4 k}{(2\pi)^4} \pi \delta(|\mathbf{k}|^2 - \omega^2) \sum_\lambda \left| \epsilon^{*\mu\nu}_\lambda(\mathbf{n}) T_{\mu\nu}(k) \right|^2 \\
&= \frac{8\pi G}{\hbar c^2} \int \frac{d^3 k}{(2\pi)^3 2\omega} \sum_\lambda \left| \epsilon^{*\mu\nu}_\lambda(\mathbf{n}) T_{\mu\nu}(k) \right|^2
\end{aligned}
\tag{3.161}
$$

with $\omega = |\mathbf{k}|$.

---

[2] In a Poisson distribution, the probability of producing n gravitons is given by

$$
P(n) = \frac{\lambda^n}{n!} e^{-\lambda},
\tag{3.159}
$$

where $\lambda = \langle n \rangle$ is the mean value.

The average energy radiated is therefore obtained by multiplying the integrand by $\hbar\omega$

$$\langle E \rangle = \frac{8\pi G}{\hbar c^2} \int \frac{d^3 k}{(2\pi)^3 2\omega} \hbar\omega \sum_\lambda |\epsilon_\lambda^{*\mu\nu}(\mathbf{n}) T_{\mu\nu}(k)|^2 . \tag{3.162}$$

The Fourier transform stress tensor in $\mathbf{k}$ space can be written as a multipole expansion

$$T^{\mu\nu}(\omega, \mathbf{k}) = \int dt d^3 x \; e^{i\mathbf{k}\cdot\mathbf{x} - i\omega t} T^{\mu\nu}(t, \mathbf{x})$$

$$= \sum_{l=0}^{\infty} \frac{i^l}{l!} \int dt e^{i\omega t} d^3 x \; T^{\mu\nu}(t, \mathbf{x}) \; (x^{i_1} \cdots x^{i_l}) \; (k_{i_1} \cdots k_{i_l}). \tag{3.163}$$

Taking $l = 0$ and $l = 2$ terms, we have the mass and the angular momentum which are conserved in the binary orbital motion. The leading order contribution to gravitational radiation is made by the $l = 2$ quadrupole moment given by

$$T^{ij}(t, \mathbf{k}) = \frac{-1}{2} \int dt d^3 x \; e^{i\omega t} T^{ij}(t, \mathbf{x}) \; (x^{i_1} x^{i_2}) \; (k_{i_1} k_{i_2}). \tag{3.164}$$

Now we can use the conservation of stress tensor $\partial_\mu T^{\mu\nu} = 0$, which implies $\omega T^{00} = -k_i T^{0i}$ and $\omega T^{0i} = -k_j T{ji}$ to write

$$T^{ij}(t, \mathbf{k}) = \frac{-1}{2} \int dt e^{i\omega t} d^3 x \; T^{ij}(t, \mathbf{x}) \; (x^{i_1} x^{i_2}) \; (k_{i_1} k_{i_2})$$

$$= \frac{1}{2} \int dt e^{i\omega t} d^3 x \; k_0^2 \left(\delta_{i_1}^i \delta_{i_2}^j + \delta_{i_1}^j \delta_{i_2}^i \right) T^{00}(t, \mathbf{x}) \; (x^{i_1} x^{i_2})$$

$$= k_0^2 \int dt e^{i\omega t} d^3 x T^{00}(t, \mathbf{x})[x^i x^j]^{STF}$$

$$\equiv \omega^2 I^{ij}(\omega), \tag{3.165}$$

where $\omega = k_0$, STF stands for taking the symmetric trace-free combination and $I^{ij}(t)$ is the quadrupole moment

$$I^{ij}(t) = \int d^3 x \; T^{00}(t, \mathbf{x})[x^i x^j]^{STF}. \tag{3.166}$$

The quadrupole moment for a collection of point masses m_a is

$$I^{ij}(t) = \sum_a m_a \left(x_a^i x_a^j - \frac{1}{3} \delta^{ij} x_{ai} x_{aj} \right). \tag{3.167}$$

Going back to the expression for the energy radiated (3.162), the amplitude squared sum is

$$\sum_\lambda |T_{ij}(k)\epsilon_{ij}^{TT}(k,\lambda)|^2 = \omega^2 I_{ij}(\omega) I_{lm}(\omega) \sum_\lambda \epsilon_{ij}^{TT}(k,\lambda)\epsilon_{lm}^{*TT}(k,\lambda). \quad (3.168)$$

The polarisation sum of the TT polarisations can be calculated as

$$\sum_\lambda \epsilon_{ij}^{TT}(k,\lambda)\epsilon_{lm}^{*TT}(k,\lambda) = \Lambda_{ij,rs}(k)\Lambda_{lm,pq}(k) \sum_\lambda \epsilon_{rs}(k,\lambda)\epsilon_{pq}^*(k,\lambda). \quad (3.169)$$

The expression for the polarisation sum is

$$\sum_\lambda \epsilon_{rs}(k,\lambda)\epsilon_{pq}^*(k,\lambda) = \frac{1}{2}\left(\delta_{rp}\delta_{sq} + \delta_{rq}\delta_{sp} - \delta_{rs}\delta_{pq}\right). \quad (3.170)$$

Using these expressions, we can compute (3.168). Then substituting (3.168) in (3.162) and doing the angular integrals using (3.146) (this calculation is the same as what we did in the previous section), we obtain the expression for the rate of energy radiated given by

$$\langle \dot{E} \rangle = \frac{G}{\pi T} \int d\omega \frac{\omega^6}{5}\left|I^{ij}(\omega)\right|^2. \quad (3.171)$$

This is the quadrupole radiation formula in frequency space.

References

1. R.P. Feynman, F.B. Morinigo, W.G. Wagner, B. Hatfield, *Feynman Lectures on Gravitation* (Addison-Wesley, Reading, 1995)
2. M.J.G. Veltman, Quantum theory of gravitation, in *Methods in Field Theory*. Les Houches, Session XXVIII, ed. by R. Balian, J. Zinn- Justin (North Holland, Amsterdam, 1976), pp. 265–328
3. B.S. DeWitt, Quantum theory of gravity. 1. The canonical theory. Phys. Rev. **160**, 1113–1148 (1967)
4. B.S. DeWitt, Quantum theory of gravity. 2. The manifestly covariant theory. Phys. Rev. **162**, 1195–1239 (1967)
5. B.S. DeWitt, Quantum theory of gravity. 3. Applications of the covariant theory. Phys. Rev. **162**, 1239–1256 (1967)
6. W.D. Goldberger, I.Z. Rothstein, An effective field theory of gravity for extended objects. Phys. Rev. D **73**, 104029 (2006). [arXiv:hep-th/0409156 [hep-th]]
7. W.D. Goldberger, Les Houches lectures on effective field theories and gravitational radiation [arXiv:hep-ph/0701129 [hep-ph]]
8. R.A. Porto, The effective field theorist's approach to gravitational dynamics. Phys. Rep. **633**, 1–104 (2016). [arXiv:1601.04914 [hep-th]]
9. A. Einstein, L. Infeld, B. Hoffmann, The Gravitational equations and the problem of motion. Ann. Math. **39**, 65–100 (1938)

10. N.E.J. Bjerrum-Bohr, J.F. Donoghue, B.R. Holstein, L. Planté, P. Vanhove, Bending of light in quantum gravity. Phys. Rev. Lett. **114**(6), 061301 (2015). [arXiv:1410.7590 [hep-th]]
11. M. Fierz, W. Pauli, On relativistic wave equations for particles of arbitrary spin in an electromagnetic field. Proc. Roy. Soc. Lond. A **173**, 211–232 (1939). https://doi.org/10.1098/rspa.1939.0140
12. H. van Dam, M.J.G. Veltman, Massive and massless Yang-Mills and gravitational fields. Nucl. Phys. B **22**, 397–411 (1970)
13. V.I. Zakharov, Linearized gravitation theory and the graviton mass. JETP Lett. **12**, 312 (1970)
14. C. Talmadge, J.P. Berthias, R.W. Hellings, E.M. Standish, Model Independent constraints on possible modifications of Newtonian gravity. Phys. Rev. Lett. **61**, 1159–1162 (1988)
15. E. Fomalont, S. Kopeikin, G. Lanyi, J. Benson, Progress in measurements of the gravitational bending of radio waves using the VLBA. Astrophys. J. **699**, 1395–1402 (2009). arXiv:0904.3992 [astro-ph.CO]
16. A.I. Vainshtein, To the problem of nonvanishing gravitation mass. Phys. Lett. B **39**, 393–394 (1972)
17. E. Babichev, C. Deffayet, An introduction to the Vainshtein mechanism. Class. Quant. Grav. **30**, 184001 (2013). [arXiv:1304.7240 [gr-qc]]
18. M. Scadron, *Advanced Quantum Theory and its Applications through Feynman Diagrams* (Springer, New York, 1979)
19. N.E.J. Bjerrum-Bohr, P.H. Damgaard, G. Festuccia, L. Planté, P. Vanhove, General relativity from scattering amplitudes. Phys. Rev. Lett. **121**(17), 171601 (2018). [arXiv:1806.04920 [hep-th]]
20. N.E.J. Bjerrum-Bohr, J.F. Donoghue, B.R. Holstein, L. Plante, P. Vanhove, Light-like scattering in quantum gravity. JHEP **11**, 117 (2016). [arXiv:1609.07477 [hep-th]]
21. N.E.J. Bjerrum-Bohr, B.R. Holstein, J.F. Donoghue, L. Planté, P. Vanhove, Illuminating light bending. PoS CORFU2016, 077 (2017). https://doi.org/10.22323/1.292.0077. [arXiv:1704.01624 [gr-qc]]
22. N.E.J. Bjerrum-Bohr, J.F. Donoghue, B.R. Holstein, Quantum gravitational corrections to the nonrelativistic scattering potential of two masses. Phys. Rev. D **67**, 084033 (2003). [erratum: Phys. Rev. D **71**, 069903 (2005)], [arXiv:hep-th/0211072 [hep-th]]
23. M. Accettulli Huber, A. Brandhuber, S. De Angelis, G. Travaglini, Eikonal phase matrix, deflection angle and time delay in effective field theories of gravity. Phys. Rev. D **102**(4), 046014 (2020). [arXiv:2006.02375 [hep-th]]
24. F. Bastianelli, F. Comberiati, L. de la Cruz, Light bending from eikonal in worldline quantum field theory [arXiv:2112.05013 [hep-th]]
25. J. Schwinger, *Particles, Sources, and Fields*, vol. I (CRC Press, Boca Raton, 1988)
26. J. Schwinger, W.Y. Tsai, T. Erber, Classical and quantum theory of synergic synchrotron—Cherenkov radiation. Ann. Phys. **96**, 303 (1976)
27. E.B. Manoukian, A quantum viewpoint of gravitational radiation. Gen. Relativ. Gravit. **22**, 501 (1990)

Gravitational Wave Radiation from Compact Binaries

4

Abstract

In this chapter we calculate the gravitational radiation as a graviton emission process from an interaction vertex using the field theory methods. We derive the Peter and Mathews formula for the energy loss from compact binaries in elliptical orbits. Comparison of the prediction of period loss with observations from the Hulse–Taylor binary gave the first experimental evidence of existence of gravitational waves. The field theory method is then also used to calculate radiation of ultralight scalars and vector fields from orbiting binaries. Comparison with observations of time period loss from the Hulse–Taylor and other binary pulsar timings enables us to constrain the couplings of ultralight scalars and vectors to matter. We also study gravitational waves from hyperbolic orbits and show how in this case we get the memory signal which is the zero frequency component of gravitational waves.

4.1   Introduction

Gravitational waves radiation from macroscopic objects like binary compact stars and blackholes is traditionally calculated using the quadrupole radiation formula. In this chapter we use the effective field theory and calculate the gravitational radiation as a graviton emission process. Using the field theory method we recover the Peter and Mathews formula for the energy loss from compact binaries in elliptical orbits. Comparison of the prediction of period loss with observations from the Hulse–Taylor binary gave the first experimental evidence of existence of gravitational waves [1–3]. The field theory method is then also used to calculate radiation of ultralight scalars and vector fields from orbiting binaries. Comparison with observations of time period loss from the Hulse–Taylor and other binary pulsar timings enables us to constrain the couplings of ultralight scalars and vectors

S. Mohanty, *Gravitational Waves from a Quantum Field Theory Perspective*,
Lecture Notes in Physics 1013, https://doi.org/10.1007/978-3-031-23770-6_4

to matter and test theories of gravity beyond Einstein's GR [4]. We also study gravitational waves from hyperbolic orbits and show how in this case we get the memory signal which is the zero frequency component of gravitational waves.

4.2   Gravitational Wave Radiation by Binaries in Circular Orbit

We first calculate the gravitational waves from binary stars in circular orbits. In the initial stages of the orbit binaries are highly eccentric and the circular orbit approximation underestimates the energy loss, for example, in the case of binary pulsars with $e \sim 0.5$ by a factor of 10. The orbit loses its eccentricity and in the coalescing stage from which we get the direct signal circular orbit assumption is a good approximation.

$$T_{\mu\nu}(\mathbf{x}', t) = \mu \delta^3(\mathbf{x}' - \mathbf{x}(t)) u_\mu u_\nu, \tag{4.1}$$

where $\mu = \frac{m_1 m_2}{m_1 + m_2}$ is the reduced mass of the binary system, m_1 and m_2 are the masses of the two stars and $u_\mu = (1, \dot{x}, \dot{y}, \dot{z})$ is the non-relativistic four-velocity of the reduced mass. We can choose the coordinates so that orbit is in the x-y plane with the origin at the centre. We can write the Keplerian orbit in the parametric form as,

$$\mathbf{x} = (a \cos \Omega t, a \sin \Omega t, 0), \tag{4.2}$$

where a is the separation between the two stars. The angular velocity in a circular orbit is $\Omega = \left(\frac{G(m_1 + m_2)}{a^3} \right)^{\frac{1}{2}}$, and we can write the Fourier transform of the stress tensor in terms the fundamental frequency Ω.

The stress tensor of this system can be written as

$$T_{ij}(\mathbf{x}', t) = \mu u_i u_j \delta^3(\mathbf{x}' - \mathbf{x}(t)), \tag{4.3}$$

where for the circular orbit (4.2) the 3-velocity $\mathbf{u}$ is

$$\mathbf{u} = \Omega a(- \sin \Omega t, \cos \Omega t, 0). \tag{4.4}$$

In Fourier space the stress tensor components are

$$T_{ij}(\mathbf{k}', \omega') = \mu \int d^3 x' dt e^{i(\omega t - \mathbf{k}' \cdot \mathbf{x}')} u_i u_j \delta^3(\mathbf{x}' - \mathbf{x}(t))$$

$$= \mu \int dt e^{i(\omega' t - \mathbf{k}' \cdot \mathbf{x})} u_i u_j. \tag{4.5}$$

Expressing exponential term in terms of the orbital parameters (4.2) we have

$$e^{i\omega' t - \mathbf{k}' \cdot \mathbf{x}} = e^{i(\omega' t - k_x' a \cos \Omega t - k_y' a \cos \Omega t)} = e^{i\omega' t (1 - \mathbf{k} \cdot \mathbf{v})} \simeq e^{i\omega' t} \tag{4.6}$$

as $|\mathbf{v}| \ll 1$. We can therefore do the time integral in (4.5) by neglecting the $e^{-i\mathbf{k}\cdot\mathbf{x}}$ contribution and obtain

$$T_{ij}(\mathbf{k}', \omega') = \mu e^{-i\mathbf{k}'\cdot\mathbf{x}} \int dt e^{i\omega't} u_i u_j$$

$$= e^{-i\mathbf{k}'\cdot\mathbf{x}} T_{ij}(\omega'). \tag{4.7}$$

In the calculation of the S-matrix the exponential factor gives a momentum conservation delta function which is eliminated in the volume integral as we have seen in (3.134). What enters the final expression of the energy loss rate is the stress tensor in frequency space

$$T_{ij}(\omega') = \int dt\, e^{i\omega't} u_i u_j. \tag{4.8}$$

Using (4.4) we have the stress tensor components

$$T_{xx}(t) = \mu a^2 \Omega^2 \sin^2 \Omega t = \frac{\mu a^2 \Omega^2}{2}(-\cos 2\Omega t + 1),$$

$$T_{yy}(t) = \mu a^2 \Omega^2 \cos^2 \Omega t = \frac{\mu a^2 \Omega^2}{2}(\cos 2\Omega t + 1),$$

$$T_{xy}(t) = -\mu a^2 \Omega^2 \sin \Omega t \cos \Omega t = -\frac{\mu a^2 \Omega^2}{2}(\sin 2\Omega t), \tag{4.9}$$

with the corresponding Fourier transforms given by

$$T_{xx}(\omega') = \frac{\mu a^2 \Omega^2}{2} \int dt e^{-i\omega't}(\cos 2\Omega t + 1)$$

$$= \frac{\mu a^2 \Omega^2}{2} 2\pi \left(2\delta(\omega') + \delta(\omega' + 2\Omega) + \delta(\omega' - 2\Omega) \right)$$

$$T_{yy}(\omega') = \frac{\mu a^2 \Omega^2}{2} 2\pi \left(2\delta(\omega') - \delta(\omega' + 2\Omega) - \delta(\omega' - 2\Omega) \right)$$

$$T_{xy}(\omega') = i \frac{\mu a^2 \Omega^2}{2} 2\pi \left(\delta(\omega' + 2\Omega) - \delta(\omega' - 2\Omega) \right). \tag{4.10}$$

The first proportional to $\delta(\omega')$ signifies emission of a zero energy graviton and does not contribute to the emission rate. The term with $\delta(\omega' + 2\Omega)$ contributes to the process of graviton absorption and we will drop it for calculating the energy radiated. The term with $\delta(\omega' - 2\Omega)$ implies that a binary system with orbital angular frequency Ω will radiate gravitons of angular frequency 2Ω. This is due to the quadrupole nature of the source in gravitational radiation. We will only retain the $\delta(\omega' - 2\Omega)$ terms which corresponds to the graviton emission process. We define

$$T_{ij}(\omega') \equiv \bar{T}_{ij}(2\pi)\delta(\omega' - 2\Omega). \tag{4.11}$$

The components of stress tensor of the binary orbit which contribute to graviton emission are therefore

$$\bar{T}_{xx} = \frac{\mu a^2 \Omega^2}{2}, \quad \bar{T}_{yy} = -\frac{\mu a^2 \Omega^2}{2}, \quad \bar{T}_{xy} = -i\frac{\mu a^2 \Omega^2}{2}. \tag{4.12}$$

The energy radiated is given by (3.148), where the sum over the initial harmonics ω'_n gets a contribution only from the single term $\omega' = 2\Omega$, and we obtain

$$\frac{dE_{gw}}{dt} = \frac{\kappa^2}{4}\frac{8\pi}{5} \int \left(\bar{T}_{ij}(\omega')\bar{T}^*_{ji}(\omega') - \frac{1}{3}|\bar{T}^i{}_i(\omega')|^2\right)\omega^3 2\pi \delta(\omega - 2\Omega)\frac{d\omega}{(2\pi)^3 2\omega}. \tag{4.13}$$

Using (4.12) we compute the tensor-product

$$\left(\bar{T}_{ij}\bar{T}^*_{ji} - \frac{1}{3}|\bar{T}^i{}_i|^2\right) = \left(\bar{T}_{xx}\bar{T}^*_{xx} + \bar{T}_{yy}\bar{T}^*_{yy} + \bar{T}_{xy}\bar{T}^*_{yx} + \bar{T}_{yx}\bar{T}^*_{xy}\right)$$

$$= \mu^2 a^4 \Omega^4 \tag{4.14}$$

as $\bar{T}^i{}_i = \bar{T}^x{}_x + \bar{T}^x{}_x = 0$. Now substituting (4.14) in (4.13) we obtain the expression for the energy radiated by binary stars in a circular orbit with orbital angular velocity Ω

$$\frac{dE_{gw}}{dt} = \frac{32G}{5}\mu^2 a^4 \Omega^6$$

$$= \frac{32G^4}{5}\frac{m_1^2 m_2^2(m_1 + m_2)}{a^5}. \tag{4.15}$$

In the next section we calculate the energy radiated by binaries in elliptical orbits. This is the same result one obtains using the quadrupole radiation formula of GR. The tree level quantum field theory calculation gives the same result as the classical calculation. This equivalence of tree level process with the classical result changes if the background is not a vacuum state of gravitons but has a non-zero graviton distribution. We explore this case detail in the next section.

4.3 Gravitational Wave Radiation from Compact Binary Stars in Elliptical Orbits

We now calculate the energy loss due to the radiation of massless graviton from compact binary systems following [5, 6].

We derived the expression for energy radiated as gravitational waves in (3.148) for a general source with stress tensor in Fourier space $T_{ij}(\omega')$. The rate of energy radiated is

$$\frac{dE_{gw}}{dt} = \frac{\kappa^2}{4} \sum_{\omega'} \int \frac{8\pi}{5} \left(T_{ij}(\omega') T_{ji}^*(\omega') - \frac{1}{3} |T^i{}_i(\omega')|^2 \right) \omega^3 \, 2\pi \delta(\omega' - \omega) \frac{d\omega}{(2\pi)^3 2\omega}.$$

(4.16)

In the elliptical orbit the orbital speed varies in the course of the orbit and we compute the average rate of energy radiated in one orbit. We can write the elliptical Keplerian orbit in the parametric form as [7],

$$x = a(\cos\xi - e), \quad y = a\sqrt{(1 - e^2)} \sin\xi, \quad z = 0, \quad \Omega t = \xi - e\sin\xi,$$

(4.17)

where a and e are the semi-major axis and eccentricity, respectively, of the elliptical orbit and $\xi \in (0, 2\pi)$.

The stress tensor or the current density for this compact binary system is

$$T_{\mu\nu}(x') = \mu \delta^3(\mathbf{x}' - \mathbf{x}(t)) u_\mu u_\nu,$$

(4.18)

where $\mu = \frac{m_1 m_2}{m_1 + m_2}$ is the reduced mass of the binary system, m_1 and m_2 are the masses of the two stars and $u_\mu = (1, \dot{x}, \dot{y}, 0)$ is the non-relativistic four-velocity of the reduced mass taking the Keplerian orbit in the x-y plane. Since the angular velocity of an eccentric orbit is not constant, we can write the Fourier transform of the current density in terms of the n harmonics of the average angular velocity $\Omega = \left(G \frac{(m_1 + m_2)}{a^3} \right)^{\frac{1}{2}}$.

Now we will calculate the Fourier transform of the stress tensor components $T_{ij}(\mathbf{x}, t)$. In the elliptical orbit the time angular velocity varies in the course of the orbit, so we calculate the time average of the energy loss over the time period T. The Fourier coefficient for the n-th harmonic $\omega'_n = n\Omega$ of the stress tensor is given by

$$T_{ij}(\mathbf{k}', \omega'_n) = \frac{1}{T} \int_0^T \int T_{ij}(\mathbf{x}', t) e^{-i(\mathbf{k}' \cdot \mathbf{x}' - \omega'_n t)} d^3 x' dt$$

$$= \frac{1}{T} \int_0^T \int \mu \, \dot{x}'_i \dot{x}'_j \delta(\mathbf{x}' - \mathbf{x}(t)) e^{-i(\mathbf{k}' \cdot \mathbf{x}' - \omega'_n t)} d^3 x' dt$$

$$= \frac{1}{T} \int_0^T \int \mu \, \dot{x}_i \dot{x}_j e^{-i(\mathbf{k}' \cdot \mathbf{x} - \omega'_n t)} dt .$$

(4.19)

For non-relativistic motion $k' \cdot \mathbf{x} \ll \omega'_n t$ where $\mathbf{x}$ is the orbit of the binary described in (4.17). The $e^{-i\mathbf{k}' \cdot \mathbf{x}}$ can be taken out of the time integral. This exponential factor integrated over $d^3 x$ gives a momentum-conserving delta function. The Fourier transform of the stress tensor in frequency space determines the gravitational wave radiation from the binary. For periodic orbits the Fourier transform can be described as a sum over the discrete harmonics of the fundamental frequency $\omega'_n = n\Omega$.

The xx-component of stress tensor in the Fourier space is

$$
\begin{aligned}
T_{xx}(\omega'_n) &= \frac{\mu}{T} \int_0^T \dot{x}^2(t) e^{i\omega'_n t} dt \\
&= \frac{-i\mu\omega'_n{}^2}{T} \int_0^T \dot{x}(t) x(t) e^{i\omega'_n t} dt,
\end{aligned}
\tag{4.20}
$$

where we used integration by parts. Using the orbit relations (4.17) we can write

$$
\dot{x} dt = \frac{dx}{d\xi} d\xi = -a \sin \xi \, d\xi
\tag{4.21}
$$

and we obtain

$$
\begin{aligned}
T_{xx}(\omega'_n) &= \frac{i\mu a^2 \omega'_n{}^2}{2\pi n} \int_0^{2\pi} \sin \xi (\cos \xi - e) e^{in(\xi - e\sin \xi)} d\xi \\
&= \frac{i\mu a^2 \omega'_n{}^2}{8\pi n} \int_0^{2\pi} \left[(e^{2i\xi} - e^{-2i\xi}) - 2e(e^{i\xi} - e^{-i\xi}) \right] e^{in(\xi - e\sin \xi)} d\xi \\
&= \frac{\mu a^2 \omega'_n{}^2}{4n} \left[(J_{n-2}(ne) - J_{n+2}(ne)) - 2e(J_{n-1}(ne) - J_{n+1}(ne)) \right],
\end{aligned}
\tag{4.22}
$$

where we used $T = 2\pi/\Omega$, $\omega'_n = n\Omega$ and in the last steps we used the integral representation of the Bessel function [7]

$$
J_n(z) = \frac{1}{2\pi} \int_0^{2\pi} e^{i(n\xi - z\sin \xi)} d\xi .
\tag{4.23}
$$

By using the Bessel function recurrence relations

$$
J_{n-1}(z) - J_{n+1}(z) = 2J'_n(z),
$$

$$
J_{n-1}(z) + J_{n+1}(z) = \frac{2n}{z} J_n(z) ,
$$

$$
J_{n-2}(z) - J_{n+2}(z) = \frac{4n}{z} J'_n(z) - \frac{4n}{z^2} J_n(z),
$$

$$
J_{n-2}(z) + J_{n+2}(z) = \left(\frac{4n^2}{z^2} - 2 \right) J_n(z) - \frac{4}{z} J'_n(z)
\tag{4.24}
$$

we can write (4.22) as

$$T_{xx}(\omega'_n) = \frac{-\mu a^2 \omega'^2_n}{n}\left[\left(\frac{1-e^2}{e}\right)J'_n(ne) - \frac{1}{ne^2}J_n(ne)\right].\qquad(4.25)$$

Similarly,

$$
\begin{aligned}
T_{yy}(\omega'_n) &= \frac{\mu}{T}\int_0^T \dot{y}^2(t)e^{i\omega'_n t}\,dt \\
&= \frac{-i\mu\omega'_n}{T}\int_0^T \dot{y}(t)y(t)e^{i\omega'_n t}\,dt \\
&= \frac{\mu\omega'^2_n a^2(1-e^2)}{2\pi i n}\int_0^{2\pi} \sin\xi\cos\xi\, e^{in(\xi - e\sin\xi)}\,d\xi \\
&= \frac{-\mu\omega'^2_n a^2(1-e^2)}{8n\pi}\int_0^{2\pi}\left(e^{2\xi} - e^{-2\xi}\right)d\xi \\
&= \frac{\mu\omega'^2_n a^2(1-e^2)}{4n}[J_{n-2}(ne) - J_{n+2}(ne)] \\
&= \frac{\mu\omega'^2_n a^2(1-e^2)}{n}\left[\frac{1}{e}J'_n(ne) - \frac{1}{ne^2}J_n(ne)\right].\qquad(4.26)
\end{aligned}
$$

The remaining non-zero component of the stress tensor is

$$
\begin{aligned}
T_{xy}(\omega'_n) &= \frac{\mu}{T}\int_0^T \dot{x}(t)\dot{y}(t)e^{i\omega'_n t}\,dt \\
&= \frac{-i\mu\omega'_n}{T}\int_0^T \dot{x}(t)y(t)e^{i\omega'_n t}\,dt \\
&= \frac{i\mu\omega'^2_n a^2\sqrt{1-e^2}}{2\pi n}\int_0^{2\pi} \sin^2\xi\, e^{in(\xi - e\sin\xi)}\,d\xi \\
&= \frac{i\mu\omega'^2_n a^2\sqrt{1-e^2}}{4\pi n}\int_0^{2\pi}(1-\cos 2\xi)\,e^{in(\xi - e\sin\xi)}\,d\xi \\
&= \frac{i\mu\omega'^2_n a^2\sqrt{1-e^2}}{2n}\left[J_n(ne) - \frac{1}{2}(J_{n-2}(ne) + J_{n+2}(ne))\right] \\
&= \frac{i\mu\omega'^2_n a^2\sqrt{1-e^2}}{n}\left[-\frac{(1-e^2)}{e^2}J_n(ne) + \frac{1}{ne}J'_n(ne)\right].
\end{aligned}
$$

$$(4.27)$$

Using (4.25) and (4.26) we see that the trace of the stress tensor $T_{ij}(\omega_n')$ is

$$T^i{}_i(\omega_n') = T_{xx}(\omega_n') + T_{yy}(\omega_n') = \frac{\mu \omega_n'^2 a^2}{n^2} J_n(ne). \tag{4.28}$$

Using (4.25)–(4.28) we compute the tensor-product which occurs in the energy loss expression

$$\left(T_{ij} T_{ji}^* - \frac{1}{3}|T^i{}_i|^2\right) = \left(T_{xx}T_{xx}^* + T_{yy}T_{yy}^* + T_{xy}T_{yx}^* + T_{yx}T_{xy}^* - \frac{1}{3}|T^i{}_i|^2\right)$$

$$= \mu^2 a^4 \Omega^4 n^2 \left\{ J_n(ne)^2 \left[\frac{2}{e^4}(e^2 - 1)^3 + \frac{6 + 2e^4 - 6e^2}{3n^2 e^4} \right] \right.$$

$$+ J_n'(ne)^2 \left[\frac{2(1 - e^2)^2}{e^2} + \frac{2(1 - e^2)}{n^2 e^2} \right]$$

$$\left. + J_n(ne) J_n'(ne) \left[\frac{-8 + 14e^2 - 6e^4}{ne^3} \right] \right\}$$

$$\equiv \mu^2 a^4 \Omega^4 g(n, e), \tag{4.29}$$

where we have defined $g(n, e)$ as

$$g(n, e) = \left\{ J_n(ne)^2 \left[\frac{2n^2}{e^4}(e^2 - 1)^3 + \frac{6 + 2e^4 - 6e^2}{3e^4} \right] \right.$$

$$+ J_n'(ne)^2 \left[\frac{2n^2(1 - e^2)^2}{e^2} + \frac{2(1 - e^2)}{e^2} \right]$$

$$\left. + J_n(ne) J_n'(ne) \left[\frac{(-8 + 14e^2 - 6e^4)n}{e^3} \right] \right\}. \tag{4.30}$$

Thus the energy loss due to gravitational radiation (4.16) becomes

$$\frac{dE_{gw}}{dt} = \frac{\kappa^2}{8(2\pi)^2} \sum_{n=0}^{\infty} \frac{8\pi}{5} \left[T_{ij}(\omega_n') T_{ji}^*(\omega_n') - \frac{1}{3}|T^i{}_i(\omega_n')|^2 \right] \omega_n'^2$$

$$= \frac{32G}{20} \Omega^6 \mu^2 a^4 \sum_{n=0}^{\infty} n^2 g(n, e). \tag{4.31}$$

We can sum the series of the Bessel functions by using the following identities given in the Appendix of [8],

$$\sum_{n=0}^{\infty} n^2 \left[J_n(ne) \right]^2 = \frac{e^2}{4(1 - e^2)^{7/2}} \left(1 + \frac{e^2}{4} \right),$$

$$\sum_{n=0}^{\infty} n^2 \left[J_n'(ne) \right]^2 = \frac{1}{4(1 - e^2)^{5/2}} \left(1 + \frac{3}{4} e^2 \right),$$

$$\sum_{n=0}^{\infty} n^3 J_n(ne) J_n'(ne) = \frac{e}{4(1 - e^2)^{9/2}} \left(1 + 3e^2 + \frac{3}{8} e^4 \right),$$

$$\sum_{n=0}^{\infty} n^4 \left[J_n'(ne) \right]^2 = \frac{1}{4(1 - e^2)^{11/2}} \left(1 + \frac{39}{4} e^2 + \frac{79}{8} e^4 + \frac{45}{64} e^6 \right),$$

$$\sum_{n=0}^{\infty} n^4 \left[J_n(ne) \right]^2 = \frac{e^2}{4(1 - e^2)^{13/2}} \left(1 + \frac{37}{4} e^2 + \frac{59}{8} e^4 + \frac{27}{64} e^6 \right).$$

$$(4.32)$$

Using (4.32) we find that the eccentricity factor

$$f(e) \equiv \sum_{n=1}^{\infty} n^2 g(n, e) = \frac{4}{(1 - e^2)^{7/2}} \left(1 + \frac{73}{24} e^2 + \frac{37}{96} e^4 \right) \qquad (4.33)$$

and the energy radiated as gravitational waves by binaries from (4.31) is given by

$$\frac{dE_{gw}}{dt} = \frac{32G}{5} \Omega^6 \left(\frac{m_1 m_2}{m_1 + m_2} \right)^2 a^4 \frac{1}{(1 - e^2)^{7/2}} \left(1 + \frac{73}{24} e^2 + \frac{37}{96} e^4 \right). \qquad (4.34)$$

This expression matches with the Peters-Mathews calculation [8] which is derived using Einstein's quadrupole formula. From the energy loss formula we can calculate the change in time period ($P_b = 2\pi / \Omega$). From Kepler's law $\Omega^2 a^3 = G(m_1 + m_2)$ we have $\dot{a}/a = (2/3)(\dot{P_b}/P_b)$. The total energy is $E = -Gm_1 m_2 / 2a$ which implies $\dot{a}/a = -(\dot{E}/E)$. Using these two relations we get $\dot{P_b}/P_b = -(3/2)(\dot{E}/E)$ which gives us the expression for the time period loss

$$\frac{dP_b}{dt} = -6\pi G^{-3/2} (m_1 m_2)^{-1} (m_1 + m_2)^{-1/2} a^{5/2} \left(\frac{dE}{dt} \right)$$

$$= -\frac{192\pi}{5} G^{5/3} \Omega^{5/3} \frac{m_1 m_2}{(m_1 + m_2)^{1/3}} \frac{1}{(1 - e^2)^{7/2}} \left(1 + \frac{73}{24} e^2 + \frac{37}{96} e^4 \right),$$

$$(4.35)$$

which is the Peters and Mathews formula [8] for the period loss due gravitational radiation from binary systems.

In a circular orbit the angular velocity of the binary is a constant over the orbital period and the Fourier expansion of the orbit $x(\omega)$ and $y(\omega)$ would be only term, the $\omega = 2\Omega$. In an eccentric orbit the angular velocity is not constant and this means that the Fourier expansion must sum over the harmonics $n\Omega$ of the fundamental. This makes the calculation tedious, but for binary pulsars with large eccentricity the enhancement factor $f(e)$ to the energy loss can be large and must be retained. For the Hulse–Taylor binary where $e = 0.617$ the enhancement factor $f(e) = 12$.

For the Hulse–Taylor binary, the expression (4.35) we have $\Omega = 0.2251 \times 10^{-3}$ sec^{-1}, the pulsar mass $m_1 = 1.4414 \pm 0.0002 M_\odot$, mass of the companion $m_2 = 1.3867 \pm 0.0002 M_\odot$, $e = 0.6171338(4)$ and for the parameters of the H-T binary, yields the energy loss is $dE/dt = 3.2 \times 10^{33}$ erg/sec and orbital period loss due to the gravitational radiation $\dot{P}_b = -2.40263 \pm 0.0005 \times 10^{-12}$ which can be compared with the observed value from the Hulse–Taylor binary [9, 10] (after correcting for the galactic motion) $\dot{P}_b(observed) = -2.40262 \pm 0.00005 \times 10^{-12}$. The cumulative time period loss over a period of 40 years of observations is shown in Fig. 4.1 with the predictions from general relativity. The time-period loss by gravitational emission has been observed in many more neutron star and neuron star -white dwarf binaries [12–14] since the first observation from the Hulse-Taylor binary.

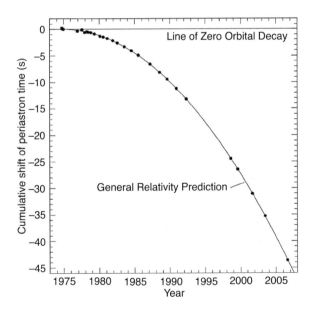

Fig. 4.1 Cumulative period loss over observation time of Hulse–Taylor binary pulsar shows agreement with GR prediction of gravitational wave radiation. Reprinted from [11]. Figure credit: ©The Author(s) 2008. Reproduced under CC-BY-4.0 license

In the Kepler orbit the there are two conserved quantities the total energy $E = -Gm_1m_2/2a$ and the angular momentum which has a magnitude

$$L = \frac{1}{(2E)^{1/2}}G\mu^{3/2}(m_1 + m_2)(1 - e^2)^{1/2}. \tag{4.36}$$

Due to the energy loss by gravitational radiation the magnitude of the angular momentum also decreases. The angular momentum loss associated with the energy loss (4.34) is [15],

$$\frac{dL}{dt} = -\frac{32G}{5}G\Omega^5\mu^2a^4\frac{1}{(1-e^2)}\left(1 + \frac{7}{8}e^2\right) \tag{4.37}$$

The change in angular momentum and energy results in a adiabatic change in time of the orbital parameters, the semi-major

$$\frac{da}{dt} = -\frac{64}{5}G\Omega^4a^3\mu\frac{1}{(1-e^2)^{7/2}}\left(1 + \frac{73}{24}e^2 + \frac{37}{96}e^4\right), \tag{4.38}$$

and eccentricity

$$\frac{de}{dt} = -\frac{304}{15}G\Omega^4a^2\mu\frac{e}{(1-e^2)^{7/2}}\left(1 + \frac{121}{304}e^2\right). \tag{4.39}$$

Due the secular decrease in eccentricity the orbits, by the time of coalescence the binary orbits can be treated as circular.

4.4 Waveforms of Elliptical Orbit Binaries

Using the frequency space computation of the stress tensor components ((4.25), (4.26), (4.27)) for elliptical orbits with eccentricity e,

$$T_{xx}(\omega'_n) = \frac{-\mu a^2\omega'^2_n}{n}\left[\left(\frac{1-e^2}{e}\right)J'_n(ne) - \frac{1}{ne^2}J_n(ne)\right],$$

$$T_{yy}(\omega'_n) = \frac{\mu\omega'^2_na^2(1-e^2)}{n}\left[\frac{1}{e}J'_n(ne) - \frac{1}{ne^2}J_n(ne)\right],$$

$$T_{xy}(\omega'_n) = \frac{i\mu\omega'^2_na^2\sqrt{1-e^2}}{n}\left[-\frac{(1-e^2)}{e^2}J_n(ne) + \frac{1}{ne}J'_n(ne)\right] \tag{4.40}$$

(where $\omega'_n = n\Omega$) we can calculate the waveform-from elliptic binaries follows.

The gravitational wave amplitude of polarisation λ measured by a detector at a distance r is given in the frequency space by

$$h_\lambda(\omega', r) = \frac{4G}{r} \epsilon_\lambda^{ij}(\mathbf{n}) T_{ij}(n, \omega'). \tag{4.41}$$

For the observer in the radial direction $\hat{n} = (\sin\theta\cos\phi, \sin\theta\sin\phi, \cos\theta)$ in the binary coordinates, the amplitudes of the $+$ and $\times$ wave-functions in the frequency space is given by

$$
\begin{aligned}
h_+(\omega', r) &= i\frac{4G}{r}\epsilon_+^{ij}(\mathbf{n}) T_{ij}(\mathbf{n}, \omega') = i\frac{4G}{r}\left(\mathbf{e}_{\theta i}\mathbf{e}_{\theta j} - \mathbf{e}_{\phi i}\mathbf{e}_{\phi j}\right) T_{ij}(k) \\
&= i\frac{4G}{r}\Big(T_{xx}(\cos^2\phi - \sin^2\phi\cos^2\theta) + T_{yy}(\sin^2\phi - \cos^2\phi\cos^2\theta) \\
&\quad - T_{xy}\sin 2\phi(1 + \cos^2\theta)\Big)
\end{aligned}
\tag{4.42}
$$

and the waveform of the $\times$ polarisation gravitational wave is

$$
\begin{aligned}
h_\times(\omega', r) &= i\frac{4G}{r}\epsilon_\times^{ij}(\mathbf{n}) T_{ij}(\mathbf{n}, \omega') = i\frac{4G}{r}\left(\mathbf{e}_{\theta i}\mathbf{e}_{\phi j} + \mathbf{e}_{\phi i}\mathbf{e}_{\theta j}\right) T_{ij}(k) \\
&= i\frac{4G}{r}\Big((T_{xx} - T_{yy})\sin 2\phi\cos\theta + 2T_{xy}\cos 2\phi\cos\theta\Big). \quad \tag{4.43}
\end{aligned}
$$

4.5   Gravitational Radiation from Binaries in the Non-relativistic Effective Theory

The non-relativistic effective theory of gravity (called NRGR) is velocity expansion of the action of linearized gravity with extended objects.

For the binary systems we treat each star as a point mass and the action is the sum of the point particle action (3.53) summed over all the particles. The interaction Lagrangian density is to the leading order in gravitational waves h_{ij} from (3.53) is

$$\mathcal{L}_{int}(\mathbf{x}') = -\frac{\kappa}{2}\sum_{a=1}^{2} m_a h_{ij} v_{ai} v_{aj} \delta^3(\mathbf{x}_a(t) - \mathbf{x}') = -\sum_a \frac{\kappa}{2} T_a^{ij}(x') h_{ij}(x'). \tag{4.44}$$

For a two-body system with masses m_1 and m_2 at positions $\mathbf{x}_1$ and $\mathbf{x}_2$ we transform to the coordinates $\mathbf{R} = (m_1\mathbf{x}_1 + m_2\mathbf{x}_2)/(m_1 + m_2)$ and $\mathbf{r} = \mathbf{x}_1 - \mathbf{x}_2$. The velocities of the two bodies can be written as

$$\mathbf{v}_1 = \mathbf{V} + \frac{m_2\mathbf{v}}{m_1 + m_2} \qquad \mathbf{v}_2 = \mathbf{V} - \frac{m_2\mathbf{r}}{m_1 + m_2}, \tag{4.45}$$

where $V = \dot{R}$ the c.m velocity and $\mathbf{v} = \mathbf{v}_1 - \mathbf{v}_2$ the relative velocity of the two bodies. The two-body interaction terms (4.44) in terms of V and v turn out to be

$$L_{int} = -\frac{\kappa}{2} \left[M V_i V_j + \mu v_i v_j \right] h_{ij}, \tag{4.46}$$

where $M = (m_1 + m_2)$ is the total mass and $\mu = \frac{m_1 m_2}{m_1 + m_2}$ is the reduced mass. By choosing the coordinates system fixed to the centre of mass we $R = V = 0$ and the interaction term reduces to one body interaction. The reduction of the two-body problem is reduced to a one body problem is simple in Newtonian orbits with central potential. However, the two-body orbit can also be reduced to the orbit of one body with reduced mass in full GR was first shown in [16] and the applicability of the effective one body theory in quantum theory is discussed in [17, 18].

The Lagrangian density for the binary star interaction with gravitational waves can therefore be written as

$$\mathcal{L}_{int}(x) = -\frac{\kappa}{2} T^{ij}(x) h_{ij}(x), \tag{4.47}$$

where the stress tensor is for a single body of mass μ and velocity v

$$T_{ij}(\mathbf{x}', t) = \mu v_i v_j \delta^3(\mathbf{x}' - \mathbf{r}(t)), \tag{4.48}$$

where we have introduced the delta function to convert the point particle Lagrangian to a Lagrangian density.

In the covariant calculation of Sect. 4.3 we used the covariant interaction vertex $T^{\mu\nu} h_{\mu\nu}$ and used the conservation of stress tensor $k_\mu T^{\mu\nu} = 0$ to eliminate the redundant components of $h_{\mu\nu}$. In the non-relativistic effective theory we have the non-covariant interaction term $T_{ij} h_{ij}$. To calculate the physical graviton emission we have to restrict the degrees of freedom of h_{ij} from 6 to 2 by imposing the 3 transverse conditions $k_i h^{ij}(\mathbf{k}) = k_j h^{ij}(\mathbf{k}) = 0$ and 1 traceless condition $h_{ii} = 0$. This is done as follows. The transverse projection operator is constructed as $P_{ij} = \delta_{ij} - \hat{k}^i \hat{k}^j$ where $\hat{k}^i$ is the component of the unit vector along the propagation direction $\mathbf{k}$. The transverse projection operator has the properties $P_{ij} = P_{ji}$, $\hat{k}^i P_{ij} = 0$, $P_{ij} P_{jk} = P_{ik}$ and $P_{ii} = 2$. The transverse–traceless projection operator is constructed from P_{ij} as follows

$$\Lambda_{ij,kl}(k) = \left(P_{ik} P_{jl} - \frac{1}{2} P_{ij} P_{kl} \right)$$

$$= \left(\delta_{ik} - \hat{k}_i \hat{k}_k \right) \left(\delta_{jl} - \hat{k}_j \hat{k}_l \right) - \frac{1}{2} \left(\delta_{ij} - \hat{k}_i \hat{k}_j \right) \left(\delta_{kl} - \hat{k}_k \hat{k}_l \right)$$

$$= \delta_{ik} \delta_{jl} - \hat{k}_i \hat{k}_l \delta_{ik} - \hat{k}_i \hat{k}_k \delta_{jl} + \frac{1}{2} \hat{k}_k \hat{k}_l \delta_{ij} + \frac{1}{2} \hat{k}_i \hat{k}_j \delta_{kl} + \frac{1}{2} \hat{k}_i \hat{k}_j \hat{k}_k \hat{k}_l,$$

$$\tag{4.49}$$

where we see that taking the trace of any of the indices gives zero. The transverse–traceless part of the graviton can be obtained by taking the projection

$$h_{ij}^{TT}(\mathbf{k}) = \Lambda_{ij,kl}(k)h_{kl}(\mathbf{k}) \tag{4.50}$$

such that $h_{ii}^{TT} = 0$ and $\hat{k}^i h_{ij}^{TT} = 0$.

The rate of graviton emission from binary stars is given by

$$d\Gamma = \frac{\kappa^2}{4} \sum_{\lambda=1}^{2} |T_{ij}(k')\epsilon_{ij}^{TT}(k,\lambda)|^2 2\pi\delta(\omega-\omega')\frac{d^3k}{(2\pi)^3}\frac{1}{2\omega} \tag{4.51}$$

and the rate of energy loss by gravitational radiation is

$$\frac{dE}{dt} = \frac{\kappa^2}{4} \int \sum_{\lambda=1}^{2} \omega|T_{ij}(k')\epsilon_{ij}^{TT}(k,\lambda)|^2 2\pi\delta(\omega-\omega')\frac{d^3k}{(2\pi)^3}\frac{1}{2\omega}. \tag{4.52}$$

Here the amplitude squared sum is

$$\sum_{\lambda} |T_{ij}(k')\epsilon_{ij}^{TT}(k,\lambda)|^2 = T_{ij}(k')T_{lm}^*(k') \sum_{\lambda} \epsilon_{ij}^{TT}(k,\lambda)\epsilon_{lm}^{*TT}(k,\lambda). \tag{4.53}$$

The polarisation sum of the TT polarisations can be calculated as

$$\sum_{\lambda} \epsilon_{ij}^{TT}(k,\lambda)\epsilon_{lm}^{*TT}(k,\lambda) = \Lambda_{ij,rs}(k)\Lambda_{lm,pq}(k) \sum_{\lambda} \epsilon_{rs}(k,\lambda)\epsilon_{pq}^*(k,\lambda). \tag{4.54}$$

Using expression (3.23) for the polarisation sum

$$\sum_{\lambda} \epsilon_{rs}(k,\lambda)\epsilon_{pq}^*(k,\lambda) = \frac{1}{2}\left(\delta_{rp}\delta_{sq} + \delta_{rq}\delta_{sp} - \delta_{rs}\delta_{pq}\right) \tag{4.55}$$

we obtain

$$\sum_{\lambda} \epsilon_{ij}^{TT}(k,\lambda)\epsilon_{lm}^{*TT}(k,\lambda) = \frac{1}{2}\Big[(\delta_{il}\delta_{jm} + \delta_{im}\delta_{lj} - \delta_{ij}\delta_{lm})$$
$$- (\delta_{il}\hat{k}_j\hat{k}_m + \delta_{im}\hat{k}_j\hat{k}_l - \delta_{ij}\hat{k}_l\hat{k}_m)$$
$$- (\delta_{jm}\hat{k}_i\hat{k}_l + \delta_{jl}\hat{k}_i\hat{k}_m - \delta_{lm}\hat{k}_i\hat{k}_j) + \hat{k}_i\hat{k}_j\hat{k}_l\hat{k}_m\Big] \tag{4.56}$$

Using (4.56) in (4.53) we obtain

$$
\sum_\lambda |T_{ij}(k')\epsilon_{ij}^{TT}(k,\lambda)|^2 = \frac{1}{2}T_{ij}(k')T_{lm}^*(k')\Lambda_{ij,rs}(k)\Lambda_{lm,pq}(k)
$$

$$
\times (\delta_{rp}\delta_{sq} + \delta_{rq}\delta_{sp} - \delta_{rs}\delta_{pq})
$$

$$
= \frac{1}{2}T_{ij}(k')T_{lm}^*(k')\Big[(\delta_{il}\delta_{jm} + \delta_{im}\delta_{lj} - \delta_{ij}\delta_{lm})
$$

$$
-(\delta_{il}\hat{k}_j\hat{k}_m + \delta_{im}\hat{k}_j\hat{k}_l - \delta_{ij}\hat{k}_l\hat{k}_m)
$$

$$
-(\delta_{jm}\hat{k}_i\hat{k}_l + \delta_{jl}\hat{k}_i\hat{k}_m - \delta_{lm}\hat{k}_i\hat{k}_j) + \hat{k}_i\hat{k}_j\hat{k}_l\hat{k}_m\Big]. \quad (4.57)
$$

We do the angular integration in (4.52) using the relations (3.146) to obtain

$$
\int d\Omega_k \sum_\lambda |T_{ij}(k')\epsilon_{ij}^{TT}(k,\lambda)|^2
$$

$$
= T_{ij}(k')T_{lm}^*(k')\frac{8\pi}{5}\left[\frac{1}{2}(\delta_{il}\delta_{jm} + \delta_{im}\delta_{jl}) - \frac{1}{3}\delta_{ij}\delta_{lm}\right]
$$

$$
= \frac{8\pi}{5}\left(T_{ij}(\omega')T_{ji}^*(\omega') - \frac{1}{3}|T^i{}_i(\omega')|^2\right). \quad (4.58)
$$

This expression is the same as what we obtain from the covariant calculation in Eq. (3.147) and from this point on the calculation becomes identical to that of Sect. 4.3 and the result for the energy loss is the same as in (4.31) which is the Peter-Mathews formula.

4.6 Multipole Expansion

The quadrupole formula for gravitational radiation can be improved by taking higher order term in the multipole moments of the position and velocity distribution of the source [19–23]. The main idea is to decompose the source terms into the 'electric' and 'magnetic' components of the Riemann tensor and taking the symmetric and trace-free term in order to calculate the energy rate and waveforms of gravitational wave as a series in the multipole expansion. Each order of multipole is a 1PN correction over the previous term in the energy loss rate and a 0.5 PN correction in the waveform. In this section we will outline the formalism of the multipole method following and calculate the higher order terms for binary orbits.

We start with the graviton–matter interaction vertex

$$
S_{int} = -\frac{\kappa}{2}\int dt d^3x\, T^{\mu\nu}(t,\mathbf{x})\, h_{\mu\nu}(t,\mathbf{x}). \quad (4.59)
$$

To find the multipole source moments do a Taylor expansion of the field near $\mathbf{x} = 0$,

$$
\begin{aligned}
h_{\mu\nu}(t, \mathbf{x}) &= \sum_{n=0}^{\infty} \frac{1}{n!} x^{i_1} \cdots x^{i_n} \, \partial_{i_1} \cdots \partial_{i_n} h_{\mu\nu}(t, \mathbf{x}) \Big|_{t=t_r, \mathbf{x}=0} \\
&\equiv \sum_{n=0}^{\infty} \frac{1}{n!} x^N \, \partial_N h_{\mu\nu}(t, \mathbf{x}) \Big|_{t=t_R, \mathbf{x}=0},
\end{aligned}
\tag{4.60}
$$

where we used the notation [24] that upper case letters denote a series of indices, for example, $x^N = x^{i_1} \cdots x^{i_n}$ or $x^N = x^i x^j x^{r_1} \cdots x^{r_{n-2}}$.

In momentum space the interaction vertex will appear as

$$
\Gamma = i \frac{\kappa}{2} T^{\mu\nu}(k) \epsilon_{\mu\nu}^{\lambda}(k),
\tag{4.61}
$$

where $\epsilon_{\mu\nu}^{\lambda}(k)$ is the polarisation tensor for gravitons with polarisation $\lambda = +, \times$. Introducing the Fourier transform in position space of $\mathbf{x}$ of $T^{\mu\nu}(t, \mathbf{x})$,

$$
T(t, \mathbf{k}) = \int d^3x \, e^{i\mathbf{x}\cdot\mathbf{k}} T(t, \mathbf{x})
\tag{4.62}
$$

and expanding the exponential in a series we have $T^{\mu\nu}(t, \mathbf{k})$ with the multipole moment form-factor

$$
T^{\mu\nu}(t, \mathbf{k}) = \sum_{l=0}^{\infty} \frac{(i^l)}{l!} \left(\int d^3x \, T^{\mu\nu}(t, \mathbf{x}) \, x^{i_1} \cdots x^{i_l} \right) (k_{i_1} \cdots k_{i_l}).
\tag{4.63}
$$

Each order l of the multipole expansion represents a factor $R\omega$ where R is the size of the system and $\omega = |\mathbf{k}|$ the angular frequency of the gravitational waves.

To compute the radiation part of the graviton in the amplitude (4.61) we will take only the $\epsilon_{ij}(k)^{\lambda}$ terms. The probability of a graviton emission in time T is given by

$$
d\Gamma = \sum_{\lambda} \frac{1}{T} \int \frac{d^3k}{(2\pi)^3 2|\mathbf{k}|} \left| T^{ij}(k) \, \epsilon_{ij}^{\lambda *}(k) \right|^2.
\tag{4.64}
$$

In the source stress tensor multipole expansion (4.63), $T^{ij}(k)$ is a second rank tensor of the rotation group $O(3)$. The only invariants of the rotation group are δ_{ij} and ϵ_{ijk}. We can therefore decompose T^{ij} into sum of irreducible representations as

$$
T^{ij}(k) = \frac{1}{3} S_i^i + \frac{1}{2} \epsilon^{ijk} A_k + \left(\delta_{ij} - \frac{1}{3} S_i^i \right),
\tag{4.65}
$$

where we have written T^{ij} as a sum of a trace, an antisymmetric tensor and a symmetric traceless tensor. Here $S_{ij} = \frac{1}{2}T_{(ij)}$ is the symmetric part, $A_{ij} = \frac{1}{2}T_{[ij]}$ the antisymmetric part of the stress tensor and $A_i = \frac{1}{2}\epsilon_{ijk}A_{jk}$.

The symmetric part of the multipole expansion generates the 'electric' part of the Riemann tensor E_{ij} and the ant-symmetric terms generate the 'magnetic part B_{ij}. We can write the terms $\partial_N h_{\mu\nu}(t, \mathbf{x})$ as a sum of the 'electric' and 'magnetic' of the Weyl tensor $C_{\mu\nu\alpha\beta}$,

$$E_{\mu\nu} = C_{\mu\nu\alpha\beta}u^\alpha u^\beta , \quad B_{\mu\nu} = \frac{1}{2}\varepsilon_{\mu\rho\sigma\alpha}C^{\rho\sigma}{}_{\nu\beta}u^\alpha u^\beta . \tag{4.66}$$

In vacuum where $R_{\mu\nu} = 0$ the Weyl tensor is equal to the Riemann tensor. We choose the reference which is the rest frame of the source centre of mass for doing these calculations. In that case $u^\alpha = (1, 0, 0, 0)$ and the non-zero components of the electric and magnetic Weyl tensor are

$$E_{ij} = R_{0i0j} = \frac{1}{2}\left(\partial_0\partial_i h_{0j} + \partial_0\partial_j h_{0i} - \partial_i\partial_j h_{00} - \partial_0^2 h_{ij}\right) ,$$

$$B_{ij} = \frac{1}{2}\varepsilon_{ilr}R_{0j}{}^{lr} = \frac{1}{2}\varepsilon_{ilr}\left(\partial_0\partial^l h_j{}^r + \partial_j\partial^r h_0{}^l\right). \tag{4.67}$$

The electric and magnetic parts of the Riemann tensor are symmetric $E_{ij} = E_{ji}, B_{ij} = B_{ji}$, and traceless $E_{ii} = 0$, $B_{ii} = 0$. The electric and magnetic components of the Riemann tensor obey the e.o.m is vacuum

$$\partial_i E_{ij} = 0 , \quad \partial_i B_{ij} = 0 , \quad \epsilon_{imn}\partial_m E_{jn} = \dot{B}_{ij} , \quad \epsilon_{imn}\partial_m B_{jn} = -\dot{E}_{ij}. \tag{4.68}$$

Combining these we see that they satisfy the wave equations $\Box E_{ij} = 0$ and $\Box B_{ij} = 0$.

The amplitude for graviton emission as a sum of multipoles l can be written in general as [21],

$$A_\lambda(k) = \frac{i\kappa}{2}\sum_{l=2}^{\infty}\left\{\left(\frac{(-i)^{l-2}}{l!}\right)|\mathbf{k}|^2\ I^L(|\mathbf{k}|)\ k_{L-2}\ \epsilon^{*\lambda}_{i_{l-1}i_l}(k)\right.$$

$$\left.-\left(\frac{2l}{(l+1)!}\right)|\mathbf{k}|\ J^L(|\mathbf{k}|)\ \epsilon_{i_{l-1}mn}\ k_{L-2}\ k^n\ \epsilon^{*\lambda m}{}_{i_l}(k)\right\}.$$

$$\tag{4.69}$$

The general expressions for the mass distribution $I^L(k)$ and current distribution $J^L(k)$ in the position space can be found in [21]. Here we will derive these for the first few multipoles for the binary system.

In terms of the amplitude the power radiated is

$$P = \int |\mathbf{k}| d\Gamma = \sum_\lambda \frac{1}{T} \int \frac{d^3 k}{(2\pi)^3 2|\mathbf{k}|} \, |\mathbf{k}| \, |A_\lambda(k)|^2. \tag{4.70}$$

In the amplitude squared the electric and magnetic amplitudes do not have any cross terms. The polarisation sum is done using (4.56).

$$\sum_\lambda \epsilon_{ij}^{TT}(\hat{k}, \lambda) \epsilon_{lm}^{*TT}(\hat{k}, \lambda) = \frac{1}{2} \Big[(\delta_{il}\delta_{jm} + \delta_{im}\delta_{lj} - \delta_{ij}\delta_{lm})$$

$$- (\delta_{il}\hat{k}_j\hat{k}_m + \delta_{im}\hat{k}_j\hat{k}_l - \delta_{ij}\hat{k}_l\hat{k}_m)$$

$$- (\delta_{jm}\hat{k}_i\hat{k}_l + \delta_{jl}\hat{k}_i\hat{k}_m - \delta_{lm}\hat{k}_i k_j) + \hat{k}_i\hat{k}_j\hat{k}_l\hat{k}_m \Big]. \tag{4.71}$$

The angular integrals of the $k^i k^j \dots$ can be performed using

$$\int d\Omega \, \hat{k}_{i_1} \dots \hat{k}_{i_l} = \frac{4\pi}{(l+1)!!} \left(\delta_{i_1 i_1} \dots \delta_{i_{l-1} i_l} + \text{symmetric permutations} \right). \tag{4.72}$$

The angular integral of the product of odd number of $\hat{k}_i$'s is zero. Using these results, the expression for the power radiated in gravitational waves (with $\omega = |\mathbf{k}|$) is given by Ross [21]

$$P = \frac{G}{\pi T} \int d\omega \sum_{l=2}^{\infty} \left\{ \left(\frac{(l+1)(l+2)}{l(l-1) \, l!(2l+1)!!} \omega^{2(l+1)} |I^L(\omega)|^2 \right) \right.$$

$$\left. + \left(\frac{4l(l+2)}{(l-1)(l+1) \, l!(2l+1)!!} \omega^{2(l+1)} |J^L(\omega)|^2 \right) \right\}$$

$$= \frac{G}{\pi T} \int d\omega \left\{ \frac{\omega^6}{5} |I^{ij}(\omega)|^2 + \frac{16\omega^6}{45} |J^{ij}(\omega)|^2 \right.$$

$$\left. + \frac{\omega^8}{189} |I^{ijk}(\omega)|^2 + \frac{\omega^8}{84} |J^{ijk}(\omega)|^2 + \cdots \right\}. \tag{4.73}$$

The waveforms at each multipole order are

$$h_{ij}(|\mathbf{x}|) = \frac{4G}{|\mathbf{x}|} \sum_{l=2}^{\infty} \Lambda_{ij,k_{l-1}k_l} \left\{ \left(\frac{1}{l!} \partial_t^l I^L(t_{\text{ret}}) n^{L-2} \right) \right.$$

$$\left. - \left(\frac{2l}{(l+1)!} \epsilon^{pq(k_l} \partial_t^l J^{k_{l-1})q L-2}(t_{\text{ret}}) n^{aL-2} \right) \right\}, \tag{4.74}$$

where $n^i = x^i/|\mathbf{x}|$ is the source observer direction, $t_{\text{ret}} = t - |\mathbf{x}|$ is the retarded time and $\Lambda_{ij,k_{l-1}k_l}$ is the transverse–traceless projection operator for each multipole l given by

$$\Lambda_{ij,k_{l-1}k_l} = \left(\delta_{ik_{l-1}} - n_i n_{k_{l-1}}\right)\left(\delta_{jk_l} - n_j n_{k_l}\right) - \frac{1}{2}\left(\delta_{ij} - n_i n_j\right)\left(\delta_{k_{l-1}k_l} - n_{k_{l-1}} n_{k_l}\right).$$

(4.75)

4.7   Radiation of Ultralight-Scalars from Binary Stars

In this section for the energy and period loss of binary orbits by emission of ultralight scalars ($m < \Omega \sim 10^{-19}$eV) following [5, 25]. Consider a coupling between scalar fields ϕ and the baryons of the form

$$\mathcal{L}_s = g_s \, \phi \, \bar{\psi} \, \psi$$

(4.76)

which for non-relativistic fermions can be written as

$$\mathcal{L}_s = g_s \, \phi \, n(x),$$

(4.77)

where $n(x) = \psi^\dagger \psi$ is the number density of the non-relativistic fermions which couple to ϕ. In the case of neutron stars we will take $n(x)$ to be the number density of nucleons. The number density $n(x)$ for the binary stars (denoted by $a = 1, 2$) may be written as

$$n(x) = \sum_{a=1,2} N_a \, \delta^3(\mathbf{x} - \mathbf{x}_a(t)),$$

(4.78)

where $N_a \sim 10^{57}$ is the total number of nucleons in the neutron star and $\mathbf{x}_a(t)$ represents the Keplerian orbit of the binary stars. For the coupling (4.77) and the source (4.78) the rate of scalar particles emitted from the neutron star in orbit with frequency Ω is

$$d\Gamma = g_s^2 |n(\omega')|^2 (2\pi) \, \delta(\omega - \omega') \frac{d^3\omega}{(2\pi)^3 \, 2\omega}$$

(4.79)

the rate of energy loss by massless scalar radiation is

$$\frac{dE_s}{dt} = \int g_s^2 |n(\omega')|^2 \, \omega \, (2\pi) \, \delta(\omega - \omega') \frac{d^3\omega}{(2\pi)^3 \, 2\omega},$$

(4.80)

where $n(\omega)$ is the Fourier expansion of the source number density (4.78)

$$
n(\omega') = \frac{1}{T} \int_0^T dt \int d^3\mathbf{x}\, e^{-i\mathbf{k'}\cdot\mathbf{x}}\, e^{i\omega't} \sum_{a=1,2} N_a \delta^3(\mathbf{x} - \mathbf{x}_a(t))
$$

$$
= \sum_{a=1,2} N_a \frac{1}{T} \int_0^T dt\, e^{-i\mathbf{k'}\cdot\mathbf{x}_a}\, e^{i\omega't}, \tag{4.81}
$$

where $\omega' = n\Omega = 2\pi/T$ and we sum over the harmonics n in the end for obtaining the total energy radiated. For non-relativistic motion $k x_a \ll \omega't$ as $v_a \ll 1$, and we can expand $e^{i\mathbf{k'}\cdot\mathbf{x}_a}$ in a series and keep terms up to linear order in $\mathbf{k}\cdot\mathbf{x}_a$ to obtain

$$
n(\omega') = \sum_{a=1,2} N_a \frac{1}{T} \int_0^T \left(1 - i\mathbf{k'}\cdot\mathbf{x}_a + O(\mathbf{k'}\cdot\mathbf{x}_a)^2\right) e^{i\omega't}\, dt
$$

$$
= \frac{N_1}{T} \int_0^T \left(1 - i\mathbf{k'}\cdot\mathbf{x}_1\right) e^{i\omega't}\, dt + \frac{N_2}{T} \int_0^T \left(1 - i\mathbf{k'}\cdot\mathbf{x}_2\right) e^{i\omega't}\, dt. \tag{4.82}
$$

In the c.o.m coordinates we have $x_1^i = \frac{m_2 x^i}{m_1+m_2} = \frac{\mu}{m_1}x^i$ and $x_2^i = -\frac{m_1 x^i}{m_1+m_2} = -\frac{\mu}{m_2}x^i$. Therefore

$$
n(\omega') = \left(\frac{N_1}{m_1} - \frac{N_2}{m_2}\right) \mu \left(-ik'_x\, x(\omega') - ik'_y\, y(\omega')\right) + O(\mathbf{k'}\cdot\mathbf{x}_a)^2, \tag{4.83}
$$

where $(x(\omega'), y(\omega'))$ are the Fourier components of the Kepler orbit of the reduced mass μ in the c.m. frame given by (4.17). We now need to calculate the $x(\omega')$ and $y(\omega')$ for the Kepler orbit (4.17). Using $\omega' = n\Omega$, we can write

$$
x(\omega') = \frac{1}{T} \int_0^T dt\, e^{in\Omega t} x(t). \tag{4.84}
$$

Integrating by parts we get

$$
x(\omega') = \frac{1}{in\Omega T} \int_0^T dt\, e^{in\Omega t} \dot{x}(t). \tag{4.85}
$$

Using the orbit equations we write $dt\,dx/dt = d\xi\,dx/d\xi = -a\sin\xi\,d\xi$ and we obtain

$$
\begin{aligned}
x(\omega') &= \frac{-a}{in2\pi} \int_0^{2\pi} d\xi\, e^{in(\xi - e\sin\xi)} \sin\xi \\
&= \frac{a}{2n} (J_{n+1}(ne) - J_{n+1}(ne)) \\
&= \frac{a}{n} J_n'(ne),
\end{aligned}
\tag{4.86}
$$

where we used the integral representation of the Bessel function (4.23) and used the recurrence relations (4.24).

Similarly,

$$
\begin{aligned}
y(\omega') &= \frac{1}{T} \int_0^T dt\, e^{in\Omega t} y(t), \\
&= \frac{1}{in\Omega T} \int_0^T dt\, e^{in\Omega t} \dot{y}(t), \\
&= \frac{a\sqrt{1-e^2}}{in2\pi} \int_0^{2\pi} d\xi\, e^{in(\xi - e\sin\xi)} \cos\xi, \\
&= \frac{-i\,a\,\sqrt{1-e^2}}{ne} J_n(ne).
\end{aligned}
\tag{4.87}
$$

Substituting

$$
x(\omega') = \frac{a}{n} J_n'(ne), \qquad y(\omega') = \frac{-i\,a\,\sqrt{1-e^2}}{ne} J_n(ne)
\tag{4.88}
$$

in (4.83) we obtain the expression for $|n(\omega')|^2$ given by

$$
|n(\omega')|^2 = \frac{1}{12} \left(\frac{N_1}{m_1} - \frac{N_2}{m_2} \right)^2 \mu^2 a^2 \Omega^2 \, [J_n'^2(ne) + \frac{(1-e^2)}{e^2} J_n^2(ne)],
\tag{4.89}
$$

where we have used the angular average $\langle k_x'^2 \rangle = \langle k_y'^2 \rangle = \frac{1}{3}(n\Omega)^2$. Substituting (4.89) in (4.80) we have the rate of energy loss by massless scalars

$$
\frac{dE_s}{dt} = \frac{1}{6\pi} \left(\frac{N_1}{m_1} - \frac{N_2}{m_2} \right)^2 \mu^2 \, g_s^2 a^2 \Omega^4 \sum_{n=1}^{\infty} n^2 \left[J_n'^2(ne) + \frac{(1-e^2)}{e^2} J_n^2(ne) \right].
\tag{4.90}
$$

The mode sum can be carried out using the Bessel function series formulas (4.32) to obtain

$$\sum_{n=1}^{\infty} n^2 \left[J'^2_n(ne) + (1 - e^2)e^{-2} J^2_n(ne) \right] = \frac{1}{4}(2 + e^2)(1 - e^2)^{-5/2}. \qquad (4.91)$$

The energy loss (4.90) in terms of the orbital parameters Ω, a and e is given b

$$\frac{dE_s}{dt} = \frac{1}{24\pi} \left(\frac{N_1}{m_1} - \frac{N_2}{m_2} \right)^2 \mu^2 \, g_s^2 \, \Omega^4 \, a^2 \, \frac{(1 + e^2/2)}{(1 - e^2)^{5/2}}. \qquad (4.92)$$

Since $N_a \, m_n = m_a - \epsilon_a$ where $\epsilon_a = \frac{Gm_a^2}{R_a}$ is the gravitational binding energy and m_n the neutron mass, the factor $\left(\frac{N_1}{m_1} - \frac{N_2}{m_2} \right) = G\left(\frac{m_1}{R_1} - \frac{m_2}{R_2} \right)$. For the H-T binary $m_1 - m_2 \simeq 0.02 M_\odot$ and $R_a \sim 10$ km, therefore $\left(\frac{N_1}{m_1} - \frac{N_2}{m_2} \right) \simeq 3 \times 10^{-3}$ GeV^{-1}.

The rate of change of the orbital period due to energy loss of scalar bosons as well as gravitons is

$$\frac{dP_b}{dt} = -6\pi G^{-3/2}(m_1 m_2)^{-1}(m_1 + m_2)^{-1/2} a^{5/2} \left(\frac{dE_s}{dt} + \frac{dE_{GW}}{dt} \right), \qquad (4.93)$$

where $\frac{dE_{GW}}{dt}$ is the rate of energy loss due to quadrupole formula for the gravitational radiation and is given in Eq (4.31).

For the H-T binary the rate of energy loss turns out to be

$$\frac{dE_s}{dt} = g_s^2 \times 9.62 \times 10^{67} \text{ ergs/sec}. \qquad (4.94)$$

Assuming that this is less than 1% of the gravitational energy loss, i.e. $dE/dt \leq 10^{31}$ ergs/sec, we obtain gives an upper bound on scalar nucleon coupling $g_s < 3 \times 10^{-19}$. This bound holds for models where the mass of the scalar boson is smaller than the frequency $\Omega \simeq 10^{-19}$ eV of the binary orbit.

The period loss in the H-T system has been determined by measuring the time of periastron over many decades [10]. The accuracy of the measured value of period loss increases quadratically with time. If in the course of observation one finds a significant discrepancy between the observed value of period loss and the prediction of the gravitational quadrupole formula, it would be a compelling signal of physics beyond standard model.

Ultralight scalar fields are candidates for being dark matter [26, 27]. Axion-like particles with coupling $f \simeq 10^{16} - 10^{18}$ GeV and mass in the range $m \sim 10^{-22} - 10^{-21}$ eV can solve the core-cusp problem of galactic dark matter and at the same time provide the closure density of the universe to be the dark matter at cosmological scales. Ultralight scalars can be axion-like Nambu–Goldstone bosons. These couple to the pseudo-scalar bilinear $\frac{\phi}{f} \bar{\psi} \gamma_5 \psi$ which gives a spin-dependent long range force [28, 29]. However, in case of a CP violation Nambu–Goldstone bosons can also couple as scalars [30] and give rise to long ranged forces in unpolarised macroscopic bodies. When the ultralight dark matter outside macroscopic bodies is an oscillating axionic field there is again a long range axionic field coupled to macroscopic bodies [31, 32].

4.8   Vector Boson Radiation by Neutron Star Binaries

Ultralight vector bosons can also be radiated from neutron star binaries if the gauge boson mass is lower than the angular frequency of the binaries. Gauge bosons which couple to baryon number can be constrained by fifth force experiments. It is more difficult to constrain vector bosons which couple only to lepton number. In the standard model the any one of the combination of leptonic charges $L_e - L_\mu$, $L_e - L_\tau$ and $L_\mu - L_\tau$ can be gauged and is anomaly free [33, 34]. Of these the $L_e - L_\mu$ and $L_e - L_\tau$ long range gauge bosons which are generated by astrophysical bodies like the Earth and the Sun can be probed by neutrino oscillations experiment [35]. In case the gauge bosons couple to the $L_\mu - L_\tau$ quantum number, they can couple to the muon charge of neutron stars and if their mass $M_{Z'} < \Omega$ the angular frequency of binary neutron stars, they can be radiated over time and result in a loss of the time period of the binary stars and hence can be probed in the binary pulsar timings. In this section we derive the formula for energy loss of binary neutron stars due to emission of ultralight vector bosons [36].

The interaction between the ultralight gauge boson Z' and fermions is described by the Lagrangian

$$\mathcal{L}_I = g J^\mu Z'_\mu + \frac{M_{Z'}^2}{2} Z'_\mu Z'^\mu. \tag{4.95}$$

The Z' boson is a U(1) gauge boson and gauge invariance demands that the current is conserved $k_\mu J^\mu = 0$. The mass term arises from a spontaneous symmetry breaking.

The rate of energy loss by vector boson emission by a classical current is

$$\frac{dE_V}{dt} = g^2 \sum_{\lambda=1}^{3} [J^\mu(k')J^{\nu*}(k')\epsilon_\mu^\lambda(k)\epsilon_\nu^{\lambda*}(k)] \, 2\pi \delta(\omega - \omega') \, \omega \, \frac{d^3k}{(2\pi)^3 2\omega}, \tag{4.96}$$

where $J^\mu(k')$ is the Fourier transform of $J^\mu(x)$ and $\epsilon_\mu^\lambda(k)$ is the polarisation vector of massive vector boson. The polarisation sum is given as,

$$\sum_{\lambda=1}^{3} \epsilon_\mu^\lambda(k)\epsilon_\nu^{\lambda*}(k) = -g_{\mu\nu} + \frac{k_\mu k_\nu}{M_{Z'}^2}. \tag{4.97}$$

If Z' boson is a gauge field then from gauge invariance $k_\mu J^\mu = 0$ and the second term in the polarisation sum Eq (4.97) will not contribute to the energy loss formula. The expression for the energy loss reduces to

$$\frac{dE_V}{dt} = \frac{g^2}{2(2\pi)^2} \int |-(J^0(\omega'))^2 + (J^i(\omega'))^2| \, \delta(\omega - \omega')\omega \left(1 - \frac{M_{Z'}^2}{\omega^2}\right)^{\frac{1}{2}} d\omega d\Omega_k.$$

(4.98)

The current density for the binary stars denoted by $a = 1, 2$ may be written as

$$J^\mu(x) = \sum_{a=1,2} Q_a \delta^3(\mathbf{x} - \mathbf{x}_a(t))u_a^\mu,$$

(4.99)

where Q_a is the total charge of the neutron star due to muons and $\mathbf{x}_a(t)$ denotes the Kepler orbit of the binaries. $u_a^\mu = (1, \dot{x}_a, \dot{y}_a, 0)$ is the non-relativistic four-velocity in the x-y plane of the neutron stars. The Fourier transform of Eq (4.99) for the spatial part with $\omega' = n\Omega$ is

$$J^i(\omega') = \sum_{a=1,2} \int \frac{1}{T} \int_0^T dt \, e^{in\Omega t} \dot{x}_a^i(t) Q_a d^3\mathbf{x}' e^{-i\mathbf{k}'\cdot\mathbf{x}'} \delta^3(\mathbf{x}' - \mathbf{x}_a(t))$$

$$= \sum_{a=1,2} \frac{1}{T} \int_0^T dt \, e^{in\Omega t} \dot{x}_a^i(t) Q_a e^{-i\mathbf{k}'\cdot\mathbf{x}_a}.$$

(4.100)

We expand the $e^{i\mathbf{k}'\cdot\mathbf{x}_a} = 1 + i\mathbf{k}'\cdot\mathbf{x}_a + \dots$ and retained the leading order term as $\mathbf{k}'.\mathbf{x}_a \sim \Omega a \ll \Omega t$ for non-relativistic velocities. Hence Eq. (4.100) becomes

$$J^i(\omega') = \frac{Q_1}{T} \int_0^T dt \, e^{in\Omega t} \dot{x}_1^i(t) + \frac{Q_2}{T} \int_0^T dt \, e^{in\Omega t} \dot{x}_2^i(t).$$

(4.101)

In the c.o.m coordinates we have $x_1^i = \frac{m_2 x^i}{m_1+m_2} = \frac{\mu}{m_1}x^i$ and $x_2^i = -\frac{m_1 x^i}{m_1+m_2} = -\frac{\mu}{m_2}x^i$. $\mu = m_1 m_2/(m_1 + m_2)$ is the reduced mass of the compact binary system. Hence we rewrite the current density as

$$J^i(\omega') = \frac{1}{T}\left(\frac{Q_1}{m_1} - \frac{Q_2}{m_2}\right)\mu \int_0^T dt \, e^{in\Omega t} \dot{x}^i(t).$$

(4.102)

For the Kepler orbit using (4.17) in (4.102) we obtain

$$J^x(\omega') = \Omega\left(\frac{Q_1}{m_1} - \frac{Q_2}{m_2}\right)\mu \frac{1}{2\pi} \int_0^T dt \, e^{in\Omega t} \dot{x}^i(t)$$

$$= -ia\Omega\left(\frac{Q_1}{m_1} - \frac{Q_2}{m_2}\right)\mu J_n'(ne)$$

(4.103)

and

$$J^y(\omega') = \Omega\left(\frac{Q_1}{m_1} - \frac{Q_2}{m_2}\right)\mu\frac{a\sqrt{1-e^2}}{e}J_n(ne). \tag{4.104}$$

Therefore

$$J^i J_i^* = J^x J^{x*} + J^y J^{y*} = a^2\Omega^2\mu^2\left(\frac{Q_1}{m_1} - \frac{Q_2}{m_2}\right)^2\left[J_n'^2(ne) + \frac{(1-e^2)}{e^2}J_n^2(ne)\right]. \tag{4.105}$$

We can solve for J^0 in terms of J^i using the current conservation $k_\mu J^\mu(k) = 0$ which gives us

$$(J^0)^2 = \hat{k}_i\hat{k}_j J^i J^j. \tag{4.106}$$

Integrating $\hat{k}_i\hat{k}_j$ over the angles $d\Omega_k$ gives us

$$\int d\Omega_k\hat{k}_i\hat{k}_j = \frac{4\pi}{3}\delta_{ij}. \tag{4.107}$$

From which we obtain

$$\int d\Omega_k(J^0)^2 = \frac{4\pi}{3}(J^i)^2. \tag{4.108}$$

Using Eqs. (4.105) and (4.108) in Eq. (4.98) we obtain the rate of energy loss

$$\frac{dE_V}{dt} = \frac{g^2}{3\pi}a^2\mu^2\left(\frac{Q_1}{m_1} - \frac{Q_2}{m_2}\right)^2\Omega^4$$

$$\times \sum_{n=1}^{\infty}n^2\left[J_n'^2(ne) + \frac{(1-e^2)}{e^2}J_n^2(ne)\right]\left(1 - \frac{n_0^2}{n^2}\right)^{\frac{1}{2}}\left(1 + \frac{1}{2}\frac{n_0^2}{n^2}\right), \tag{4.109}$$

where $n_0 = (M_{Z'}/\Omega)$. In the limit $M_{Z'} \ll \Omega$ energy loss from Eq. (4.109) can be written as a closed form expression

$$\frac{dE_V}{dt} = \frac{g^2}{6\pi}a^2\mu^2\left(\frac{Q_1}{m_1} - \frac{Q_2}{m_2}\right)^2\Omega^4\frac{(1 + \frac{e^2}{2})}{(1-e^2)^{\frac{5}{2}}}. \tag{4.110}$$

The rate of change of the orbital period due to energy loss of vector bosons is

$$\frac{dP_b}{dt} = -6\pi G^{-3/2}(m_1m_2)^{-1}(m_1+m_2)^{-1/2}a^{5/2}\left(\frac{dE_V}{dt} + \frac{dE_{GW}}{dt}\right), \tag{4.111}$$

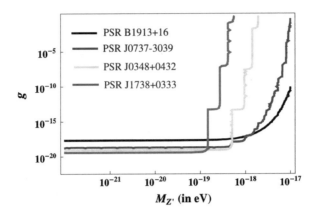

Fig. 4.2 Exclusion plot of gauge coupling g of $L_\mu - L_\tau$ gauge boson as a function of mass $M_{Z'}$ from different binary timing observations. Reprinted from [36]. Figure credit: ©The Author(s) 2021. Reproduced under CC-BY-4.0 license

where $\frac{dE_{GW}}{dt}$ is the rate of energy loss due to quadrupole formula for the gravitational radiation and is given in Eq. (4.31).

The bounds on gauge coupling g of $L_\mu - L_\tau$ gauge boson as a function of mass $M_{Z'}$ from different binary timing observations [10, 12–14] are shown in Fig. 4.2.

4.9 Gravitational Waveforms from Coalescing Binaries

Elliptical binaries lose their eccentricity by gravitational radiation and by the time they coalesce the orbit is nearly circular. The velocity increases, however, and post-Newtonian (powers of v/c) corrections are important for computing the accurate waveforms. The relation between the metric perturbation at the detector at time t and distance r from the source and the stress-tensor at an earlier time $t - r$, is derived in (3.123) for a binary orbit in the x-y plane with the observer along the z-axis ($\mathbf{n} = (0, 0, 1)$, and is given by,

$$h_{ij}^{TT}(\mathbf{x}, t) = -\frac{4G}{r}\Lambda_{ij,kl}(\mathbf{n})\int \frac{d\omega'}{2\pi} T_{kl}(\omega', \mathbf{n}\omega')e^{-i\omega'(t-r)} \qquad (4.112)$$

The stress tensor components for binaries in a circular orbit in the Newtonian approximation are derived in (4.10),

$$T_{xx}(\omega') = \frac{\mu a^2 \Omega^2}{2} 2\pi \delta(\omega' - 2\Omega)\,,$$

$$T_{yy}(\omega') = -\frac{\mu a^2 \Omega^2}{2} 2\pi \delta(\omega' - 2\Omega)\,,$$

$$T_{xy}(\omega') = -i\frac{\mu a^2 \Omega^2}{2} 2\pi \delta(\omega' - 2\Omega)\,. \qquad (4.113)$$

Using (4.113) in (4.112) we obtain the waveform of gravitational waves from binaries in a circular orbit

$$h_{xx}(t, r) = -h_{yy}(t, r) = -\frac{4G}{r}\frac{\mu a^2 \Omega^2}{2}\, \mathrm{Re}\left(e^{-i(2\Omega t - \phi_0)}\right)$$

$$= -\frac{2G}{r}\mu a^2 \Omega^2 \cos(2\Omega t - \phi_0)$$

$$h_{xy}(t, r) = -\frac{4G}{r}\frac{\mu a^2 \Omega^2}{2}\, \mathrm{Re}\left(i e^{-i(2\Omega t - \phi_0)}\right)$$

$$= -\frac{2G}{r}\mu a^2 \Omega^2 \sin(2\Omega t - \phi_0). \tag{4.114}$$

Since the stress tensor matrix is already (4.113) is already transverse–traceless the TT projection operator $\Lambda_{ij,kl}(\mathbf{n})$ in (4.112) acts as identity operator.

4.10 Gravitational Waves from Hyperbolic Orbits

At the centre of our Milky Way galaxy there is a supermassive blackhole of mass $4.15 \times 10^6 M_\odot$ called Sagittarius A*. Many large blackholes are expected to be going around this supermassive blackhole in hyperbolic orbits. These hyperbolic encounters are estimated to produce gravitational waves which may be observable at AdvLIGO and LISA [37–43].

In this section we calculate the gravitational energy radiated by binary blackholes in hyperbolic orbits. The energy radiated is then used to calculate the waveform of gravitational waves in these encounters. What is interesting in the hyperbolic case is that there is non-zero energy radiated even with zero frequency gravitational waves. There is thus a non-oscillatory component in the gravitational waves which causes a permanent separation of the mirrors at the arms of the detector. This effect is called the memory effect and will be studied in detail in Chap. 5.

Consider a blackhole of mass m_2 in a hyperbolic orbit around a larger blackhole of mass m_1 as shown in Fig. 4.3. In the centre of mass frame of the two blackholes this motion can be described by a single body with reduced mass $\mu = m_1 m_2/(m_1 + m_2)$ around the c.m as the origin. Choosing the plane of the orbit as the x-y plane, the hyperbolic orbit in Cartesian coordinates of the reduced mass can be described in the parametric form

$$x(\xi) = a(e - \cosh\xi), \quad y(\xi) = b \sinh\xi, \quad \Omega t = (e \sinh\xi - \xi) \equiv \frac{\omega'}{\nu}t \tag{4.115}$$

with $b = a\sqrt{e^2 - 1}$, $\Omega = (G(m_1 + m_2)/a^3)^{1/2}$ and $\xi \in (-\infty, \infty)$. The angular frequency $\omega' \equiv \nu\Omega$ where for the unbounded orbit ν is a real number, $\nu \in (0, \infty)$. Here a is the magnitude of the semi-major axis and $e > 1$ is the eccentricity of

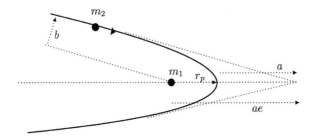

Fig. 4.3 Hyperbolic orbit of a blackhole of mass m_2 around a supermassive blackhole of mass m_1

the hyperbolic orbit. The initial conditions are the asymptotic initial velocity v_0 and the impact parameter b. These two initial conditions determine the two constants of motion, the energy $E = (1/2)\mu v_0^2$ and angular momentum $L = \mu b v_0$. The orbital parameters a and e depend upon the initial conditions as $a = G(m_1 + m_2)/v_0^2$ and $e = (1 + b^2/a^2)^{1/2}$.

We derived the expression for energy radiated as gravitational waves in (3.148) for a general source with stress tensor in Fourier space $T_{ij}(\omega')$. The rate of energy radiated in an hyperbolic encounter is

$$\frac{E_{gw}}{dt} = \frac{\kappa^2}{4} \int \frac{8\pi}{5} \left(T_{ij}(\omega') T_{ji}^*(\omega') - \frac{1}{3}|T^i{}_i(\omega')|^2 \right) \omega^3 \, 2\pi (\omega - \omega') \frac{d\omega}{(2\pi)^3 2\omega}.$$
(4.116)

The Fourier transforms for the unbounded orbits spanning the time $t \in (-\infty, \infty)$ is given by

$$T_{ij}(\omega') = \int_{-\infty}^{\infty} dt \, e^{i\omega' t} \, \mu \, \dot{x}_i \dot{x}_j \,,$$
(4.117)

where $\omega' = \nu\Omega$. The non-zero components of $T_{ij}(\omega')$ are

$$T_{xx} = \int_{-\infty}^{\infty} dt \, e^{i\omega' t} \, \mu \, \dot{x}^2 = -i\mu\omega' \int_{-\infty}^{\infty} dt \, e^{i\omega' t} \, x \, \dot{x}$$

$$= -i\mu\omega' \int_{-\infty}^{\infty} d\xi \, e^{i\nu(e\sinh\xi - \xi)} \, x \, \frac{dx}{d\xi}$$

$$= i\mu\omega' a^2 \int_{-\infty}^{\infty} d\xi \, e^{i\nu(e\sinh\xi - \xi)} \, \sinh\xi \, (e - \cosh\xi) \,,$$

(4.118)

where in the first step we integrated by parts and then used the orbit equations (4.115) to change the integration variable to ξ. To perform this integral we use the integral representation of Hankel functions

$$H_p^{(1)}(q) = \frac{1}{i\pi} \int_{\infty}^{\infty} d\xi\, e^{q\sinh\xi - p\xi} \,. \tag{4.119}$$

Using (4.119) in (4.118) we obtain

$$\begin{aligned} T_{xx}(\omega') &= \frac{i\mu\omega' a^2}{4} \int_{-\infty}^{\infty} d\xi\, e^{i\nu(e\sinh\xi - \xi)} \left(e^\xi - e^{-\xi}\right)\left(2e - \left(e^\xi + e^{-\xi}\right)\right) \\ &= \frac{-\mu\omega' a^2 \pi}{4} \left[2e\left(H_{i\nu-1}^{(1)}(iev) - H_{i\nu+1}^{(1)}(iev)\right)\right. \\ &\quad \left. - \left(H_{i\nu-2}^{(1)}(iev) + H_{i\nu+2}^{(1)}(iev)\right)\right]. \end{aligned} \tag{4.120}$$

We make use of the recurrence relations of the Hankel functions

$$H_{i\nu-1}^{(1)}(iev) + H_{i\nu+1}^{(1)}(iev) = \frac{2}{e} H_{i\nu}^{(1)}(iev)\,,$$

$$H_{i\nu-1}^{(1)}(iev) - H_{i\nu+1}^{(1)}(iev) = 2H_{i\nu}^{(1)'}(iev)\,,$$

$$H_{i\nu-2}^{(1)}(iev) + H_{i\nu+2}^{(1)}(iev) = \left(\frac{4}{e^2} - 2\right) H_{i\nu}^{(1)}(iev) + \frac{4i}{ve} H_{i\nu}^{(1)'}(iev)\,,$$

$$H_{i\nu-2}^{(1)}(iev) - H_{i\nu+2}^{(1)}(iev) = \frac{4i}{ve^2} H_{i\nu}^{(1)}(iev) + \frac{4}{e} H_{i\nu}^{(1)'}(iev)\,. \tag{4.121}$$

Using (4.121) we can write (4.120) as

$$T_{xx}(\omega') = \mu\omega' a^2 \pi \left[\frac{i}{ve} H_{i\nu}^{(1)}(iev) - \left(e - \frac{1}{e}\right) H_{i\nu}^{(1)'}(iev)\right]. \tag{4.122}$$

Similarly we can calculate the other components of the stress tensor in Fourier space

$$\begin{aligned} T_{yy}(\omega') &= \int_{-\infty}^{\infty} dt\, e^{i\omega' t}\, \mu\, \dot{y}^2 = -i\mu\omega' \int_{-\infty}^{\infty} dt\, e^{i\omega' t}\, y\, \dot{y} \\ &= -i\mu\omega' b^2 \int_{-\infty}^{\infty} d\xi\, e^{i\nu(e\sinh\xi - \xi)}\, \cosh\xi\, \sinh\xi \\ &= \frac{-i\mu\omega' b^2}{4} \int_{-\infty}^{\infty} d\xi\, e^{i\nu(e\sinh\xi - \xi)} \left(e^{2\xi} - e^{-2\xi}\right) \\ &= \frac{\mu\omega' b^2 \pi}{4} \left[H_{i\nu-2}^{(1)}(iev) - H_{i\nu+2}^{(1)}(iev)\right] \\ &= \mu\omega' a^2 (e^2 - 1)\pi \left(\frac{i}{ve^2} H_{i\nu}^{(1)}(iev) + \frac{1}{e} H_{i\nu}^{(1)'}(iev)\right) \end{aligned} \tag{4.123}$$

and the remaining non-zero component of $T_{ij}(\omega')$ is

$$
\begin{aligned}
T_{xy}(\omega') &= \int_{-\infty}^{\infty} dt\, e^{i\omega' t}\, \mu\, \dot{x}\, \dot{y} = -i\mu\omega' \int_{-\infty}^{\infty} dt\, e^{i\omega' t}\, y\dot{x} \\
&= i\mu\omega' ab \int_{-\infty}^{\infty} d\xi\, e^{i\nu(e\sinh\xi - \xi)}\, \sinh^2\xi \\
&= \frac{i\mu\omega' ab}{4} \int_{-\infty}^{\infty} d\xi\, e^{i\nu(e\sinh\xi - \xi)}\, \left(e^{2\xi} + e^{-2\xi} - 2\right) \\
&= \frac{-\mu\omega' ab\,\pi}{4} \left(H^{(1)}_{i\nu-2}(iev) + H^{(1)}_{i\nu+2}(iev) - 2H^{(1)}_{i\nu}(iev)\right) \\
&= -\mu\omega' a^2 \sqrt{e^2 - 1}\,\pi \left[\left(\frac{1}{e^2} - 1\right) H^{(1)}_{i\nu}(iev) + \frac{i}{ve} H^{(1)'}_{i\nu}(iev)\right].
\end{aligned}
$$

$$(4.124)$$

Using (4.122), (4.123) and (4.124) we calculate

$$
\begin{aligned}
\left(T_{ij}(\omega')T^*_{ji}(\omega') - \frac{1}{3}|T^i{}_i(\omega')|^2\right) &= \mu^2\omega'^2 a^4 \pi^2 \Bigg\{ \left|H^{(1)}_{i\nu}(iev)\right|^2 \\
&\times \left(\frac{2}{e^4}(e^2 - 1)^3 + \frac{6 - 6e^2 + 2e^4}{3v^2 e^4}\right) \\
&+ \left|H^{(1)'}_{i\nu}(iev)\right|^2 \frac{2(e^2 - 1)}{e^2} \left(\frac{1}{v^2} + (e^2 - 1)\right) \Bigg\}.
\end{aligned}
$$

$$(4.125)$$

The rate of energy radiated in gravitational waves in the hyperbolic orbit is given by integrating over ω using the delta function in (4.116) and using (4.125) to give

$$
\frac{dE_{gw}}{dt} = \frac{\kappa^2\,\omega'^2}{20\,\pi} \left(T_{ij}(\omega')T^*_{ji}(\omega') - \frac{1}{3}|T^i{}_i(\omega')|^2\right),
$$

$$(4.126)$$

where $\omega' = v\Omega$.

Using (4.125) in (4.126) we find that gravitational energy radiation rate in a hyperbolic orbit is given by

$$
\begin{aligned}
\frac{dE_{gw}}{dt} = \frac{\kappa^2}{20}\mu^2\Omega^4 v^4 a^4 \pi \Bigg\{ &\left|H^{(1)}_{i\nu}(iev)\right|^2 \left(\frac{2}{e^4}(e^2 - 1)^3 + \frac{6 - 6e^2 + 2e^4}{3v^2 e^4}\right) \\
&+ \left|H^{(1)'}_{i\nu}(iev)\right|^2 \frac{2(e^2 - 1)}{e^2} \left(\frac{1}{v^2} + (e^2 - 1)\right) \Bigg\}.
\end{aligned}
$$

$$(4.127)$$

4.11 Waveform of Gravitational Waves from Hyperbolic Encounters

The gravitational wave amplitude of polarisation λ measured by a detector at a distance r is given in the frequency space by

$$h_\lambda(\omega', r) = \frac{4G}{r} \epsilon_\lambda^{ij}(\mathbf{n}) T_{ij}(n, \omega'). \tag{4.128}$$

For hyperbolic encounters we gave the stress tensor components (4.122), (4.123) and (4.124),

$$T_{xx}(\omega') = \mu\omega'a^2\pi \left[\frac{i}{ve} H_{iv}^{(1)}(iev) - \left(e - \frac{1}{e}\right) H_{iv}^{(1)'}(iev) \right],$$

$$T_{yy}(\omega') = \mu\omega'a^2(e^2 - 1)\pi \left(\frac{i}{ve^2} H_{iv}^{(1)}(iev) + \frac{1}{e} H_{iv}^{(1)'}(iev) \right),$$

$$T_{xy}(\omega') = -\mu\omega'a^2\sqrt{e^2 - 1}\,\pi \left[\left(\frac{1}{e^2} - 1\right) H_{iv}^{(1)}(iev) + \frac{i}{ve} H_{iv}^{(1)'}(iev) \right]. \tag{4.129}$$

For the observer in the radial direction $\hat{n} = (\sin\theta\cos\phi, \sin\theta\sin\phi, \cos\theta)$ in the binary coordinates, using the stress tensor components (4.129), the amplitudes of the $+$ and $\times$ wave-functions in the frequency space is given by

$$h_+(\omega', r) = i\frac{4G}{r}\epsilon_+^{ij}(\mathbf{n}) T_{ij}(\mathbf{n}, \omega') = i\frac{4G}{r} \left(\mathbf{e}_{\theta i}\mathbf{e}_{\theta j} - \mathbf{e}_{\phi i}\mathbf{e}_{\phi j} \right) T_{ij}(k)$$

$$= i\frac{4G}{r} \Big(T_{xx}(\cos^2\phi - \sin^2\phi\cos^2\theta) + T_{yy}(\sin^2\phi - \cos^2\phi\cos^2\theta)$$

$$-T_{xy}\sin 2\phi(1 + \cos^2\theta) \Big) \tag{4.130}$$

and the waveform of the $\times$ polarisation gravitational wave is

$$h_\times(\omega', r) = i\frac{4G}{r}\epsilon_\times^{ij}(\mathbf{n}) T_{ij}(\mathbf{n}, \omega') = i\frac{4G}{r} \left(\mathbf{e}_{\theta i}\mathbf{e}_{\phi j} + \mathbf{e}_{\phi i}\mathbf{e}_{\theta j} \right) T_{ij}(k)$$

$$= i\frac{4G}{r} \Big((T_{xx} - T_{yy})\sin 2\phi\cos\theta + 2T_{xy}\cos 2\phi\cos\theta \Big). \tag{4.131}$$

4.12 Memory Effect in Hyperbolic Orbits

Gravitational memory is the phenomenon of a permanent change in the metric after the passage of a gravitational wave. This results in a non-oscillatory permanent shift in the location of test particles after the passage of gravitational waves. The memory component of $h_{ij}(t, \mathbf{x})$ appears as a pole in the Fourier transformed signal $\tilde{h}_{ij}(f)$. This can be seen as follows. Consider the signal in frequency space of the form

$$\tilde{h}(\omega) = \frac{A}{\omega}. \tag{4.132}$$

This will correspond to gravitational waveform in time of the form,

$$h(t) = \int_{-\infty}^{\infty} \frac{d\omega}{2\pi} e^{i\omega t} \tilde{h}(\omega) = \int_{-\infty}^{\infty} \frac{d\omega}{2\pi} e^{i\omega t} \frac{A}{f}$$

$$= A \,\Theta(t), \tag{4.133}$$

where $\Theta(t)$ is the Heaviside step function[1] This is the memory waveform.

Hyperbolic orbits have the property that even at zero angular velocity the energy radiated is non-zero. This is a signature of the gravitational wave memory effect where the waveform dominates at zero frequency.

To compute the rate of energy radiated at $\omega' \to 0$ limit we expand the stress tensor in the $v \to 0$ limit. To do this we use the asymptotic forms of the Hankel functions for small argument $ve \to 0$,

$$H_{iv}^{(1)}(iev) \simeq \frac{2i}{\pi} \ln(ve),$$

$$H_{iv}^{(1)'}(iev) \simeq \frac{2}{\pi ve}. \tag{4.135}$$

In the low frequency limit the stress tensor components (4.129) reduce to

$$T_{xx}(\omega') = \frac{-2\mu\omega' a^2}{ve^2} \left[\ln(ve) + (e^2 - 1) \right],$$

$$T_{yy}(\omega') = \frac{-2\mu\omega' a^2 (e^2 - 1)}{ve^2} \left[\ln(ve) - 1 \right],$$

$$T_{xy}(\omega') = 2i\,\mu\omega' a^2 \sqrt{e^2 - 1} \left[\frac{e^2 - 1}{e^2} \ln(ve) - \frac{1}{v^2 e^2} \right] \tag{4.136}$$

---

[1] The Heaviside function has a Fourier representation

$$\Theta(t) = \int_{\infty}^{\infty} \frac{d\omega}{2\pi} \frac{e^{i\omega t}}{\omega + i\epsilon}. \tag{4.134}$$

with $\omega' = \nu\Omega$. In the $\nu \to 0$ limit only one term in (4.136) is non-zero which is

$$T_{xy}(\omega' \to 0) = 4i\mu\Omega a^2 \sqrt{e^2 - 1}\frac{1}{\nu e^2}. \tag{4.137}$$

Using this in the waveforms (4.130) and (4.131) we find that the zero frequency memory waveforms for hyperbolic orbits are of the form

$$\Delta h_+ = \frac{8G}{r}\mu\Omega a^2 \frac{\sqrt{e^2 - 1}}{\nu e^2} \sin 2\phi \, (1 + \cos^2\theta)$$

$$\Delta h_\times = -\frac{16}{r}G\mu\Omega a^2 \frac{\sqrt{e^2 - 1}}{\nu e^2} \cos 2\phi \, \cos\theta. \tag{4.138}$$

The zero frequency memory waveform vanishes in the limit $e \to 1$.

We can compute the energy radiated in zero frequency gravitational waves by starting from the expression (4.127) for energy rate in hyperbolic orbits. From the limits of Hankel functions (4.135) it is clear that only the $H_{i\nu}^{(1)'}(ie\nu)(1/\nu^2)$ term in (4.127) will survive in the $\nu \to 0$ limit. The energy rate at zero frequency for hyperbolic orbits is given by

$$\lim_{\nu \to 0} \frac{dE_{gw}}{dt} = \frac{32G}{5}\mu^2\Omega^4 a^4 \frac{2(e^2 - 1)}{e^4}. \tag{4.139}$$

References

1. R.A. Hulse, J.H. Taylor, Discovery of a pulsar in a binary system. Astrophys. J. **195**, L51–L53 (1975)
2. J.H. Taylor, J.M. Weisberg, A new test of general relativity: gravitational radiation and the binary pulsar PSR 1913+16. Astrophys. J. **253**, 908–920 (1982)
3. J.M. Weisberg, J.H. Taylor, Observations of post-newtonian timing effects in the binary pulsar PSR 1913+16. Phys. Rev. Lett. **52**,1348–1350 (1984)
4. N. Yunes, K. Yagi, F. Pretorius, Theoretical physics implications of the binary black-hole mergers GW150914 and GW151226. Phys. Rev. D **94**(8), 084002 (2016). [arXiv:1603.08955[gr-qc]]
5. S. Mohanty, P. Kumar Panda, Particle physics bounds from the Hulse-Taylor binary. Phys. Rev. D **53**, 5723 (1996)
6. T.K. Poddar, S. Mohanty, S. Jana, Gravitational radiation from binary systems in massive graviton theories (2021). [arXiv:2105.13335 [gr-qc]]
7. L.D. Landau, E.M. Lifschitz, *The Classical Theory of Fields*. Course of Theoretical Physics, vol. 2 (Pergamon Press, Oxford, 1975), p. 181
8. P.C. Peters, J. Mathews, Gravitational radiation from point masses in a Keplerian orbit. Phys. Rev. **131**, 435 (1963)
9. J.H. Taylor, J.M. Weisberg, A new test of general relativity: Gravitational radiation and the binary pulsar PS R 1913+16. Astrophys. J. **253**, 908 (1982)
10. J.M. Weisberg, Y. Huang, Relativistic measurements from timing the binary pulsar PSR B1913+16. Astrophys. J. **829**(1), 55 (2016). [arXiv:1606.02744 [astro-ph.HE]]
11. D.R. Lorimer, Binary and millisecond pulsars. Living Rev. Relativ. **11**, 8 (2008)

12. M. Kramer, et al., Tests of general relativity from timing the double pulsar. Science **314**, 97 (2006)
13. P.C.C. Freire, et al., The relativistic pulsar-white dwarf binary PSR J1738+0333 II. The most stringent test of scalar-tensor gravity. Mon. Not. Roy. Astron. Soc. **423**, 3328 (2012)
14. J. Antoniadis, et al., A massive pulsar in a compact relativistic binary. Science **340**, 6131 (2013)
15. P.C. Peters, Gravitational radiation and the motion of two point masses. Phys. Rev. **136**, B1224 (1964)
16. A. Buonanno, T. Damour, Effective one-body approach to general relativistic two-body dynamics. Phys. Rev. D **59**, 084006 (1999). [arXiv:gr-qc/9811091 [gr-qc]]
17. T. Damour, Classical and quantum scattering in post-Minkowskian gravity. Phys. Rev. D **102**(2), 024060 (2020). [arXiv:1912.02139 [gr-qc]]
18. P.H. Damgaard, P. Vanhove, Remodeling the effective one-body formalism in post-Minkowskian gravity. Phys. Rev. D **104**(10), 104029 (2021). [arXiv:2108.11248 [hep-th]]
19. K.S. Thorne, Multipole expansions of gravitational radiation. Rev. Mod. Phys. **52**, 299–339 (1980)
20. T. Damour, B.R. Iyer, Multipole analysis for electromagnetism and linearized gravity with irreducible Cartesian tensors. Phys. Rev. D **43**(10), 3259–3272 (1991)
21. A. Ross, Multipole expansion at the level of the action. Phys. Rev. D **85**, 125033 (2012). https://doi.org/10.1103/PhysRevD.85.125033. [arXiv:1202.4750 [gr-qc]]
22. W.D. Goldberger, A. Ross, Gravitational radiative corrections from effective field theory. Phys. Rev. D **81**, 124015 (2010). [arXiv:0912.4254 [gr-qc]]
23. M.M. Riva, Effective Field Theory for Gravitational Radiation in General Relativity and beyond (2021). [arXiv:2111.07433 [gr-qc]]
24. L. Blanchet, T. Damour, Radiative gravitational fields in general relativity I. General structure of the field outside the source. Philosoph. Trans. Roy. Soc. London. Ser. A Math. Phys. Sci. **320**(1555), 379–430 (1986)
25. T. Kumar Poddar, S. Mohanty, S. Jana, Constraints on ultralight axions from compact binary systems. Phys. Rev. D **101**(8), 083007 (2020). https://doi.org/10.1103/PhysRevD.101.083007. [arXiv:1906.00666 [hep-ph]]
26. W. Hu, R. Barkana, A. Gruzinov, Cold and fuzzy dark matter. Phys. Rev. Lett. **85**, 1158 (2000). [astro-ph/0003365]
27. L. Hui, J.P. Ostriker, S. Tremaine, E. Witten, Ultralight scalars as cosmological dark matter. Phys. Rev. D **95**(4), 043541 (2017). [arXiv:1610.08297 [astro-ph.CO]]
28. G. Raffelt, Limits on a CP-violating scalar axion-nucleon interaction. Phys. Rev. D **86**, 015001 (2012). [arXiv:1205.1776 [hep-ph]]
29. J.E. Moody, F. Wilczek, New macroscopic forces? Phys. Rev. D **30**, 130 (1984)
30. D. Chang, R.N. Mohapatra, S. Nussinov, Could goldstone bosons generate an observable 1/r potential? Phys. Rev. Lett. **55**, 2835 (1985)
31. A. Hook, J. Huang, Probing axions with neutron star inspirals and other stellar processes. J. High Energy Phys. **1806**, 036 (2018). [arXiv:1708.08464 [hep-ph]]
32. T. Kumar Poddar, S. Mohanty, S. Jana, Constraints on ultralight axions from compact binary systems. Phys. Rev. D **101**(8), 083007 (2020). [arXiv:1906.00666 [hep-ph]]
33. R. Foot, Charge quantization in the standard model and some of its extensions. Mod. Phys. Lett. A**6**, 527 (1991)
34. X.-G. He, G.C. Joshi, H. Lew, R.R. Volkas, New Z' phenomenology. Phys. Rev. D **44**, 2118 (1991)
35. A.S. Joshipura, S. Mohanty, Constraints on flavor dependent long range forces from atmospheric neutrino observations at super-Kamiokande. Phys. Lett. B **584**, 103–108 (2004). [arXiv:hep-ph/0310210 [hep-ph]]
36. T. Kumar Poddar, S. Mohanty, S. Jana, Constraints on long range force from perihelion precession of planets in a gauged $L_e - L_{\mu,\tau}$ scenario. Eur. Phys. J. C **81**(4), 286 (2021). [arXiv:2002.02935 [hep-ph]]
37. Turner, M.A., Gravitational radiation from point-masses in unbound orbits: Newtonian results. Ap. J., **216**, 610–619 (1977)

38. S. Capozziello, M. De Laurentis, F. De Paolis, G. Ingrosso, A. Nucita, Gravitational waves from hyperbolic encounters," Mod. Phys. Lett. A **23**, 99–107 (2008). [arXiv:0801.0122 [gr-qc]]
39. L. De Vittori, P. Jetzer, A. Klein, Gravitational wave energy spectrum of hyperbolic encounters. Phys. Rev. D **86**, 044017 (2012). [arXiv:1207.5359 [gr-qc]]
40. L. De Vittori, A. Gopakumar, A. Gupta, P. Jetzer, Gravitational waves from spinning compact binaries in hyperbolic orbits. Phys. Rev. D **90**(12), 124066 (2014). [arXiv:1410.6311 [gr-qc]]
41. J. García-Bellido, S. Nesseris, Gravitational wave energy emission and detection rates of Primordial Black Hole hyperbolic encounters. Phys. Dark Univ. **21**, 61–69 (2018). [arXiv:1711.09702 [astro-ph.HE]]
42. M. Gröbner, P. Jetzer, M. Haney, S. Tiwari, W. Ishibashi, A note on the gravitational wave energy spectrum of parabolic and hyperbolic encounters. Class. Quant. Grav. **37**(6), 067002 (2020). [arXiv:2001.05187 [gr-qc]]
43. D. Bini, A. Geralico, Frequency domain analysis of the gravitational wave energy loss in hyperbolic encounters. Phys. Rev. D **104**(10), 104019 (2021). [arXiv:2108.02472 [gr-qc]]

Gravitational Memory and Soft-Graviton Theorem

5

Abstract

Gravitational memory effect is the phenomenon where test particles are permanently displaced after the passage of a gravitational wave. The memory signals arise as zero frequency poles in the scattering amplitudes of soft-graviton theorems. Memory signal also arises from the secondary gravitational waves from a sudden burst of primary gravitational waves which can arise in supernovae explosions. In this chapter we study soft-graviton theorems and show how they can be used for predicting the memory signals from diverse astrophysical events, like supernova explosions and hyperbolic encounters of stars orbiting past super-massive black holes. We also calculate the energy radiated in collisional scatterings of a gas of particles and test the viability of measuring such gravitational waves as stochastic GW signals.

5.1 Introduction

Gravitational memory effect is the phenomenon where test particles are permanently displaced after the passage of a gravitational wave [1]. Test particles initially separated by ξ_i will have an extra separation given by

$$\Delta \xi_i = \frac{1}{2} \Delta h^{ij} \xi_j,$$

(5.1)

where

$$\Delta h_{ij}(\mathbf{x}) = h_{ij}(t \to \infty, \mathbf{x}) - h_{ij}(t \to -\infty, \mathbf{x}).$$

(5.2)

Gravitational memory is the phenomenon of a permanent change in the metric after the passage of a gravitational wave. This results in a non-oscillatory permanent shift

© The Author(s), under exclusive license to Springer Nature Switzerland AG 2023
S. Mohanty, *Gravitational Waves from a Quantum Field Theory Perspective*,
Lecture Notes in Physics 1013, https://doi.org/10.1007/978-3-031-23770-6_5

in the location of test particles after the passage of gravitational waves. The memory component of $h_{ij}(t, \mathbf{x})$ appears as a pole in the Fourier transformed signal $\tilde{h}_{ij}(f)$. This can be seen as follows. Consider the signal in frequency space of the form

$$\tilde{h}(f) = \frac{h}{if}. \tag{5.3}$$

This will correspond to gravitational waveform in time of the form

$$h(t) = \int_{-\infty}^{\infty} \frac{df}{2\pi} e^{2\pi ft} \tilde{h}(f) = \int_{-\infty}^{\infty} \frac{df}{2\pi} e^{2\pi ft} \frac{h}{if}$$
$$= h \, \Theta(t), \tag{5.4}$$

where $\Theta(t)$ is the Heaviside step function. This is the memory waveform.

The memory component of the waveforms in the Frequency space dominates at zero frequency. The amplitude for the memory component of gravitational waves is therefore conveniently derived using he soft-graviton theorems, which derive the amplitude of emission of a zero frequency graviton from the amplitude of hard particle scattering by multiplication of a kinematical factor.

5.2   Soft-Graviton Amplitudes

Consider a general n-particle scattering process as shown in Fig. 5.1a. We use the conventions that the four-momenta of particles $a = 1, 2, \ldots, n$ are outgoing. Let us denote the scattering amplitude of the n-particle scattering process by $\mathcal{A}_n$. Now consider another diagram Fig. 5.1b where a soft graviton of four-momentum q is emitted from any of the external legs.

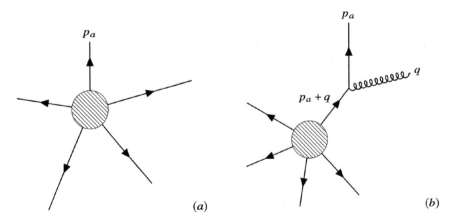

Fig. 5.1 (a) n-particle scattering and (b) n-particle scattering with soft-graviton emission

The graviton emission has the vertex factor $\frac{-i\kappa_a}{2}T_{\mu\nu}\epsilon_\lambda^{*\mu\nu}$ given by

$$p_1 \longleftrightarrow p_2 \qquad = \frac{-i\kappa}{2}\epsilon_\lambda^{*\mu\nu}(q)\left[p_{1\mu}p_{2\nu} + p_{1\nu}p_{2\mu} - \eta_{\mu\nu}(p_1 \cdot p_2 - m^2)\right].$$

$$q$$

$$(5.5)$$

So for each graviton emission from external leg with momenta p_a, we have to multiply $\mathcal{A}_n$ by the factor

$$\frac{-i\kappa}{2} \frac{\epsilon_\lambda^{*\mu\nu}(q)}{(p_a + q)^2 - m_a^2}$$

$$\times\left[(p_{a\mu} + q_\mu)p_{a\nu} + (p_{a\nu} + q_\nu)p_{a\mu} - \eta_{\mu\nu}\big((p_a + q)\cdot p_a - m_a^2\big)\right]$$

$$= \frac{-i\kappa}{2} \frac{\epsilon_\lambda^{*\mu\nu}(q)}{2p_a \cdot q}\left[2p_{a\mu}p_{a\nu} + q_\mu p_{a\nu} + q_\nu p_{a\mu} - \eta_{\mu\nu}p_a \cdot q\right] \qquad (5.6)$$

and sum over the external legs. So the amplitude of a single graviton emission from the n-particle amplitude can be written as

$$\mathcal{A}_{n+1} = \frac{-i\kappa}{2}\epsilon_\lambda^{*\mu\nu}(q)\sum_a \frac{1}{2p_a \cdot q}\left[2p_{a\mu}p_{a\nu} + q_\mu p_{a\nu} + q_\nu p_{a\mu} - \eta_{\mu\nu}p_a \cdot q\right]\mathcal{A}_n.$$

$$(5.7)$$

We take only the graviton emission to the external legs as graviton emission from the loops will not give the poles in the $q \to 0$ limit.

Keeping the leading order terms in q in (5.7), we see that the n-particle amplitude can be related to the single soft-graviton emission amplitude by a factor

$$\mathcal{A}_{n+1}(p_a, q) = \frac{-i\kappa}{2}\epsilon_\lambda^{*\mu\nu}(q)\sum_{a=1}^n \frac{p_{a\mu}p_{a\nu}}{p_a \cdot q}\mathcal{A}_n(p_a). \qquad (5.8)$$

The $n + 1$ particle amplitude is factorised into the amplitude for the soft-graviton emission $\mathcal{A}_\lambda$ given to the leading order in graviton momenta q by

$$\mathcal{A}_\lambda = \frac{-i\kappa}{2}\epsilon_\lambda^{*\mu\nu}(q)\sum_{a=1}^n \frac{p_{a\mu}p_{a\nu}}{p_a \cdot q} \qquad (5.9)$$

times the amplitude for the hard scattering $\mathcal{A}_n(p_1, p_2 \cdots p_n)$. This is Weinberg's soft-graviton theorem to the leading order in graviton momenta [2–5].

We will now show that the universality of the gravitational coupling κ for all matter (and gravity) follows from gauge transformation [3,4]. GR is invariant under the general coordinate transformation $x^\mu \to x^\mu + \xi^\mu$, which implies $h^{\mu\nu}(x) \to h^{\mu\nu} + (\partial^\mu \xi^\nu + \partial^\nu \xi^\mu)$ and in momentum space

$$\epsilon^{\mu\nu}(q) \to \epsilon^{\mu\nu}(q) + (q^\mu \xi^\nu + q^\mu \xi^\nu), \qquad (5.10)$$

which implies that contracting the amplitude (5.7) with q^μ or q^ν we should get zero. We will assume that κ in (5.7) is not the same for all particles and we take the gravitational coupling for particle 'a' in the external leg as κ_a. Operating on the terms in the summation in (5.7) with q^μ, we obtain

$$\sum_a \kappa_a \frac{1}{2p_a \cdot q} \Big[(2p_{a\mu} \cdot q) \, p_{a\nu} \Big] = \sum_a \kappa_a p_{a\nu} = 0. \qquad (5.11)$$

Now, by Lorentz invariance of the Minkowski background, we must conservation of momentum $\sum_a p_{a\nu} = 0$. Therefore Lorentz invariance and gravitational gauge invariance (5.11) imply that $\kappa_a = \kappa$ is identical for all particles. Therefore we have a derivation of the equivalence principle in the Minkowski space linearised gravity theory [2].

5.3 Soft-Graviton Amplitudes in the NLO

The soft factor up to order q^2 can be written as [6, 7]

$$\mathscr{A}_{n+1}(p_a, q) = \Big(S_0 + S_1 + S_2 \Big) \mathscr{A}_n(p_a) \qquad (5.12)$$

with

$$S_0 = \frac{\kappa}{2} \sum_{a=1}^{n} \frac{\epsilon_\lambda^{*\mu\nu} p_{a\mu} p_{a\nu}}{p_a \cdot q},$$

$$S_1 = -i \frac{\kappa}{2} \sum_{a=1}^{n} \frac{\epsilon_{\lambda\,\mu\nu}^* p_a^\mu q_\beta J_a^{\beta\nu}}{p_a \cdot q},$$

$$S_2 = -\frac{\kappa}{2} \sum_{a=1}^{n} \frac{\epsilon_{\lambda\mu\nu}^* q_\alpha q_\beta J_a^{\alpha\mu} J_a^{\beta\nu}}{p_a \cdot q}, \qquad (5.13)$$

where $J_a^{\alpha\beta} = x_a^\alpha p_a^\beta - x_a^\beta p_a^\alpha + S_a^{\alpha\beta}$ is the total angular momentum of particle 'a'. The series of the soft factors are the same order in the gravitational coupling but are increasing powers of the graviton frequency $q_0 = \omega$. The leading term $S_0 \sim \omega^{-1}$, while the sub-leading terms go as $S_1 \sim \omega^0$ and $S_2 \sim \omega^1$. The gauge invariance of the amplitude (5.12) under the transformation (5.10) follows from the conservation of the total momentum and angular momentum. The sub-leading order soft terms (5.13) can in fact be derived using gauge invariance and the conservation of momentum and angular momentum [8]. To see this, define the tensor $\mathcal{M}_{\mu\nu}$ from the relation

$$\mathcal{A}_{n+1} \equiv \frac{\kappa}{2} \epsilon_\lambda^{*\mu\nu} \mathcal{M}_{\mu\nu}. \tag{5.14}$$

At the leading order in q^μ, we have from (5.8)

$$\mathcal{M}_{\mu\nu}^{(0)} = \sum_{a=1}^n \frac{p_{a\mu} p_{a\nu}}{p_a \cdot q} \mathcal{A}_n(p_1, ..p_a + q, \dots p_n). \tag{5.15}$$

We would like to derive corrections to $\mathcal{M}_{\mu\nu}^0$ up to order q^2

$$\mathcal{M}_{\mu\nu} \equiv \sum_{a=1}^n \frac{p_{a\mu} p_{a\nu}}{p_a \cdot q} \mathcal{A}_n(p_1, ..p_a + q, \dots p_n) + N_{\mu\nu}(q, p_1, \dots, p_n), \tag{5.16}$$

where $N_{\mu\nu}(q, p_1, \dots, p_n)$ is symmetric under $\mu \leftrightarrow \nu$.

From gauge invariance, we must have

$$q_\mu \mathcal{M}_{\mu\nu} = 0, \tag{5.17}$$

which gives us the relation

$$\sum_{a=1}^n p_{a\nu} \mathcal{A}_n + q_\mu N_{\mu\nu} = 0. \tag{5.18}$$

We Taylor expand $\mathcal{A}_n$ and $N_{\mu\nu}$ in powers of q to obtain from (5.18)

$$\sum_{a=1}^n \left(p_{a\nu} + p_{a\nu} q_\mu \frac{\partial}{\partial p_{a\mu}} + p_{a\nu} q_\mu q_\rho \frac{\partial^2}{\partial p_{a\mu} \partial p_{a\rho}} \right) \mathcal{A}_n$$

$$+ q_\mu N_{\mu\nu} + q_\mu q_\rho \left(\frac{\partial N_{\mu\nu}}{\partial q^\rho} + \frac{\partial N_{\rho\nu}}{\partial q^\mu} \right) = 0. \tag{5.19}$$

Taking terms order by order in q, we obtain the following. To the zeroeth order in q, we obtain from (5.19) the relation

$$\sum_{a=1}^n p_{a\nu} \mathcal{A}_n = 0, \tag{5.20}$$

which is just the conservation of momentum. Equating order q^1 terms, we have

$$\sum_{a=1}^{n} p_{av} \frac{\partial \mathscr{A}_n}{\partial p_{a\mu}} + N_{\mu v} = 0, \tag{5.21}$$

and from order q^2 terms, we have

$$\sum_{a=1}^{n} p_{av} \frac{\partial^2 \mathscr{A}_n}{\partial p_{a\mu} \partial p_{a\rho}} + \left(\frac{\partial N_{\mu v}}{\partial q^\rho} + \frac{\partial N_{\rho v}}{\partial q^\mu} \right) = 0. \tag{5.22}$$

We introduce the angular operator of particle 'a' given by the operator

$$J_a^{\mu v} = i \left(p_a^\mu \frac{\partial}{\partial p_{av}} - p_a^v \frac{\partial}{\partial p_{a\mu}} \right) + S_a^{\mu v}. \tag{5.23}$$

Using (5.23) and (5.22), we obtain

$$\sum_{a=1}^{n} -i J_a^{\mu\rho} \frac{\partial \mathscr{A}_n}{\partial p_{a\sigma}} = \left(\frac{\partial N^{\rho\sigma}}{\partial q_\mu} - \frac{\partial N^{\mu\sigma}}{\partial q_\rho} \right). \tag{5.24}$$

To get the order q^2 correction terms of $\mathcal{M}_{\mu v}$, expand $\mathscr{A}_n$ and $N_{\mu v}$ in (5.16) in a Taylor series

$$\mathcal{M}_{\mu v} = \sum_{a=1}^{n} \frac{p_{a\mu} p_{av}}{p_a \cdot q} \left[\left(\mathscr{A}_n + q_\rho \frac{\partial \mathscr{A}_n}{\partial p_{a\rho}} + q_\sigma q_\rho \frac{\partial^2 \mathscr{A}_n}{\partial p_{a\sigma} \partial p_{a\rho}} \right) + N_{\mu v} \right.$$
$$\left. + q_\rho \left(\frac{\partial N_{\mu v}}{\partial q^\rho} + \frac{\partial N_{\rho v}}{\partial q^\mu} \right) \right]. \tag{5.25}$$

Using (5.21), (5.22) and (5.24) in (5.25), we obtain after some algebra

$$M^{\mu v} = \sum_{a=1}^{n} \frac{1}{p_a \cdot q} \left[p_a^\mu p_a^v - i p_a^\mu q_\rho J^{v\rho} - q_\rho q_\sigma J^{\mu\rho} J^{v\sigma} \right] \mathscr{A}_n. \tag{5.26}$$

Since $M^{\mu v}$ is contracted with the polarisation $\epsilon^{*\mu v}$ to obtain the graviton emission amplitude (5.14), in deriving we can drop terms proportional to q_μ, q_v and $\eta_{\mu v}$ as we take the polarisation tensor to be transverse and traceless.

From (5.26) and (5.14), we have the soft-graviton amplitude up to order q^2 as given in (5.12) and (5.13). Soft graviton amplitude calculations also give memory terms which are logarithmic in frequency [7, 9, 10].

5.4   Memory Effect from Soft-Graviton Amplitudes

We have seen (3.120) that the amplitude of a graviton emission from a scattering process can be related to the waveform of the gravitational wave measured by a detector at a distance r as

$$\tilde{h}_{\alpha\beta}(\mathbf{n}r, t) = \frac{\kappa}{4\pi r} \int \frac{dk_0}{(2\pi)} \sum_{\lambda=1}^{2} \epsilon_{\alpha\beta}^{\lambda}(\mathbf{n}) \mathcal{A}_{\lambda}(k_0, \mathbf{n}k_0) e^{-ik_0(t-r)}; \qquad (5.27)$$

here $k_{\mu} = (k_0, \mathbf{n}k_0)$ is the momentum of the emitted graviton and $\mathbf{n}$ is the unit vector pointing from the source to the detector located at a distance r.

To obtain the gravitational wave from a general n-body scattering, we use the expression for the amplitude (5.9) in (5.27) to obtain

$$\tilde{h}_{\alpha\beta}(\mathbf{n}r, t) = \frac{\kappa}{4\pi r} \int \frac{dk_0}{(2\pi)} \frac{-i\kappa}{2} \sum_{\lambda=1}^{2} \epsilon_{\alpha\beta}^{\lambda}(\mathbf{n}) \epsilon_{\lambda}^{*\mu\nu}(\mathbf{n}) \sum_{a=1}^{n} \frac{p_{a\mu} p_{a\nu}}{p_a \cdot k} e^{-ik_0(t-r)}. \quad (5.28)$$

Summing over the polarisations using (3.23), we have

$$\sum_{\lambda=1}^{2} \epsilon_{\mu\nu}^{\lambda}(k) \epsilon_{\alpha\beta}^{*\lambda}(k) = \frac{1}{2}(\eta_{\mu\alpha}\eta_{\nu\beta} + \eta_{\mu\beta}\eta_{\nu\alpha}) - \frac{1}{2}\eta_{\mu\nu}\eta_{\alpha\beta}. \qquad (5.29)$$

Using (5.29) in (5.28), we obtain

$$\tilde{h}_{\alpha\beta}(\mathbf{n}r, t) = \frac{\kappa^2}{8\pi r} \int \frac{dk_0}{(2\pi i)} \sum_{a=1}^{n} \frac{(p_{a\alpha} p_{a\beta} - (1/2)\eta_{\alpha\beta} p_a^2)}{p_a \cdot k} e^{-ik_0(t-r)}. \quad (5.30)$$

Using the traceless–transverse gauge, we see that gravitational waves from scattering amplitude (5.8) will be given by

$$\tilde{h}_{ij}^{\mathrm{TT}}(\mathbf{n}r, t) = \frac{\kappa^2}{8\pi r} \int \frac{dk_0}{(2\pi i)} \sum_{a=1}^{n} \left[\frac{(p_{ai} p_{aj} + (1/2)\delta_{ij} m_a^2)}{p_a \cdot k} \right]^{\mathrm{TT}} e^{-ik_0(t-r)} \quad (5.31)$$

where we have used the on-shell condition for the particles in the external legs of the amplitude $p_a^2 = m_a^2$. The TT projection is obtained by operating with the projection operator (4.49) on (5.31). The TT projector eliminates the $\delta_{ij} m_a^2$ term in (5.31). We express the particle momenta in terms of their velocities $p_{a\mu} = m_a \gamma_a(1, \mathbf{v}_a)$, where $\gamma_a = (1 - v_a^2)^{-1/2}$ is the Lorentz factor. The graviton 4-momentum is $k_\mu = (k_0, k_0\mathbf{n})$. Using these relations, the expression (5.31) then simplifies to

$$\tilde{h}_{ij}^{TT}(\mathbf{n}r, t) = \frac{4G}{r} \int \frac{dk_0}{2\pi i k_0} \sum_{a=1}^{n} \frac{m_a}{\sqrt{1 - v_a^2}} \left[\frac{v_{ai} v_{aj}}{(1 - \mathbf{v}_a \cdot \mathbf{n})} \right]^{TT} e^{-ik_0(t-r)}. \tag{5.32}$$

We can perform the integral over the graviton frequency using the relation

$$\int \frac{dk_0}{2\pi i k_0} e^{ik_0(t-r)} = \Theta(t - r), \tag{5.33}$$

where $\Theta(t - r)$ is the Heaviside step function.[1] Suppose the gravitational detector consists of two inertial masses separated by distance L at a distance r_0 from the source of the gravitational waves. There will be a change in the separation of the two masses after $t > r$, with the strain $\Delta h_{ij} = h_{ij}(t \gg r_0) - h_{ij}(t \ll r_0)$ given by

$$\Delta h^{TT} = \frac{4G}{r} \sum_{a=1}^{n} \frac{m_a}{\sqrt{1 - v_a^2}} \left[\frac{v_{ai} v_{aj}}{(1 - \mathbf{v}_a \cdot \mathbf{n})} \right]^{TT}. \tag{5.34}$$

This permanent change in the separation of two masses after the passage of a gravitational wave is called the memory effect.

5.5   Gravitational Energy Radiated in Particle Collisions

We can use the soft-graviton theorem to calculate the gravitational energy radiated in a multiparticle scattering [3, 11–13] as follows.

The total number of gravitons radiated per volume per time can be written in terms of $\mathcal{A}_{n+1}$ (Eq. (5.8)) as

$$\dot{N}_{gr} = \int \frac{d^3q}{2q_0(2\pi)^3} |\mathcal{A}_{n+1}|^2. \tag{5.35}$$

The spectrum of graviton number density radiated is

$$\frac{d\dot{N}_{gr}}{dq_0} = \int \frac{d^3q}{2q_0(2\pi)^3} \delta(q_0 - |\mathbf{q}|) \sum_{\lambda} |\mathcal{A}_{n+1}|^2, \tag{5.36}$$

---

[1] Taking the time derivative of both sides results in the integral representation of $\delta(t - r)$.

and the radiated energy density spectrum of the gravitons is

$$\frac{d\dot{E}_{gr}}{dq_0} = \int \frac{d^3q}{2q_0(2\pi)^3} q_0 \,\delta(q_0 - |\mathbf{q}|) \sum_\lambda |\mathcal{A}_{n+1}|^2. \tag{5.37}$$

Taking the leading order term for the soft-graviton amplitude,

$$\mathcal{A}_{n+1} = \frac{-i\kappa}{2} \epsilon_\lambda^{*\mu\nu}(q) \sum_{a=1}^n \eta_a \frac{p_{a\mu} p_{a\nu}}{p_a \cdot q} \mathcal{A}_n(p_a), \tag{5.38}$$

where $\mathcal{A}_n(p_a)$ is the hard scattering of particles with four-momenta p_a and $\eta_a = 1(-1)$ for incoming (outgoing) particles. The amplitude squared summed over the graviton polarisation is

$$\sum_\lambda |\mathcal{A}_{n+1}|^2 = 8\pi G \sum_{a,b} \eta_a \eta_b \frac{p_{a\mu} p_{a\nu}}{p_a \cdot q} \frac{1}{2} \left(\eta^{\mu\alpha} \eta^{\nu\beta} + \eta^{\mu\beta} \eta^{\nu\alpha} - \eta^{\mu\nu} \eta^{\alpha\beta} \right) \frac{p_{b\alpha} p_{b\beta}}{p_b \cdot q}$$

$$= 8\pi G \sum_{a,b} \eta_a \eta_b \frac{(p_a \cdot p_b)^2 - \frac{1}{2} m_a^2 m_b^2}{(p_a \cdot q)(p_b \cdot q)} |\mathcal{A}_n|^2 \tag{5.39}$$

where we have used the relation for the polarisation sum $\sum_\lambda \epsilon_\lambda^{\mu\nu} \epsilon_\lambda^{*\alpha\beta}$. We have used the on-shell relation $p_a^2 = m_a^2$ and then used the conservation of momenta $\sum_a p_a^\mu = 0$ of the hard particles. Using (5.39) in (5.36), we find the graviton emission rate spectrum

$$\frac{d\dot{N}_{gr}}{dq_0} = \frac{G}{2\pi^2} \sum_{a,b} \eta_a \eta_b \int d\Omega_\mathbf{n} \frac{(p_a \cdot p_b)^2 - \frac{1}{2} m_a^2 m_b^2}{q_0 E_a E_b (1 - \mathbf{v}_a \cdot \mathbf{n})(1 - \mathbf{v}_b \cdot \mathbf{n})} |\mathcal{A}_n|^2 \tag{5.40}$$

where we have taken the graviton four-momentum as $q = q_0(1, \mathbf{n})$ and $v_i = p_i / E_i$. The integration over the angles of the graviton emission gives [11]

$$\int d\Omega_\mathbf{n} \frac{1}{(1 - \mathbf{v}_a \cdot \mathbf{n})(1 - \mathbf{v}_b \cdot \mathbf{n})} = \frac{2\pi}{\beta_{ab}} \log \frac{1 + \beta_{ab}}{1 - \beta_{ab}}, \tag{5.41}$$

where β_{ab} is the relative velocity given by

$$\beta_{ab} = \left(1 - \frac{(m_a m_b)^2}{(p_a \cdot p_b)^2} \right)^{1/2}. \tag{5.42}$$

The remaining factor in (5.40) can be written in terms of β_{ab} as

$$\frac{(p_a \cdot p_b)^2 - \frac{1}{2} m_a^2 m_b^2}{E_a E_b} = \frac{m_a m_b (1 + \beta_{ab}^2)}{2(1 - \beta_{ab}^2)^{1/2}}. \tag{5.43}$$

Using (5.41) and (5.43), we obtain the expression for graviton rate spectrum (5.40) given by

$$\frac{d\dot{N}_{gr}}{dq_0} = \frac{G}{2\pi} \frac{1}{q_0} \sum_{a,b} \eta_a \eta_b \frac{m_a m_b (1 + \beta_{ab}^2)}{\beta_{ab}(1 - \beta_{ab}^2)^{1/2}} \log\left(\frac{1 + \beta_{ab}}{1 - \beta_{ab}}\right) |\mathcal{A}_n|^2. \tag{5.44}$$

The energy rate spectrum is given by

$$\frac{d\dot{E}_{gr}}{dq_0} \equiv B|\mathcal{A}|_n^2 = \frac{G}{2\pi} \sum_{a,b} \eta_a \eta_b \frac{m_a m_b (1 + \beta_{ab}^2)}{\beta_{ab}(1 - \beta_{ab}^2)^{1/2}} \log\left(\frac{1 + \beta_{ab}}{1 - \beta_{ab}}\right) |\mathcal{A}_n|^2, \tag{5.45}$$

which defines the B factor [3] between the energy radiated and the rate of hard scattering $\Gamma_0 = |\mathcal{A}_n|^2$. The graviton radiation rate has infrared and ultraviolet divergences

$$\dot{N}_{gr} = \frac{G}{2\pi} \int_\lambda^\Lambda \frac{dq_0}{q_0} \sum_{a,b} \eta_a \eta_b \frac{m_a m_b (1 + \beta_{ab}^2)}{\beta_{ab}(1 - \beta_{ab}^2)^{1/2}} \log\left(\frac{1 + \beta_{ab}}{1 - \beta_{ab}}\right) \Gamma_0$$

$$= \frac{G}{2\pi} \log\left(\frac{\Lambda}{\lambda}\right) \sum_{a,b} \eta_a \eta_b \frac{m_a m_b (1 + \beta_{ab}^2)}{\beta_{ab}(1 - \beta_{ab}^2)^{1/2}} \log\left(\frac{1 + \beta_{ab}}{1 - \beta_{ab}}\right) \Gamma_0, \tag{5.46}$$

where λ and Λ are the infrared and ultraviolet cutoffs, respectively. The radiated energy rate is ultraviolet divergent

$$\dot{E}_{gr} = \Lambda B \Gamma_0. \tag{5.47}$$

5.5.1 Graviton Emission from Non-relativistic Particle Scattering

Consider the scattering of non-relativistic particles with a graviton bremsstrahlung from an external leg [3]. The B factor

$$B = \frac{G}{2\pi} \sum_{a,b} \eta_a \eta_b \frac{m_a m_b (1 + \beta_{ab}^2)}{\beta_{ab}(1 - \beta_{ab}^2)^{1/2}} \log\left(\frac{1 + \beta_{ab}}{1 - \beta_{ab}}\right) \tag{5.48}$$

can be expanded in powers of the relative velocity β_{ab} as

$$B = \frac{G}{\pi} \sum_{a,b} \eta_a \eta_b \left(1 + \frac{11}{6}\beta_{ab}^2 + \frac{63}{40}\beta_{ab}^4 + \cdots\right) \tag{5.49}$$

and then by taking the non-relativistic expansion of β_{ab} given by

$$\beta_{ab}^2 = v_a^2 + v_b^2 - 2\mathbf{v}_a \cdot \mathbf{v}_b - v_a^2 v_b^2 - 3(\mathbf{v}_a \cdot \mathbf{v}_b)^2 + 2(v_a^2 + v_b^2)(\mathbf{v}_a \cdot \mathbf{v}_b) + \cdots . \quad (5.50)$$

Using (5.50) in (5.49) and using the conservation energy and momentum conservation equations

$$\sum_a \eta_a m_a \left(1 + \frac{1}{2}v_a^2 + \frac{3}{8}v_a^4 + \cdots\right) = 0,$$

$$\sum_a \eta_a m_a \mathbf{v}_a \left(1 + \frac{1}{2}v_a^2 + \cdots\right) = 0, \quad (5.51)$$

we have the non-relativistic expression for B

$$B = \frac{G}{\pi}\left[\frac{16}{5}Q_{ij}Q_{ij} + \frac{94}{15}Q_{ii}^2\right], \quad (5.52)$$

where

$$Q_{ij} = \frac{1}{2}\sum_a \eta_a m_a \, v_{ai} v_{aj}, \qquad Q_{ii} = -\sum_a \eta_a m_a. \quad (5.53)$$

Q_{ij} is invariant under the Galilean transformation $\mathbf{v}_a \rightarrow \mathbf{v}_a + \mathbf{V}$. So it can be computed in any reference frame like the centre of mass (CM) frame. Consider a $2 \rightarrow 2$ elastic scattering with bremsstrahlung of gravitons. In the CM frame,

$$Q_{ij}Q_{ij} = \frac{1}{2}\mu^2 v^4 \sin^2\theta_c, \qquad Q_{ii} = 0, \quad (5.54)$$

where $\mu = m_1 m_2/(m_a + m_2)$ is the reduced mass, $v = |\mathbf{v}_1 - \mathbf{v}_2|$ is the relative velocity and θ_c is the scattering angle in the CM frame. For this system, the B factor is

$$B = \frac{8G}{5\pi}\mu^2 v^4 \sin^2\theta_c. \quad (5.55)$$

If the number densities of the two particles are n_1 and n_2, then their collision rate is

$$\frac{d\Gamma}{d\Omega} = V n_1 n_2 v \frac{d\sigma}{d\Omega}, \quad (5.56)$$

where V is the volume of the source. The rate of gravitational energy radiated by a source in $2 \rightarrow 2$ collisions is therefore

$$\dot{E}_{gw} = \frac{8G}{5\pi}\mu^2 v^5 V n_1 n_2 \Lambda \int d\Omega \frac{d\sigma}{d\Omega} \sin^2\theta_c. \quad (5.57)$$

The cutoff in the energy spectrum can be taken half the KE of the particles $\Lambda = (1/4)\mu v^2$. Now consider the gravitational radiation from a typical star like the Sun. In the Sun, the scatterings take place due to Coulomb interaction between the electrons and electron and protons. Here $\mu = m_e$, $v = (3T/m_e)^{1/2}$, $n_1 = n_2$, $n_2 = n_e + n_p = 2n_e$. The scattering cross section for Coulomb interactions is

$$\int d\Omega \frac{d\sigma}{d\Omega} \sin^2 \theta_c = \frac{8\pi e^2}{(3T)^2} \ln \Lambda_D, \tag{5.58}$$

where Λ_D is the Debye cutoff. The gravitational power radiated by the Sun is therefore

$$\dot{E}_{gw} = P_\odot = \frac{32}{5} G(3T)^{3/2} m_e^{-1/2} n_e^2 V_\odot e^4 \ln \Lambda_D. \tag{5.59}$$

For the Sun, the parameters are $T \simeq 10^7 K$, $n_e = 3 \times 10^{25} \, cm^{-3}$, $V_\odot = 2 \times 10^{31} \, cm^3$ and $\ln \Lambda_D = 4$. With these parameters, the power radiated by the Sun in gravitational waves is

$$P_\odot = 6 \times 10^{14} \frac{erg}{sec}. \tag{5.60}$$

The power radiated by the Hulse–Taylor in gravitational waves is 7.35×10^{33} ergsec.

5.5.2 Graviton Emission from Massless Particle Collisions

To determine the gravitons emission from ultra-relativistic or massless particles, we start with the relation (5.39) and set the masses to zero to obtain

$$\sum_\lambda |\mathcal{A}_{n+1}|^2 = 8\pi G \sum_{a,b} \eta_a \eta_b \frac{(p_a \cdot p_b)^2}{(p_a \cdot q)(p_b \cdot q)} |\mathcal{A}_n|^2. \tag{5.61}$$

The energy rate spectrum is given by

$$\frac{d\dot{E}_{gr}}{dq_0} \equiv B|\mathcal{A}|_n^2 = \int \frac{d^3 q}{2q_0(2\pi)^3} q_0 \, \delta(q_0 - |\mathbf{q}|) \sum_\lambda |\mathcal{A}_{n+1}|^2$$

$$= 8\pi G \int \frac{d^3 q}{2q_0(2\pi)^3} q_0 \, \delta(q_0 - |\mathbf{q}|) \sum_{a,b} \eta_a \eta_b \frac{(p_a \cdot p_b)^2}{(p_a \cdot q)(p_b \cdot q)} |\mathcal{A}_n|^2,$$

$$\tag{5.62}$$

which gives the B factor for massless particles as [11]

$$B = 8\pi G \int \frac{d^3q}{2(2\pi)^3} \delta(q_0 - |\mathbf{q}|) \sum_{a,b} \eta_a \eta_b \frac{(p_a \cdot p_b)^2}{(p_a \cdot q)(p_b \cdot q)}$$

$$= -\frac{2G}{\pi} \sum_{a,b} \eta_a \eta_b (p_a \cdot p_b) \log \frac{|p_a \cdot p_b|}{\mu^2}, \tag{5.63}$$

where μ^2 is an ultraviolet cutoff. Note that owing to conservation of energy momentum,

$$\sum_{a,b} \eta_a \eta_b (p_a \cdot p_b) = 0, \tag{5.64}$$

and the expression B is independent of μ.

The graviton number density radiated is both ultra-violet and infrared divergent

$$\dot{N}_{gr} = 8\pi G \int_\lambda^\Lambda dq^0 \int \frac{d^3q}{2q_0(2\pi)^3} \delta(q_0 - |\mathbf{q}|) \sum_{a,b} \eta_a \eta_b \frac{(p_a \cdot p_b)^2}{(p_a \cdot q)(p_b \cdot q)} |\mathcal{A}_n|^2$$

$$= -\frac{2G}{\pi} \ln \frac{\Lambda}{\lambda} \sum_{a,b} \eta_a \eta_b (p_a \cdot p_b) \log \frac{|p_a \cdot p_b|}{\mu^2} |\mathcal{A}_n|^2, \tag{5.65}$$

and the energy density radiated in gravitons by massless or ultra-relativistic particle collisions is linearly dependent on the ultraviolet cutoff

$$\dot{E}_{gr} = 8\pi G \int_\lambda^\Lambda dq^0 q_0 \int \frac{d^3q}{2q_0(2\pi)^3} \delta(q_0 - |\mathbf{q}|) \sum_{a,b} \eta_a \eta_b \frac{(p_a \cdot p_b)^2}{(p_a \cdot q)(p_b \cdot q)} |\mathcal{A}_n|^2$$

$$= -\frac{2G}{\pi} \Lambda \sum_{a,b} \eta_a \eta_b (p_a \cdot p_b) \log \frac{|p_a \cdot p_b|}{\mu^2} |\mathcal{A}_n|^2$$

$$= \Lambda B \Gamma_0. \tag{5.66}$$

5.6   Gravitational Wave from Sudden Impulse

In a scattering process with a single graviton emission given by (3.118), the energy carried away by gravitational waves is

$$E_{gw} = \int \frac{d^3k}{(2\pi)^3 2k_0} \sum_\lambda |\mathcal{A}_\lambda(k_0, \mathbf{n}k_0)|^2 k_0, \tag{5.67}$$

which is the amplitude squared integrated over the phase space for a graviton multiplied by the energy of a single graviton. The amplitude squared summed over the polarisations is given by

$$
\sum_\lambda \left| \mathcal{A}_\lambda(k_0, \mathbf{n}k_0) \right|^2 = \frac{\kappa^2}{4} \sum_\lambda \epsilon^{*\lambda}_{\mu\nu}(\mathbf{n}) \tilde{T}^{\mu\nu}(k_0, \mathbf{n}k_0) \times \epsilon^\lambda_{\alpha\beta}(\mathbf{n}) \tilde{T}^{*\alpha\beta}(k_0, \mathbf{n}k_0)
$$

$$
= \frac{\kappa^2}{4} \left(\tilde{T}_{\alpha\beta}(k_0, \mathbf{n}k_0) \tilde{T}^{*\alpha\beta}(k_0, \mathbf{n}k_0) - \frac{1}{2} \eta_{\alpha\beta} \left| \tilde{T}^\mu_\mu(k_0, \mathbf{n}k_0) \right|^2 \right)
$$

$$
(5.68)
$$

where we used the completeness relation (3.23). Using (5.68) in (5.67), we obtain the expression for the energy radiated as gravitational waves in terms of the stress tensor in momentum space

$$
E_{gw} = \frac{2G}{(2\pi)^2} \int d\Omega_n dk_0 \, k_0^2 \left(\tilde{T}_{\alpha\beta}(k_0, \mathbf{n}k_0) \tilde{T}^{*\alpha\beta}(k_0, \mathbf{n}k_0) - \frac{1}{2} \eta_{\alpha\beta} \left| \tilde{T}^\mu_\mu(k_0, \mathbf{n}k_0) \right|^2 \right),
$$

$$
(5.69)
$$

which is also written in the differential form to represent the energy spectrum in a given direction

$$
\frac{d^2 E_{gw}}{dk_0 d\Omega_n} = \frac{2G}{(2\pi)^2} k_0^2 \left(\tilde{T}_{\alpha\beta}(k_0, \mathbf{n}k_0) \tilde{T}^{*\alpha\beta}(k_0, \mathbf{n}k_0) - \frac{1}{2} \eta_{\alpha\beta} \left| \tilde{T}^\mu_\mu(k_0, \mathbf{n}k_0) \right|^2 \right).
$$

$$
(5.70)
$$

We will calculate the gravitational radiation from a collection of point particles which scatter of each other and can radiate gravitational waves in the process. Consider a particle 'a' which has an initial velocity $\mathbf{V}_a$ and at time $t = 0$ is scattered to the velocity $\mathbf{V}'_a$. The energy momenta of particles which get a sudden impulse at $t = 0$ can be written as

$$
T^{\mu\nu}(\mathbf{x}, t) = \sum_a \frac{P_a^\mu P_a^\nu}{E_a} \delta^3(\mathbf{x} - \mathbf{V}_a t) \Theta(-t) + \frac{P_a'^\mu P_a'^\nu}{E_a'} \delta^3(\mathbf{x} - \mathbf{V}'_a t) \Theta(t). \quad (5.71)
$$

In Fourier space, the energy–momentum tensor is

$$
\tilde{T}^{\mu\nu}(k_0, \mathbf{n}k_0) = \int d^3x \int_{-\infty}^\infty dt \, e^{-i(k_0 t - k_0 \mathbf{n} \cdot \mathbf{x})} T^{\mu\nu}(\mathbf{x}, t)
$$

$$
= \sum_a \frac{P_a^\mu P_a^\nu}{E_a} \int_{-\infty}^0 dt \, e^{-i(1 - \mathbf{n} \cdot \mathbf{V}_a)k_0 t} + \frac{P_a'^\mu P_a'^\nu}{E_a'} \int_0^\infty dt \, e^{-i(1 - \mathbf{n} \cdot \mathbf{V}'_a)k_0 t}
$$

$$
= i \sum_a \frac{P_a^\mu P_a^\nu}{E_a} \frac{1}{k_0(1 - \mathbf{n} \cdot \mathbf{V}_a)} - \frac{P_a'^\mu P_a'^\nu}{E_a'} \frac{1}{k_0(1 - \mathbf{n} \cdot \mathbf{V}'_a)}
$$

$$
= i \sum_a \eta_a \frac{P_a^\mu P_a^\nu}{k \cdot P_a} \quad (5.72)
$$

with $\eta_a = 1(-1)$ for incoming (outgoing) particles and $k = (k_0, k_0\mathbf{n})$. Now consider a process with where two masses collide and form a single body. This includes processes like a test body falling into a black hole. We can calculate the gravitational radiation emitted in this process using a stress tensor of the form (5.72) in the expression (5.70).

Consider particles with masses m_1 and m_2 with four-momenta P_1 and P_2, respectively, which collide and come to rest. The gravitational wave emitted in the collision has four-momentum $k = P_1 + P_2 - P'$. Without loss of generality, we choose the initial velocities of the two particles along the z-axis. The direction of the gravitational wave is $\mathbf{n} = (\sin\theta\cos\phi, \sin\theta\sin\phi, \cos\theta)$ and the velocities of the particles are $\mathbf{V}_1 = (0, 0, v_1)$, $\mathbf{V}_2 = (0, 0, -v_2)$ and $\mathbf{V}'_{1,2} = (0, 0, 0)$. We choose the four-momenta as follows:

$$P_1 = \gamma_1 m_1(1, 0, 0, v_1), \quad P_2 = \gamma_2 m_2(1, 0, 0, -v_2), \quad P'_1 = (E'_1, 0, 0, 0),$$

$$P'_2 = (E'_2, 0, 0, 0), \quad k = (k_0, k_0 \sin\theta\cos\phi, k_0\sin\theta\sin\phi, k_0\cos\theta). \quad (5.73)$$

Using these in calculating the stress tensor (5.72), we obtain the expression for the energy spectrum radiated in gravitational waves (5.70) given by

$$\frac{d^2 E_{gw}}{dk_0 d\Omega_n} = \frac{G}{(2\pi)^2} \frac{(\gamma_1 m_1 v_1)(\gamma_2 m_2 v_2)(v_1^2 + v_2^2)\sin^4\theta}{(1 - v_1\cos\theta)^2(1 + v_2\cos\theta)^2}. \quad (5.74)$$

Note that by conservation of momenta $\gamma_1 m_1 v_1 = \gamma_2 m_2 v_2$. Consider the case of a high energy particle of mass m_1 and velocity v_1 falling on a black hole with mass m_2 and velocity v_2. Let $m_1 \equiv m \ll m_2 \equiv M$, $v_1 \equiv v \gg v_2$ and $m\gamma \ll M$. We can use the momentum conservation relation to write $(\gamma_1 m_1 v_1)(\gamma_2 m_2 v_2) = (\gamma m v)^2$ and the expression for the energy spectrum of gravitational wave becomes

$$\frac{d^2 E_{gw}}{dk_0 d\Omega_n} = \frac{G}{4\pi^2} \frac{(\gamma^2 m^2 v^4)\sin^4\theta}{(1 - v\cos\theta)^2}. \quad (5.75)$$

Integrating over the angles, we have

$$\frac{dE_{gw}}{dk_0} = \frac{G}{2\pi v}(\gamma^2 m^2)\left(8v - \frac{16}{3}v^3 - 4(1 - v^2)\ln\left(\frac{1+v}{1-v}\right)\right). \quad (5.76)$$

The energy spectrum is flat, but there is a cutoff at $k_{0c} = 0.613/GM$, when the wavelength approaches the size of the horizon [14].

Now consider two identical particles which have a collision and come to rest in the cm frame. We have $m_1 = m_2 \equiv m$ and $v_1 = v_2 = v$. The expression (5.74) now becomes

$$\frac{d^2 E_{gw}}{dk_0 d\Omega_n} = \frac{G}{\pi^2} \frac{(\gamma^2 m^2 v^4)\sin^4\theta}{(1 - v^2\cos^2\theta)^2}. \quad (5.77)$$

Integrating over the angles, we have

$$\frac{dE_{gw}}{dk_0} = \frac{G}{\pi}(2\gamma^2 m^2)\left(2 + (1 - v^2)\left[1 - \frac{1}{2v}(3 + v^2)\ln\left(\frac{1+v}{1-v}\right)\right]\right). \quad (5.78)$$

5.7 Gravitational Waves from Relativistic Bremsstrahlung

We now compute the waveform of gravitational waves radiated by relativistic colliding particles we discussed in the previous section. The amplitude for radiating a graviton of polarisation λ is

$$\mathcal{A}_\lambda = i\frac{\kappa}{2}\epsilon_\lambda^{ij}(\mathbf{n})T_{ij}. \quad (5.79)$$

The amplitude of radiation of the $+$ polarisation graviton is

$$\begin{aligned}
\mathcal{A}_+ &= i\frac{\kappa}{2}\epsilon_+^{ij}(\mathbf{n})T_{ij}(k) = i\frac{\kappa}{2}\left(\mathbf{e}_{\theta i}\mathbf{e}_{\theta j} - \mathbf{e}_{\phi i}\mathbf{e}_{\phi j}\right)T_{ij}(k) \\
&= i\frac{\kappa}{2}\Big(T_{11}(\cos^2\phi - \sin^2\phi\cos^2\theta) + T_{22}(\sin^2\phi - \cos^2\phi\cos^2\theta) \\
&\quad - T_{33}\sin^2\theta - T_{12}\sin 2\phi(1 + \cos^2\theta) + T_{13}\sin\phi\sin 2\theta + T_{23}\cos\phi\sin 2\theta\Big),
\end{aligned}$$
$$(5.80)$$

and the amplitude of emission of graviton of $\times$ polarisation is

$$\begin{aligned}
\mathcal{A}_\times &= i\frac{\kappa}{2}\epsilon_\times^{ij}(\mathbf{n})T_{ij}(k) = i\frac{\kappa}{2}\left(\mathbf{e}_{\theta i}\mathbf{e}_{\phi j} + \mathbf{e}_{\phi i}\mathbf{e}_{\theta j}\right)T_{ij}(k) \\
&= i\frac{\kappa}{2}\Big((T_{11} - T_{22})\sin 2\phi\cos\theta + 2T_{12}\cos 2\phi\cos\theta \\
&\quad - 2T_{13}\cos\phi\sin\theta + 2T_{23}\sin\phi\sin\theta\Big).
\end{aligned}$$
$$(5.81)$$

The gravitational wave amplitude of polarisation λ measured by a detector at a distance r is given in the frequency space by

$$h_\lambda(k_0, r) = \frac{4G}{r}\epsilon_\lambda^{ij}(\mathbf{n})T_{ij}(k). \quad (5.82)$$

Consider the case of a relativistic collision of two particles which come to rest after collision. For the two particles with four-momenta given in (5.73), the stress tensor $T_{ij}(k)$ calculated using (5.72) has only one non-zero component

$$
\begin{aligned}
T_{33}(k) &= \frac{i}{\gamma_1 m_1} \frac{(\gamma_1 m_1 v_1)^2}{(1 - v_1 \cos\theta)} + \frac{i}{\gamma_2 m_2} \frac{(\gamma_2 m_2 v_2)^2}{(1 + v_2 \cos\theta)} \\
&= i \frac{(\gamma_1 m_1 v_1)(v_1 + v_2)}{(1 - v_1 \cos\theta)(1 + v_2 \cos\theta)}.
\end{aligned}
\tag{5.83}
$$

The amplitudes of emitting $+$ and $\times$ polarised gravitons using (5.80) and (5.81) are

$$
\begin{aligned}
\mathcal{A}_+ &= i \frac{\kappa}{2} \frac{(\gamma_1 m_1 v_1)(v_1 + v_2)}{(1 - v_1 \cos\theta)(1 + v_2 \cos\theta)} \sin^2\theta \\
\mathcal{A}_\times &= 0.
\end{aligned}
\tag{5.84}
$$

The gravitational wave amplitude of $+$ and $\times$ polarisations measured at the detector at a distance r is

$$
\begin{aligned}
h_+(k_0, r) &= \frac{4G}{r} \frac{(\gamma_1 m_1 v_1)(v_1 + v_2)}{(1 - v_1 \cos\theta)(1 + v_2 \cos\theta)} \sin^2\theta \\
h_\times(k_0, r) &= 0.
\end{aligned}
\tag{5.85}
$$

Due to the azimuthal symmetry of rotation by angle ϕ around the z-axis, the $\times$ polarisation is absent.

5.8 Non-linear Memory

The non-linear memory is due to the secondary gravitational waves which are emitted by the primary gravitational waves from an oscillating source like a coalescing binary [15–21]. The stress tensor of gravitational waves is related to the energy radiated as

$$
\tau_{ij}^{gw} = \frac{dE_{gw}}{dt d\Omega} n_i n_j,
\tag{5.86}
$$

and the non-linear memory waveform is given by Braginskii and Thorne[16], Christodoulou [17] and Wiseman and Will [18]

$$
h_{ij}(t, r\hat{\mathbf{n}}) = \frac{4G}{r} \int_{-\infty}^{t-r} dt' d\Omega' \frac{dE_{gw}}{dt' d\Omega'} \left[\frac{n_i' n_j'}{1 - \mathbf{n}' \cdot \mathbf{n}} \right]^{TT}.
\tag{5.87}
$$

Here $\mathbf{n}'$ is the unit vector from the source to the solid angle denoted by $d\Omega'$ and $\mathbf{n}$ is the unit vector along the line of sight from the source to the detector.

We derive these results in this section. The source term of the non-linear gravitational waves is the Landau–Lifshitz stress tensor of the primary gravitational waves given by

$$\tau_{ij}^{gw} = \frac{1}{32\pi G} \langle \partial_i h^{ab} \partial_j h_{ab} \rangle, \tag{5.88}$$

where the angular brackets denote averaging over time longer than the time period and volumes larger than the wavelengths of the source gravitational waves h_{ab}. The source gravitational waves travel outward radially with speed of light and are functions of $t - r$, i.e. $h_{ab}(t, \mathbf{x}) = h_{ab}(t - r, \Omega)$. This implies we can relate their spatial and time derivatives as $\partial_i h_{ab}(t - r) = -n_i \partial_0 h_{ab}(t - r)$, where $n_i = x_i/r$. Therefore we can write

$$\tau_{ij}^{gw} = \frac{1}{32\pi G} \langle \partial_i h^{ab} \partial_j h_{ab} \rangle = n_i n_j \tau_{00}^{gw}. \tag{5.89}$$

We can model the energy density τ_{00}^{gw} produced from a source and propagating radially outward on null rays as

$$\tau_{00}^{gw}(t, \mathbf{x}) = \frac{1}{r^2} \frac{dE_{gw}(t - r, \Omega)}{dt d\Omega}, \tag{5.90}$$

where dE_{gw}/dt is the luminosity of the source in gravitational waves and $dE_{gw}/d\Omega$ denotes the angular distribution of the source luminosity. Therefore we can write τ_{ij}^{gw} in terms of the energy flux as

$$\tau_{ij}^{gw} = n_i n_j \tau_{00}^{gw} = n_i n_j \frac{1}{r^2} \frac{dE_{gw}(t - r, \Omega)}{dt d\Omega}. \tag{5.91}$$

The secondary gravitational waves from the sourced by the gravitational wave stress tensor will obey the inhomogeneous wave equation

$$\Box h_{ij} = -16\pi G \tau_{ij}^{gw} \tag{5.92}$$

where on the rhs we take the transverse–traceless projection of τ_{ij}^{gw}. The solution of (5.92) is of the form

$$h_{ij}(t, \mathbf{x}) = 4G \int dt' d^3x' \, \tau_{ij}^{gw}(t', \mathbf{x}') \frac{\delta \left(t' - (t - |\mathbf{x} - \mathbf{x}'|)\right)}{|\mathbf{x} - \mathbf{x}'|}. \tag{5.93}$$

We can express the source (5.89) in terms of the null coordinate $u = t' - r'$ as follows:

$$\tau_{ij}^{gw}(t', \mathbf{x}') = \frac{n_i' n_j'}{r'^2} \frac{dE_{gw}(t' - r', \Omega')}{dt' d\Omega'}$$

$$= \int du \, \frac{n_i' n_j'}{r'^2} \delta(u - (t' - r')) \frac{dE_{gw}(u, \Omega)}{dt' d\Omega'}. \tag{5.94}$$

Substituting (5.94) in (5.93), we obtain

$$h_{ij}(t, \mathbf{x}) = 4G \int du \, dt' dr' \, r'^2 d\Omega' \frac{n_i' n_j'}{r'^2} \frac{\delta\left(t' - (t - |\mathbf{x} - \mathbf{x}'|)\right)}{|\mathbf{x} - \mathbf{x}'|}$$

$$\times \delta(u - (t' - r')) \frac{dE_{gw}(u, \Omega)}{dt' d\Omega'}. \tag{5.95}$$

The distance of the observer is much larger than the source size, $r \gg r'$, and we take the approximations

$$\frac{1}{|\mathbf{x} - \mathbf{x}'|} \simeq \frac{1}{r(1 - \mathbf{n}' \cdot \mathbf{n})}, \quad \delta\left(t' - (t - |\mathbf{x} - \mathbf{x}'|)\right) \simeq \delta(t' - (t - r)). \tag{5.96}$$

Now we can perform the integral over r' using the second delta function in (5.95) and then do the t' integration using the remaining delta function to obtain

$$h_{ij}(t, \mathbf{x}) = \frac{4G}{r} \int_{-\infty}^{t-r} du \int_{4\pi} d\Omega' \frac{dE_{gw}(u, \Omega)}{du \, d\Omega'} \frac{n_i' n_j'}{(1 - \mathbf{n}' \cdot \mathbf{n})}. \tag{5.97}$$

5.9 Non-linear Memory from Binary Orbits

We will now compute the non-linear memory from binary orbits. The case of elliptical orbits is important from the point of observations as the non-linear memory signal arises at the same Newtonian order as the oscillatory signals. For the case of hyperbolic orbits, the non-linear term is suppressed by a factor $(v/c)^5$ compared to the quadrupole signal [18].

We will start by computing the non-linear memory from binaries in a circular orbit. The expression for rate of energy radiated in the direction $d\Omega'$ is given by

$$\frac{dE_{gw}}{dt' d\Omega'} = \frac{\kappa^2}{4} \int \left(T_{ij}(\omega') T_{kl}^*(\omega') \Lambda_{ij,kl}(\mathbf{n}')\right) \omega^3 2\pi \delta(\omega - 2\omega_0) \frac{d\omega}{(2\pi)^3 2\omega}, \tag{5.98}$$

where $\omega_0 = \sqrt{G(m_1 + m_2)/a^3}$ is the angular frequency of the Kepler orbit and $\Lambda_{ij,kl}(\mathbf{n}')$ is the transverse–traceless projection operator defined w.r.t. the direction of the gravitational wave emitted $\hat{n}'$. The explicit form of the TT projection operator is

$$\Lambda_{ij,kl}(\hat{n}') = P_{ik}(\hat{n}')P_{jl}(\hat{n}') - \frac{1}{2}P_{ij}(\hat{n}')P_{kl}(\hat{n}')$$

$$= \left(\delta_{ik} - \hat{n}'_i\hat{n}'_k\right)\left(\delta_{jl} - \hat{n}'_j\hat{n}'_l\right) - \frac{1}{2}\left(\delta_{ij} - \hat{n}'_i\hat{n}'_j\right)\left(\delta_{kl} - \hat{n}'_k\hat{n}'_l\right). \quad (5.99)$$

Using this, we can write

$$T_{ij}T^*_{kl}\Lambda_{ij,kl}(\mathbf{n}') = \left(T_{ij}T^*_{ji} - 2T_{ij}T^*_{jl}\hat{n}'_i\hat{n}'_l + \frac{1}{2}T_{ij}T^*_{kl}\hat{n}'_i\hat{n}'_j\hat{n}'_k\hat{n}'_l\right). \quad (5.100)$$

We will choose the orbital plane of the binary to be the x–y plane, see Fig.5.2. The observer is located at $\mathbf{r} = r\hat{n} = r(0, \sin i, \cos i)$. The primary graviton emits the secondary graviton at $\mathbf{r}' = r'\hat{n}' = r'(\sin\theta'\cos\phi', \sin\theta'\sin\phi', \cos\theta')$, which is integrated over all space. As the energy density of the primary gravitons falls off as $1/r'^2$, taking the approximation $r' \ll r$, it is a valid assumption.

For the circular orbit non-zero, the components of the stress tensor in Fourier space are given by

$$T_{xx}(\omega') = \frac{\mu a^2 \omega_0^2}{2}, \quad T_{yy}(\omega') = -\frac{\mu a^2 \omega_0^2}{2}, \quad T_{xy}(\omega') = i\frac{\mu a^2 \omega_0^2}{2}. \quad (5.101)$$

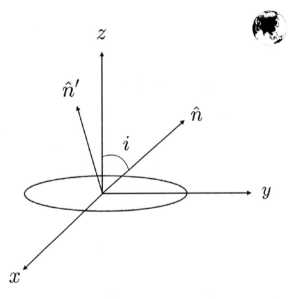

Fig. 5.2 The primary graviton emits a secondary graviton at $\mathbf{r}' = r'\hat{n}' = r'(\sin\theta'\cos\phi', \sin\theta'\sin\phi', \cos\theta')$ and the secondary graviton travels to the located at $\mathbf{r} = r\hat{n} = (0, \sin i, \cos i)$

Using (5.101) in (5.100), we obtain for the circular binary orbit.

$$T_{ij}T_{kl}^*\Lambda_{ij,kl}(\mathbf{n'}) = \frac{\mu^2 a^4 \omega_0^4}{4}\frac{1}{2}\left(1 + 6\cos^2\theta'\sin^2\phi' + \cos^4\theta'\cos^2\phi'\right). \quad (5.102)$$

Using (5.102) in (5.98), we obtain

$$\frac{dE_{gw}}{dt'd\Omega'} = \frac{G}{2\pi}\omega_0^6\mu^2 a^4\left(1 + 6\cos^2\theta'\sin^2\phi' + \cos^4\theta'\cos^2\phi'\right). \quad (5.103)$$

Using the expression $\omega_0 = (GM/a^3)^{1/2}$ for the Kepler orbital frequency (where $M = m_1 + m_2$), we can have the expression for the rate of energy radiated in terms of the semi-major axis a

$$\frac{dE_{gw}}{dt'd\Omega'} = \frac{G^4}{2\pi}\frac{\mu^2 M^3}{a^5}\left(1 + 6\cos^2\theta'\sin^2\phi' + \cos^4\theta'\cos^2\phi'\right). \quad (5.104)$$

Using this expression in (5.97), we obtain the expression for the memory signal

$$\begin{aligned} h_{ij}(t,\mathbf{x}) &= \frac{4G}{r}\int_{-\infty}^{t-r}dt'\int_{4\pi}d\Omega'\frac{dE_{gw}(t',\Omega)}{dt'd\Omega'}\frac{n_i'n_j'}{(1-\hat{n}'\cdot\hat{n})} \\ &= \frac{4G}{r}\int_{-\infty}^{t-r}dt'\frac{G^4}{2\pi}\frac{\mu^2 M^3}{a^5}\int_{4\pi}d\Omega'\left(1 + 6\cos^2\theta'\sin^2\phi' + \cos^4\theta'\cos^2\phi'\right) \\ &\quad \times \frac{n_i'n_j'}{(1-\hat{n}'\cdot\hat{n})} \end{aligned} \quad (5.105)$$

where we have written $u = t' - r' \simeq t'$ since the gravitational wave energy which is the source falls of as $1/r'^2$ while $t' \in (0, t-r)$ and we can take $t' \gg r'$ and replace u by t'. To obtain a transverse–traceless part of h_{ij}, we apply the TT projection operator

$$\begin{aligned} h_{ij}^{TT}(t,\mathbf{x}) &= \frac{G^5}{2\pi r}\mu^2 M^3\int_{-\infty}^{t-r}dt'\frac{1}{a^5} \\ &\quad \times \int_{4\pi}d\Omega'\left(1 + 6\cos^2\theta'\sin^2\phi' + \cos^4\theta'\cos^2\phi'\right)\frac{\Lambda_{ij,kl}(\hat{n})n_k'n_l'}{(1-\hat{n}'\cdot\hat{n})}. \end{aligned} \quad (5.106)$$

The angular integral can be now performed by substituting

$$\hat{n}'\cdot\hat{n} = \sin i\,\sin\theta'\cos\phi' + \cos i\,\cos\theta' \quad (5.107)$$

in the denominator of the angular part. The angular integration over θ' and ϕ' gives

$$\int_{4\pi} d\Omega' \left(1 + 6\cos^2\theta'\sin^2\phi' + \cos^4\theta'\cos^2\phi'\right) \frac{\Lambda_{ij,kl}(\hat{n})n'_k n'_l}{(1 - \hat{n}'\cdot\hat{n})}$$

$$= \frac{1}{96}\sin^2 i \left(17 + \cos^2 i\right)\epsilon_{ij}^+. \quad (5.108)$$

The time dependence of the integrand in (5.106) is due to the change in the radius of the orbit which occurs due to the energy loss of the orbit by the primary gravitational waves. There change in the radius is given $da/dt = (dE/dt)(da/dE)$ which using $E = -(1/2)G\mu M/a$ is

$$\frac{da}{dt} = -\frac{64}{5}G^3\frac{\mu M^2}{a^3}. \quad (5.109)$$

Solving this equation gives us the time dependence of the separation distance $a(t)$

$$a(t) = \left(\frac{256}{5}G^3\mu M^2\right)^{1/4}(t_c - t)^{1/4}, \quad (5.110)$$

where t_c is the time of coalescence of the black holes.

The growth of memory signal stops at the time of the coalescence t_c when the black hole separation becomes smaller than innermost stable circular orbit radius $r_c = 6GM$. The frequency $f = \omega_0/\pi = ((GM)/a^3)^{1/2}/pi$ increases till the time of coalescence as

$$f(t') \simeq \left(\frac{5}{256}\right)^{3/8}\frac{1}{\pi}(G\mathcal{M}_c)^{-5/8}|t' - t_c|^{-3/8}, \quad (5.111)$$

where $\mathcal{M}_c = \mu^{3/5}M^{2/5}$ is the chirp mass of the binary pair.

Using (5.108) and (5.110) in (5.106), we obtain the expression for the non-linear memory signal of binaries with rotation axis at an angle i to the Earth-source direction given by

$$h_+(t) = \frac{2G}{r}\left(\frac{5G\mu^3 M}{r_c^4(t - t_c)}\right)^{1/4}\frac{1}{96}\sin^2 i \left(17 + \cos^2 i\right). \quad (5.112)$$

The memory effect will be strongest in edge on binaries $i = \pi/2$ and will be zero in the face on binaries $i = 0$. In Fig. 5.3, the simulated memory signal for the first observed black hole binary merger event GW150914 is shown.

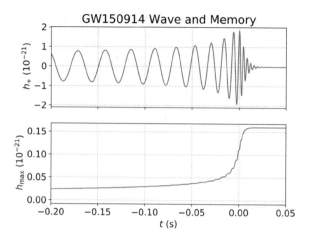

Fig. 5.3 Simulated non-linear memory signal for the black hole merger event GW150914. Reprinted from Johnson et al. [22]. Figure credit: ©2019 American Physical Society. Reproduced with permissions. All rights reserved

5.10 Gravitational Wave Signal from Supernova Neutrino Burst

In a supernova burst, the neutrino emission during core-collapse is a transient event which lasts about 1 sec, and the neutrinos carry away the gravitational binding energy of the proto-star which could be of the order $\mathcal{L}_\nu \simeq 10^{52}$ ergs/sec. The gravitational wave signal from a supernova event can be significant if the neutrino emission is asymmetric and results in a change in the quadrupole moment of the mass distribution. Thus supernovae explosions can be a source of the gravitational memory signal [23–25].

For an observer located at $\mathbf{x}$, the gravitational wave from a supernova arrives at a time t for a supernova collapse at time $t' = t - r$. The gravitational wave signal is

$$h_{ij}(t, \mathbf{x}) = 4G \int \frac{T_{ij}(t' - (t - r), \mathbf{x}')}{|\mathbf{x} - \mathbf{x}'|} d^3 x'. \tag{5.113}$$

The transient neutrino emission which propagates radially with the speed of light can be modelled by the stress tensor [26]

$$T_{ij}(t', \mathbf{x}') = \frac{n_i' n_j'}{r'^2} \mathcal{L}_\nu(t' - r') f(t' - r', \Omega'), \tag{5.114}$$

where $\mathcal{L}_\nu(t' - r')$ is the neutrino luminosity which is the energy carried away as neutrinos per unit time. The function $f(t' - r', \Omega')$, with the property $\int d\Omega' f = 1$, determines the angular anisotropy of neutrino emission. The neutrino emission should be asymmetric and have an angular dependence in order for gravitational waves to be generated. We write the neutrino stress tensor (5.114) as

$$
T_{ij}(t', \mathbf{x}') = \frac{n'_i n'_j}{r'^2} \int_{-\infty}^{\infty} dt'' \delta(t'' - (t' - r')) \mathcal{L}_v(t'') f(t'', \Omega') .
\tag{5.115}
$$

Inserting (5.115) in (5.116), we can perform the integration over r' using the delta function to obtain

$$
\begin{aligned}
h_{ij}(t, \mathbf{x}) &= 4G \int_{-\infty}^{t-r} dt'' \int_{4\pi} d\Omega' \, (n'_i n'_j)^{TT} \mathcal{L}_v(t'') f(t'', \Omega') \frac{1}{t - t'' - r \cos \theta'} \\
&\simeq \frac{4G}{r} \int_{-\infty}^{t-r} dt'' \int_{4\pi} d\Omega' \, (n'_i n'_j)^{TT} \mathcal{L}_v(t'') f(t'', \Omega') \frac{1}{(1 - \hat{n}' \cdot \hat{n})} ,
\end{aligned}
\tag{5.116}
$$

where $\hat{n} = \mathbf{r}/r$ is the direction of the detector from the source.

Typical strain that can be expected from supernova neutrinos is

$$
h = 10^{-21} \left(\frac{10 \text{kpc}}{r} \right) \left(\frac{\alpha}{0.1} \right) \left(\frac{\mathcal{L}_v}{10^{53} \text{ ergs/sec}} \right) \left(\frac{\Delta t}{1 \text{ sec}} \right) ,
\tag{5.117}
$$

where α is the asymmetry factor for the neutrino luminosity (Fig. 5.4).

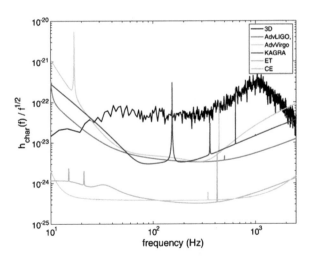

Fig. 5.4 Detection of the gravitational wave signal from a core-collapse supernova across the full spectrum of emission. Reprinted from [23]. Figure credit: ©2020 American Physical Society. Reproduced with permissions. All rights reserved

References

1. Y.B. Zel'dovich, A.G. Polnarev, Radiation of gravitational waves by a cluster of superdense stars. Soviet Astr. **18**, 17 (1974). https://adsabs.harvard.edu/full/1974SvA....18...17Z
2. S. Weinberg, Photons and gravitons in perturbation theory: derivation of Maxwell's and Einstein's equations. Phys. Rev. **138**, B988 (1965)
3. S. Weinberg, Infrared photons and gravitons. Phys. Rev. **140**, B516 (1965)
4. S. Weinberg, *The Quantum Theory of Fields: Volume 1, Foundations* (Cambridge University Press, Cambridge, 1995)
5. A. Strominger, *Lectures on the Infrared Structure of Gravity and Gauge Theory* (Princeton University Press, Princeton, 2017). [arXiv:1703.05448 [hep-th]]
6. F. Cachazo, A. Strominger, Evidence for a New Soft Graviton Theorem (2014). [arXiv:1404.4091 [hep-th]]
7. B. Sahoo, A. Sen, Classical and quantum results on logarithmic terms in the soft theorem in four dimensions. J. High Energy Phys. **02**, 086 (2019) [arXiv:1808.03288 [hep-th]]
8. Z. Bern, S. Davies, P. Di Vecchia, J. Nohle, Low-energy behavior of gluons and gravitons from gauge invariance. Phys. Rev. D **90**(8), 084035 (2014). [arXiv:1406.6987 [hep-th]]
9. A. Laddha, A. Sen, Observational signature of the logarithmic terms in the soft graviton theorem. Phys. Rev. D **100**(2), 024009 (2019). [arXiv:1806.01872 [hep-th]]
10. D. Ghosh, B. Sahoo, Spin dependent gravitational tail memory in $D = 4$ (2021). [arXiv:2106.10741 [hep-th]]
11. A. Addazi, M. Bianchi, G. Veneziano, Glimpses of black hole formation/evaporation in highly inelastic, ultra-planckian string collisions. J. High Energy Phys. **02**, 111 (2017). [arXiv:1611.03643 [hep-th]]
12. A. Addazi, M. Bianchi, G. Veneziano, Soft gravitational radiation from ultra-relativistic collisions at sub- and sub-sub-leading order. J. High Energy Phys. **05**, 050 (2019). [arXiv:1901.10986 [hep-th]]
13. A. Addazi, K.A. Zeng, Soft gravitational radiation from multi-body collisions. J. High Energy Phys. **11**, 193 (2021). [arXiv:2110.01194 [hep-th]]
14. L. Smarr, Gravitational radiation from distant encounters and from headon collisions of black holes: the zero frequency limit. Phys. Rev. D **15**, 2069–2077 (1977)
15. D. Kennefick, Prospects for detecting the Christodoulou memory of gravitational waves from a coalescing compact binary and using it to measure neutron-star radii. Phys. Rev. D **50**, 3587 (1994)
16. V.B. Braginskii, K.S. Thorne, Gravitational-wave bursts with memory and experimental prospects. Nature **327**, 123–125 (1987)
17. D. Christodoulou, Nonlinear nature of gravitation and gravitational wave experiments. Phys. Rev. Lett. **67**, 1486–1489 (1991)
18. A.G. Wiseman, C.M. Will, Christodoulou's nonlinear gravitational wave memory: Evaluation in the quadrupole approximation. Phys. Rev. **D44**(10), R2945–R2949 (1991)
19. M. Favata, Post-Newtonian corrections to the gravitational-wave memory for quasi-circular, inspiralling compact binaries. Phys. Rev. D **80**, 024002 (2009). [arXiv:0812.0069 [gr-qc]]
20. M. Favata, The gravitational-wave memory effect. Class. Quant. Grav. **27**, 084036 (2010). [arXiv:1003.3486 [gr-qc]]
21. A. Tolish, R.M. Wald, Retarded fields of null particles and the memory effect. Phys. Rev. D **89**(6), 064008 (2014). [arXiv:1401.5831 [gr-qc]]
22. A.D. Johnson, S.J. Kapadia, A. Osborne, A. Hixon, D. Kennefick, Prospects of detecting the nonlinear gravitational wave memory. Phys. Rev. D **99**(4), 044045 (2019). https://doi.org/10.1103/PhysRevD.99.044045. [arXiv:1810.09563 [gr-qc]]
23. A. Mezzacappa, P. Marronetti, R.E. Landfield, E.J. Lentz, K.N. Yakunin, S.W. Bruenn, W.R. Hix, O.E. Bronson Messer, E. Endeve, J.M. Blondin, et al., Gravitational-wave signal of a core-collapse supernova explosion of a 15 $M_\odot$ star. Phys. Rev. D **102**(2), 023027 (2020). https://doi.org/10.1103/PhysRevD.102.023027. [arXiv:2007.15099 [astro-ph.HE]]

24. K. Kotake, K. Sato, K. Takahashi, Explosion mechanism, neutrino burst, and gravitational wave in core-collapse supernovae. Rept. Prog. Phys. **69**, 971–1144 (2006). arXiv:astro-ph/0509456 [astro-ph]]

25. D. Radice, V. Morozova, A. Burrows, D. Vartanyan, H. Nagakura, Characterizing the gravitational wave signal from core-collapse supernovae. Astrophys. J. Lett. **876**(1), L9 (2019). https://doi.org/10.3847/2041-8213/ab191a. [arXiv:1812.07703 [astro-ph.HE]]

26. R. Epstein, The generation of gravitational radiation by escaping supernova neutrinos. Astrophys. J. **223**, 1037–1045 (1978)

Backreaction and Dissipation: The In-In Formalism

Abstract

Gravitational radiation causes a backreaction which is described by the dissipative self-force on the binary dynamics. This dissipative force can be obtained by going to a closed time path formulation or the in-in formulation which is the method suitable for describing non-equilibrium phenomenon. We discuss the CTP path integrals and the Keldysh formulation and including extensions to finite temperature. We illustrate these concepts by calculating the wave form and energy radiated from binaries in the CTP formalism. The Burke-Thorne dissipative potential on binaries is derived.

6.1 Introduction

We have seen in Sect. 3.14 that the imaginary part of the effective action gives the energy radiated. Due to the loss of energy there will be a backreaction on the orbit, which can be calculated by introducing the Burke-Thorne dissipative potential [1,2]

$$\Phi_{BT}(t, \mathbf{x}) = \frac{G}{5} x^i x^j \frac{d^5 I_{ij}}{dt^5}, \tag{6.1}$$

where I_{ij} is the quadrupole distribution of the mass of the source.

In Sect. 3.14 we have seen that the probability of gravitational radiation is obtained by starting with the amplitude for the 'in' vacuum at $t = -\infty$ to persist to be the 'out' vacuum at time $t \to \infty$ in the presence of sources J is given by

$$\langle 0_{\text{out}} | e^{i W(J)/\hbar} | 0_{\text{in}} \rangle \tag{6.2}$$

© The Author(s), under exclusive license to Springer Nature Switzerland AG 2023
S. Mohanty, *Gravitational Waves from a Quantum Field Theory Perspective*,
Lecture Notes in Physics 1013, https://doi.org/10.1007/978-3-031-23770-6_6

The probability of the persistence of vacuum (no particle production) is therefore

$$\left| \langle 0_{\text{out}} | e^{i W(J)/\hbar} | 0_{\text{in}} \rangle \right|^2 . \tag{6.3}$$

The departure from unity of this probability if

$$\frac{2}{\hbar} Im\, W(J) \neq 0 . \tag{6.4}$$

The graviton production probability is a Poisson distribution average number of gravitons given by

$$\langle N \rangle = \frac{2}{\hbar} Im\, W(J) . \tag{6.5}$$

Computing the imaginary part of the action $W(J)$ gives classical quadrupole formula for the gravitational energy loss as we have seen in Sect. 3.14. We expect the radiation reaction potential to arise from the real part of the action (6.7). This turns out not to be the case, however, as the real part of the action turns out to be a time integral of a total time derivative which does not contribute to the equation of motion. The reason that the dynamics of the backreaction is not obtained from the real part of the action is due to the fact that the real part of the Feynman propagator is time symmetric whereas the dissipative dynamics is not. Hence we need to generalise the formalism in order to capture the backreaction dynamics in a Lagrangian or Hamiltonian formalism.

6.2   Radiation Reaction in the In-Out Formalism

Schwinger's action integral $S_{eff}^{(2)}$ is the effective action for matter interactions at the second order in matter currents, after integrating out the graviton. This is described by the Feynman diagram

$$S_{eff}^{(2)} = \underline{\qquad\qquad\qquad} \tag{6.6}$$

is given by 3.156 (we switch back to $\hbar = c = 1$ units),

$$
\begin{aligned}
S_{eff}^{(2)} = W &= 8\pi G \frac{1}{2} \int \frac{d^4 k}{(2\pi)^4}\; T_{\mu\nu}(k)\; P_{\mu\nu\alpha\beta}\; T_{\alpha\beta}^*(k)\; \frac{1}{|\mathbf{k}| - \omega^2 - i\epsilon} \\
&= 8\pi G \frac{1}{2} \int \frac{d^4 k}{(2\pi)^4} \sum_{\lambda} \left| \epsilon_{\lambda}^{*\mu\nu}(\mathbf{n}) T_{\mu\nu}(k) \right|^2 \frac{1}{|\mathbf{k}| - \omega^2 - i\epsilon} .
\end{aligned} \tag{6.7}
$$

This action is second order in gravitational coupling and the factor of $1/2$ the symmetry factor of the diagram. The propagator is the Feynman propagator.

6.3 Propagators

The Greens function is the solution of the wave equation with delta function source. Here we consider massless particles Greens function which obey

$$\Box G(x, x') = -\delta^4(x - x').$$ (6.8)

Now for solving the wave equation in presence of a source $S(x)$

$$\Box f(x) = S(x),$$ (6.9)

the solution is given by

$$f(x) = \int d^4x \, d^4x' \, G(x, x') \, S(x').$$ (6.10)

To solve (6.8) we do a transformation to the momentum space

$$G(x, x') = \int \frac{d^4k}{(2\pi)^4} G(k) e^{ik \cdot (x-x')}$$ (6.11)

and similarly for the delta function

$$\delta^4(x - x') = \int \frac{d^4k}{(2\pi)^4} e^{ik \cdot (x-x')}.$$ (6.12)

Then substituting in (6.8), we can see that the solution $G(k)$ takes the form

$$G(k) = -\frac{1}{k^2}.$$ (6.13)

The Greens function in spacetime is then given by

$$G(x, x') = \int \frac{d^4k}{(2\pi)^4} \frac{e^{ik \cdot (x-x')}}{-k^2 \pm i\epsilon},$$ (6.14)

where ϵ is a positive constant chosen to regulate the integrals, which we will take to zero at the end of the calculation.

Greens function in spacetime $G(x, x')$ takes different forms depending upon the sign chosen of the pole term.

We have the retarded and advanced propagators

$$
\begin{aligned}
G_{R,A}(t - t', |\mathbf{x} - \mathbf{x}'|) &= \int \frac{d^4 k}{(2\pi)^4} \frac{e^{ik \cdot (x - x')}}{(\omega \pm i\epsilon)^2 - |\mathbf{k}|^2} \\
&= -\frac{1}{4\pi} \left[\frac{\delta((t - t') \mp |\mathbf{x} - \mathbf{x}'|)}{|\mathbf{x} - \mathbf{x}'|} \right],
\end{aligned}
\tag{6.15}
$$

where $k = (\omega, |\mathbf{k}|)$ and we the $-(+)$ signs in the delta functions are for the retarded(advanced) propagators. The retarded Greens function propagates signals in the past light-cone $t = t' + |\mathbf{x} - \mathbf{x}'|$. This is used in classical physics. The advanced Greens function propagates signals in the future light-cone. Under time reversal the advance and retarded propagators get interchanged,

$$
G_R(t - t', |\mathbf{x} - \mathbf{x}'|) = G_A(t' - t, |\mathbf{x} - \mathbf{x}'|).
\tag{6.16}
$$

The Feynman propagator is given by

$$
G_F(t - t', |\mathbf{x} - \mathbf{x}'|) = \int \frac{d^4 k}{(2\pi)^4} \frac{e^{-i\omega(t - t') + i\mathbf{k} \cdot (\mathbf{x} - \mathbf{x}')}}{\omega^2 - |\mathbf{k}|^2 + i\epsilon}.
\tag{6.17}
$$

In momentum space the Feynman propagator is

$$
G_F(k) = \frac{1}{\omega^2 - |\mathbf{k}|^2 + i\epsilon}.
\tag{6.18}
$$

The real and imaginary parts of the Feynman propagator are

$$
G_F(k) = P \left(\frac{1}{\omega^2 - |\mathbf{k}|^2} \right) - i\pi \delta(\omega^2 - |\mathbf{k}|^2).
\tag{6.19}
$$

The imaginary part of the Feynman propagator are for on-shell particles which obey $k^2 = 0$.

The Dyson propagator is

$$
G_F(t - t', |\mathbf{x} - \mathbf{x}'|) = \int \frac{d^4 k}{(2\pi)^4} \frac{e^{-i\omega(t - t') + i\mathbf{k} \cdot (\mathbf{x} - \mathbf{x}')}}{\omega^2 - |\mathbf{k}|^2 - i\epsilon}.
\tag{6.20}
$$

QFT calculations use the Feynman propagator as the particles at short distances from the sources can be off-shell ($k^2 \neq 0$).

Wightman functions are

$$
\Delta_{\pm}(t - t', |\mathbf{x} - \mathbf{x}'|) \equiv \int \frac{d^3 k}{(2\pi)^3 2\omega} e^{-i\omega(t-t')+i\mathbf{k}\cdot(\mathbf{x}-\mathbf{x}')}
$$

$$
= \int \frac{d^4 k}{(2\pi)^4} \, \Theta(\pm\omega) \, \delta(\omega^2 - |\mathbf{k}|^2) e^{-i\omega(t-t')+i\mathbf{k}\cdot(\mathbf{x}-\mathbf{x}')}.
$$

$$(6.21)$$

The Wightman functions support propagation on the past and future light-cones. The Hadamard propagator is the sum of Wightman functions

$$
G_H(t - t', |\mathbf{x} - \mathbf{x}'|) = \frac{1}{2} (\Delta_+ + \Delta_-).
$$

$$(6.22)$$

The Feynman propagator is related to the others as

$$
G_F(t - t', |\mathbf{x} - \mathbf{x}'|) = \frac{1}{2} (G_R + G_A) - \frac{i}{2} (\Delta_+ + \Delta_-).
$$

$$(6.23)$$

6.4 Lagrangian Dynamics of Dissipative Systems

An accelerating charge radiates energy. Due to energy–momentum conservation the charge must lose kinetic energy and there is a recoil due to the momentum of the emitted photon. This backreaction is given by the Abraham-Lorentz self-force

$$
m\dot{\mathbf{v}} = \frac{e^2}{6\pi} \ddot{v} + \mathbf{f}(t),
$$

$$(6.24)$$

where $\mathbf{f}(t)$ is the external force that accelerates the electron. This equation of motion can be obtained from the general Lagrangian equation

$$
\frac{d}{dt} \left(\frac{\partial L}{\partial \dot{x}_i} \right) - \frac{\partial L}{\partial x_i} = 0
$$

$$(6.25)$$

by choosing the Lagrangian

$$
L = \frac{1}{2} m v^2 - K, \quad K = \frac{-e^2}{6\pi} \mathbf{v} \cdot \dot{\mathbf{v}} + \mathbf{f}(t) \cdot \mathbf{x}.
$$

$$(6.26)$$

Although the Lorentz self-force can be implemented in the equation of motion by adding the dissipation potential K at the level of equation of motion, such an equation of motion cannot be derived from the usual variational principle of minimising the action. The reason is that in the usual action

$$
S = \int_{t_i}^{t_f} dt \, L(q, \dot{q}, t)
$$

$$(6.27)$$

we vary the action with the boundary conditions $q(t_i) = q_i$ and $q(t_f) = q_f$. On the other hand with the velocity dependent self-force once can solve the dynamics using the initial conditions $q(t_i) = q_i$ and $\dot{q}(t_i) = v_i$. In order to convert the initial value problem to a boundary value problem and formulate the dissipative dynamics by a Lagrangian variational principle we have double the degrees of freedom [3,4].

Consider a generalised set of coordinates $q_1 = \{q_1^I\}, I = 1 \cdots N$. Introduce a set of auxiliary coordinates $q_2 = \{q_2^I\}, I = 1 \cdots N$. We propagate q_1 forward in time $t \geq t_i$, using the initial conditions $q_1(t_i) = q_{1i}$ and $\dot{q}_1(t_i) = v_{1i}$. On the other hand we propagate the auxiliary fields back-wards in time $t \leq t_f$ using the 'initial' conditions on t_f, $q_2(t_f) = q_{2f}$ and $\dot{q}_2(t_f) = v_{2f}$. The Lagrangian for the q_1, q_2 system can then be written as

$$S(q_1, q_2) = \int_{t_i}^{t_f} dt \Big(L(q_1, \dot{q}_1, t) - L(q_2, \dot{q}_2, t) + K(q_1, \dot{q}_1, q_2, \dot{q}_2) \Big). \qquad (6.28)$$

Since

$$- \int_{t_i}^{t_f} L(q_2, \dot{q}_2, t)\, dt = \int_{t_f}^{t_i} L(q_2, \dot{q}_2, t)\, dt \qquad (6.29)$$

it is clear that $-L(q_2, \dot{q}_2, t)$ propagates the auxiliary fields from t_f to t_i. At this points both q_1 and q_2 are on the same footing the action should be invariant (up to an overall sign which does not change the equations of motions) to the interchange $q_1 \leftrightarrow q_2$. This implies that K which represents the dissipative interactions should be antisymmetric in the exchange $q_1 \leftrightarrow q_2$.

Denote the conjugate momenta as $\pi_{1,2}$, these are defined through the total Lagrangian including the dissipative term

$$\Lambda = L(q_1, \dot{q}_1, t) - L(q_2, \dot{q}_2, t) + K(q_1, \dot{q}_1, q_2, \dot{q}_2) \qquad (6.30)$$

and the conjugate momenta of q_1 and q_2 are

$$\pi_1 \equiv \frac{\partial \Lambda}{\partial \dot{q}_1} = \frac{\partial L(q_1, \dot{q}_1, t)}{\partial \dot{q}_1} + \frac{\partial K(q_1, \dot{q}_1, q_2, \dot{q}_2)}{\partial \dot{q}_1}$$

$$\pi_2 \equiv -\frac{\partial \Lambda}{\partial \dot{q}_2} = \frac{\partial L(q_2, \dot{q}_2, t)}{\partial \dot{q}_2} - \frac{\partial K(q_1, \dot{q}_1, q_2, \dot{q}_2)}{\partial \dot{q}_1}. \qquad (6.31)$$

Note that the conjugate momenta are defined in such a way that the canonical momenta without the dissipative term comes out with the usual sign,

$$p_a = \frac{\partial L(q_a, \dot{q}_a, t)}{\partial \dot{q}_a}, \quad a = 1, 2, \qquad (6.32)$$

while the momentum associated with K,

$$\kappa_a \equiv \frac{\partial K(q_1, \dot{q}_1, q_2, \dot{q}_2)}{\partial \dot{q}_a}, \qquad a = 1, 2. \tag{6.33}$$

contributes with opposite signs (this is due to the antisymmetry of K on exchange of $1 \leftrightarrow 2$. The boundary conditions that are imposed are, at t_f we have $q_1(t_f) = q_2(t_f) = q_{2f}$ and $\pi_1(t_f) = \pi_2(t_f) = \pi_{2f}$ and at t_i we have $q_1(t_i) = q_2(t_i) = q_{1i}$ and $\pi_1(t_i) = \pi_2(t_i) = \pi_{1i}$. With four variables we can solve for the dynamics of $q_1(t)$ and $q_2(t)$ for $t_i \le t \le t_f$. Here the variable $q_1(t)$ starts with the value q_{1i} at $t = t_i$ and time-evolves to q_{2f} at t_f while the variable $q_2(t)$, starts at q_{2f} at time $t = t_f$ and evolves to q_{1i} at time $t = t_i$. This is why it is called evolution on a closed time loop from t_i back to t_i although the field changes switches from q_1 to q_2 at t_f.

By varying the Lagrangian Λ w.r.t $q_1, \dot{q}_1$ and $q_2, \dot{q}_2$ we get the equations of motion

$$\frac{d}{dt} \left(\frac{\partial \Lambda}{\partial \dot{q}_A} \right) - \frac{\partial \Lambda}{\partial q_A} = 0 \quad A = 1, 2. \tag{6.34}$$

The physical path in the dissipative system is then obtained by the path which solves the two equations simultaneously $q_1(t) = q_2(t) = q(t)$ and $\dot{q}_1(t) = \dot{q}_1(t) = \dot{q}(t)$, when the forward and the reverse path coincide. To check the physical limit it is more convenient to define the average of the forward and reverse histories $q_+(t) = (q_1(t) + q_2(t))/2$ and their difference $q_-(t) = q_1(t) - q_2(t)$. The equations of motion are of the same form as (6.34) with $A = +, -$. The equation of motion for the physical $q(t)$ is obtained by taking the physical limits in the e.o.m,

$$\left(\frac{d}{dt} \left(\frac{\partial \Lambda}{\partial \dot{q}_-} \right) - \frac{\partial \Lambda}{\partial q_-} \right) \Bigg|_{q_+=q, q_-=0} = 0 . \tag{6.35}$$

In terms of the conservative part L and dissipative part K the e.o.m becomes

$$\frac{d}{dt} \left(\frac{\partial L(q, \dot{q})}{\partial \dot{q}} \right) - \frac{\partial L(q, \dot{q})}{\partial q} = \left(\frac{d}{dt} \left(\frac{\partial K(q, \dot{q}, q+, \dot{q}_-)}{\partial \dot{q}_-} \right) \right.$$

$$\left. - \frac{\partial K(q, \dot{q}, q_-, \dot{q}_-)}{\partial q_-} \right) \Bigg|_{q_+=q, q_-=0}$$

$$\equiv Q(q, \dot{q}, q_-, \dot{q}_-) \tag{6.36}$$

and $Q(q, \dot{q}, q_-, \dot{q}_-)$ is identified as the dissipative force.

The Noether's charges under symmetries are also modified by the dissipative term K. Under time translation $t \rightarrow t + \delta t$ the conservative part of the action changes as

$$0 = \delta \int_{t_i}^{t_f} L(q_I, \dot{q}_I, t) = \int_{t_i}^{t_f} dt \delta t \left[\frac{d}{dt} \left(\dot{q}^I \frac{\partial L}{\partial q^I} - L \right) + \frac{\partial L}{\partial t} - \dot{q}_I \left(\frac{\partial L}{\partial \dot{q}_I} - \frac{\partial L}{\partial q^I} \right) \right].$$
(6.37)

The Hamiltonian associated with L is

$$E(q_I, \dot{q}_I, t) = \dot{q}^I \frac{\partial L}{\partial q^I} - L.$$
(6.38)

The action being L being time translation invariant therefore implies therefore that the terms inside the square brackets in (6.37) vanish, which using (6.36) and (6.38) implies

$$\frac{dE}{dt} = -\frac{\partial L}{\partial t} + \dot{q}^I Q_I.$$
(6.39)

This is the dissipation of the energy from the conservative Lagrangian, although this is not the total energy associated with the total Lagrangian Λ. The total energy including dissipative potential can be defined by including the canonical momentum κ^I associated with K in the Hamiltonian function. The total momentum in physical coordinates is

$$\pi^I(q^I, \dot{q}^I, t) = \pi^I(q_+^i, \dot{q}_+^I, q_-^I, \dot{q}_-^I, t) \Big|_{q_+^I = q, q_-^I = 0}$$

$$= \frac{\partial \Lambda(q_+^i, \dot{q}_+^I, q_-^I, \dot{q}_-^I, t)}{\partial \dot{q}_-^I} \Big|_{q_+^I = q, q_-^I = 0}$$

$$= \frac{\partial L}{\partial \dot{q}_I} + \frac{\partial K}{\partial \dot{q}_-^I} \Big|_{q_+^I = q, q_-^I = 0}.$$
(6.40)

The second term is κ_I the momentum associated with the K which in physical coordinates is

$$\kappa^I(q^I, \dot{q}^I, t) \equiv \kappa_+ \Big|_{q_+^I = q, q_-^I = 0} = \frac{\partial K(q_+^i, \dot{q}_+^I, q_-^I, \dot{q}_-^I, t)}{\partial \dot{q}_-^I} \Big|_{q_+^I = q, q_-^I = 0}.$$
(6.41)

The Hamiltonian function including the dissipative momentum is therefore defined as

$$\mathcal{E}(q^I, \dot{q}^I, t) = \pi_I(q^I, \dot{q}^I, t) \dot{q}^I - L(q^I, \dot{q}^I, t)$$

$$= E + \dot{q}^I \kappa_I.$$
(6.42)

The change in time of the total energy $\mathcal{E}$ is then given by

$$\frac{d\mathcal{E}}{dt} = -\frac{\partial L}{\partial t} + \dot{q}^I \frac{\partial K}{\partial \dot{q}^I}\bigg|_{q_+^I = q, q_-^I = 0} + \ddot{q}^I \kappa_I \tag{6.43}$$

which can be compared with the change in the conservative energy (6.39).

For the Abraham-Lorentz self-force, we chose the conservative and dissipative Lagrangians as

$$L = \frac{1}{2}m\mathbf{v}^2, \qquad K = -\frac{e^2}{6\pi}\mathbf{v}_- \cdot \dot{\mathbf{v}}_+ + \mathbf{x}_+ \cdot \mathbf{f}(t). \tag{6.44}$$

The e.o.m using (6.36) is

$$m\dot{\mathbf{v}} = \mathbf{f}(t) + \frac{e^2}{6\pi}\ddot{\mathbf{v}}. \tag{6.45}$$

The second term is the Abraham-Lorentz radiation reaction. The energy dissipated from the conservative system from this force is from (6.42)

$$\mathcal{E} = \frac{1}{2}mv^2 - \frac{e^2}{4\pi}\mathbf{v} \cdot \mathbf{a}, \tag{6.46}$$

where the second term is the Schott term which arises from the electron self-energy from its near field. The rate of energy change is

$$\frac{dE}{dt} = -\frac{e^2}{6\pi}\mathbf{a}^2 + \mathbf{v} \cdot \mathbf{f}. \tag{6.47}$$

The first term is Larmor term for the energy lost due to acceleration and the second term is the energy pumped into the electron by the external force which is causing the acceleration. The energy radiated by the electron therefore gives rise to a self-force and a self-energy. Similar effects also arise from the gravitational radiation, there is a Burke-Thorne dissipative potential during gravitational radiation by a system of bodies.

6.5 Path Integral: In-Out Formalism

Consider the scalar field theory with the action given by

$$S = \int d^4x \frac{1}{2}\left(\partial_\mu \varphi \partial^\mu \varphi - m^2\phi^2\right) + J(x)\varphi(x), \tag{6.48}$$

where we have introduced a classical source J which has a linear coupling with φ. The transition of a vacuum state at past infinity to remain a vacuum state in the future in the presence of the source is given by the path integral

$$\langle 0_+|0_-\rangle_J = Z(J) = \int \mathcal{D}\varphi \; e^{i \int d^4x \; \frac{1}{2}(\partial_\mu\varphi\partial^\mu\varphi - m^2\phi^2) + J(x)\varphi(x)} \tag{6.49}$$

and the path integral measure

$$\mathcal{D}\varphi = \prod_{x_i} d\phi(x_i) \tag{6.50}$$

covers all field configurations connecting the past infinity and the future infinity. Take the Fourier transform of all the fields in the action. In Fourier space the action is

$$S = \frac{1}{2} \int \frac{d^4k}{(2\pi)^4} \tilde{\varphi}(k)\left(k^2 + m^2\right)\varphi(-k) + \tilde{J}(k)\tilde{\varphi}(-k) + \tilde{J}(-k)\tilde{\varphi}(k). \tag{6.51}$$

Redefine the field by a shift

$$\tilde{\chi}(k) = \tilde{\varphi}(k) - \frac{\tilde{J}(k)}{k^2 + m^2}. \tag{6.52}$$

The action in the new field is $\tilde{\chi}(k)$

$$S = \frac{1}{2} \int \frac{d^4k}{(2\pi)^4} \frac{\tilde{J}(k)\tilde{J}(-k)}{k^2 + m^2} + \tilde{\chi}(k)\left(k^2 + m^2\right)\tilde{\chi}(-k). \tag{6.53}$$

The path integral measure remains of the same form in $\tilde{\chi}$. We can write the generating function (6.54) as

$$Z(J) = e^{iW(J)} \int \mathcal{D}\chi \; e^{\frac{i}{2}\int \frac{d^4k}{(2\pi)^4} \tilde{\chi}(k)\left(k^2 + m^2 + i\epsilon\right)\tilde{\chi}(-k)}, \tag{6.54}$$

where

$$W(J) \equiv \frac{1}{2} \int \frac{d^4k}{(2\pi)^4} \frac{\tilde{J}(k)\tilde{J}(-k)}{k^2 + m^2 + i\epsilon}, \tag{6.55}$$

where we have added $i\epsilon$ with $\epsilon > 0$ to make the integral convergent. This term is important as it gives the locations of the poles of the Greens functions and leads to the fact that the correlations of the fields we obtain from $Z(J)$ are time ordered.

The path integral over χ is just the vacuum to vacuum transition without sources, this just provides a normalisation constant

$$\langle 0_+|0_-\rangle_{J=0} = Z(0) = \int \mathcal{D}\chi \ e^{\frac{i}{2}\int \frac{d^4k}{(2\pi)^4}\tilde{\chi}(k)(k^2+m^2)\tilde{\chi}(-k)}. \tag{6.56}$$

We can normalise the vacuum to vacuum transition amplitude in the presence of sources by defining it as

$$Z(J) \equiv \frac{\langle 0_+|0_-\rangle_J}{\langle 0_+|0_-\rangle_0} = \frac{\langle 0_+|0_-\rangle_J}{Z(0)}. \tag{6.57}$$

We therefore have the result for the normalised generating function

$$Z(J) = \int \mathcal{D}\varphi \ e^{i\int d^4x \ \frac{1}{2}(\partial_\mu\varphi\partial^\mu\varphi - m^2\phi^2) + J(x)\varphi(x)} = e^{\frac{i}{2}\int \frac{d^4k}{(2\pi)^4}\frac{\tilde{J}(k)\tilde{J}(-k)}{k^2+m^2}}. \tag{6.58}$$

We can write W in terms of the Feynman propagator

$$G_F(x-y) = \int \frac{d^4k}{(2\pi)^4}\frac{e^{ik\cdot(x-x')}}{k^2+m^2-i\epsilon} \tag{6.59}$$

by transforming $W(J)$ back to position space

$$W = \frac{1}{2}\int d^4x d^4y J(x)G_F(x-y)J(y). \tag{6.60}$$

The choice of the pole term $i\epsilon$ which results in the Feynman propagator is made so that the generating function can give time ordered correlation of fields.

The fact that the Feynman propagator gives the time ordered correlation can be seen from the following time order property the Feynman propagator.

$$G_F(x-x') = \int \frac{d^4k}{(2\pi)^4}\frac{e^{ik\cdot(x-y)}}{k^2+m^2-i\epsilon}$$

$$= \theta(t-t')\int \frac{d^3k}{(2\pi)^3 2\omega_k}e^{ik\cdot(x-x')} + \theta(t'-t)\int \frac{d^3k}{(2\pi)^3 2\omega_k}e^{-ik\cdot(x-x')}. \tag{6.61}$$

Therefore starting from the expression for the partition function in terms of the propagators with sources at two ends,

$$Z(J) = e^{\frac{i}{2}\int d^4x d^4y J(x)G_F(x-y)J(y)}$$

we see that the n-point correlations of the fields ϕ can be obtained by taking repeated derivatives of J of the generating function,

$$\langle T\varphi(x_1)\ldots\varphi(x_n)\rangle = \frac{-i\delta}{\delta J(x_1)}\cdots\frac{-i\delta}{\delta J(x_n)}Z(J)\Bigg|_{J=0}$$

$$= \int \mathcal{D}\varphi \; \varphi(x_1)\ldots\varphi(x_n) \; e^{i\int d^4x\,\frac{1}{2}\left(\partial_\mu\varphi\partial^\mu\varphi - m^2\phi^2\right)}. \quad (6.62)$$

The time ordering of fields in (6.62) means that $t_1 \geq t_2 \cdots t_n$. Consider the two point function

$$\langle T\varphi(x_1)\varphi(x_2)\rangle = \frac{-i\delta}{\delta J(x_1)}\frac{-i\delta}{\delta J(x_2)}e^{\frac{i}{2}\int d^4x d^4y J(x)G_F(x-y)J(y)}\Bigg|_{J=0}$$

$$= \frac{-i\delta}{\delta J(x_1)}\int d^4x\, J(x)G_F(x-x_2)\; e^{\frac{i}{2}\int d^4x d^4y J(x)G_F(x-y)J(y)}\Bigg|_{J=0}$$

$$= G_F(x_1 - x_2). \quad (6.63)$$

Functional derivatives are taken using

$$\frac{\delta J(x)}{\delta J(y)} = \delta^4(x-y). \quad (6.64)$$

In a theory of interacting fields with interactions of the type say $\mathcal{L}_{\text{int}} = \lambda\varphi^4$ the classical sources $J(x)$ play the role of book-keeping devices which are sent to zero at the end of the calculations.

In the applications which have actual interaction of fields with classical sources, these are introduces in the Lagrangian in the same way $J^c\varphi$, as terms linear in φ. These sources are not set to zero at the end and they occur in the n-point correlations.

The one-point correlation of a field in the in-out formalism in the presence of a classical source is given by

$$\langle\varphi(x)\rangle = \int d^4y \; G_F(x-y) \; J^c(y). \quad (6.65)$$

This is a problem if φ is an on-shell classical field as the Feynman propagator gives acausal propagation. What is required for propagating an on-shell field from a source is the retarded propagator. This is achieved by going over to the in-in formalism for computing correlations. The other problem is that the Feynman propagator is symmetric in time and is not capable of describing time asymmetric dissipative phenomenon. Again This is achieved in the closed time path (CTP) in-in formulation that we will discuss next.

6.6   Closed Time Path (CTP) Formulation

As we discussed in Sect. 6.4 if we want to include dissipative effects in a Lagrangian framework we have to double the variables $q- \rightarrow (q_1(t), q_2(t))$. Here q_1 evolves from $t_i = -\infty$ to $t_f = \infty$ while $q_2(t)$ evolves backward in time from $t_f = \infty$ to $t_i = -\infty$. The boundary condition is that $q_1(t_i) = q_2(t_i)$, $q_1(t_f) = q_2(t_f)$, $\pi_1(t_i) = \pi_2(t_i)$ and $\pi_1(t_f) = \pi_2(t_f)$. The time evolution is thus on a lose time path as show in Fig. 6.1. The physical solution is obtain by taking the average path $q_+ = (q_1 + q_2)/2$ and the difference in path histories $q_- = q_1 - q_2$. The physical solution is when the forward path and the reversed paths are identical. That is, the physical solution is $q = q_+$ with q_- set to zero.

The same procedure is followed in the Schwinger-Keldysh method [5–7] when addressing problems with non-equilibrium or dissipation processes. We consider scalar fields interacting with classical sources. The discussion on non-equilibrium method involves generalisation of $G(x - y) = \langle \varphi(x)\varphi(y) \rangle$ in the 'in-out' formalism to four different propagators for different bilinear combinations of φ_1 and φ_2 in the 'in-in' CTP formulation.

We double the fields from φ to φ_1 and φ_2. We evolve the fields along the contour shown in Fig. 6.2. The field φ_1 is time evolved from $t = -\infty$ to $t = \tau$ and the field φ_2 is evolved backward in time from $t = \tau$ to $t = -\infty$. At time τ the two fields must match $\varphi_1(\tau) = \varphi_2(\tau)$. The fields φ_1 and φ_2 have sources J_1 and J_2. The φ_2 fields can be evolved backward in time by choosing $\mathcal{L}(\varphi_2(t), \dot{\varphi}_2(t), t) = -\mathcal{L}(\varphi_1(t), \dot{\varphi}_1(t), t)$.

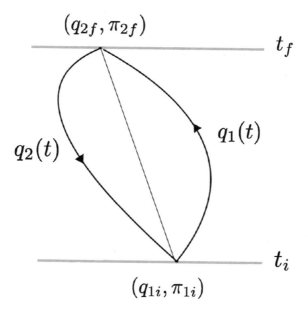

Fig. 6.1   Time evolution of q_1 and q_2

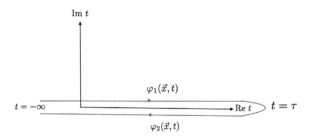

Fig. 6.2 Time evolution of φ_1 and φ_2

The generating function this theory written as a path integral over all configurations of $\varphi_{1,2}$ is

$$Z(J) = e^{i W(J_1, J_2)} = \int \mathcal{D}\varphi_1 \mathcal{D}\varphi_2 \ \delta(\varphi_1(\tau, \mathbf{x}) - \varphi_2(\tau, \mathbf{x}))$$

$$- \int_{-\infty - i\epsilon}^{\tau} dt \int d^3 y \ \left(\frac{1}{2} \partial_\mu \varphi_2(y) \partial^\mu \varphi_2(y) - \frac{1}{2} m^2 \varphi_2^2(y) + J_2(y) \varphi_2(y) \right) \Bigg\}. \tag{6.66}$$

We can write the delta function which equates φ_1 with φ_2 at $t = \tau$ as

$$\delta(\varphi_1(\tau) - \varphi_2(\tau)) = \lim_{\epsilon \to 0} \left(\frac{i}{\pi \epsilon} \right)^{1/2} e^{-\frac{1}{\epsilon}(\varphi_1(\tau) - \varphi_2(\tau))^2}. \tag{6.67}$$

Integrating out φ_1 and φ_2 we obtain the generating function as a function of the sources and propagators, which can be written as

$$Z(J_1, J_2) = e^{i W(J_1, J_2)}$$

$$= \exp \left[\frac{i}{2} \int d^4 x \, d^4 y \, J_a(x) G^{ab}(x - y) J_b(y) \right], \quad a, b = 1, 2. \tag{6.68}$$

The different propagators operating on the sources $J_a = (J_1(x), J_2(y))$ are written as 2×2 matrices

$$G^{ab}(x - y) = \begin{pmatrix} G_F(x - y) & G_-(x - y) \\ G_+(x - y) & G_D(x - y) \end{pmatrix}, \tag{6.69}$$

where G_F is the Feynman, G_D is the Dyson and $G_\pm$ are the Wightman functions. These propagators correspond to the following two point functions of the fields φ_1 and φ_2,

$$G_F(x, x') = \langle T\varphi_1(x)\varphi_1(x')\rangle = \frac{-i\delta}{\delta J_1(x)}\frac{-i\delta}{\delta J_1(x')}Z(J_1, J_2)\ \Big|_{J_{a,b}=0}$$

$$G_D(x, x') = \langle \bar{T}\varphi_2(x)\varphi_2(x')\rangle = \frac{-i\delta}{\delta J_2(x)}\frac{-i\delta}{\delta J_2(x')}Z(J_1, J_2)\ \Big|_{J_{a,b}=0}$$

$$G_-(x, x') = \langle \varphi_1(x)\varphi_2(x')\rangle = \frac{-i\delta}{\delta J_1(x)}\frac{-i\delta}{\delta J_2(x')}Z(J_1, J_2)\ \Big|_{J_{a,b}=0}$$

$$G_+(x, x') = \langle \varphi_2(x)\varphi_1(x')\rangle = \frac{-i\delta}{\delta J_2(x)}\frac{-i\delta}{\delta J_1(x')}Z(J_1, J_2)\ \Big|_{J_{a,b}=0} \tag{6.70}$$

The time ordering operator T means ordering along the contour shown in Fig. 6.2. In terms of time t the φ_1 fields are time ordered and the φ_2 fields are anti-time ordered by the operator $\bar{T}$. For the Wightman functions no time ordering is needed as the φ_2 fields always occur 'later' than the φ_1 fields along the contour.

The Feynman propagator gives the time ordered two point correlation

$$G_F(x, x') = \theta(t - t')G_+(x, x') + \theta(t' - t)G_-(x, x') \tag{6.71}$$

while the Dyson propagator gives the anti-time ordered two point correlation

$$G_D(x, x') = \theta(t' - t)G_+(x, x') + \theta(t - t')G_-(x, x'). \tag{6.72}$$

In applications involving loops, it is more convenient for regularisation to go to the momentum space. For tracking non-equilibrium properties or for application in finite temperature it is necessary to retain the time variable. We therefore do a partial Fourier transform of the Greens functions

$$G_{ab}(k; t, t') = \int d^3x d^3y\, e^{-i\mathbf{k}\cdot(\mathbf{x}-\mathbf{y})} G_{ab}(t, \mathbf{x}; t'\mathbf{y}). \tag{6.73}$$

We can write the Greens functions in terms of the mode functions $u(\mathbf{x}, t)$ which solve the homogenous wave equation

$$\left(\Box + m^2\right) u(\mathbf{x}, t) = 0. \tag{6.74}$$

The mode functions in the Minkowski background are $u(\mathbf{x}, t) = e^{-i\omega_k t + i\mathbf{k}\cdot\mathbf{x}}$ with $\omega_k = \sqrt{|\mathbf{k}|^2 + m^2}$. After the partial Fourier transform the mode functions are

$$u(k, t) = \frac{1}{\sqrt{2\omega_k}} e^{-i\omega_k t}. \tag{6.75}$$

We can express the different Greens functions in terms of the mode functions. The Wightman functions are

$$
G_+(k; t, t') = u(k, t)u^*(k, t'),
$$
$$
G_-(k; t, t') = u^*(k, t)u(k, t'). \tag{6.76}
$$

The time ordered Feynman's Greens function is

$$
G_F(k; t, t') = G_+(k; t, t')\theta(t - t') + G_-(k; t, t')\theta(t' - t). \tag{6.77}
$$

And finally the anti-time ordered Dyson Greens function is

$$
G_D(k; t, t') = G_-(k; t, t')\theta(t - t') + G_+(k; t, t')\theta(t' - t). \tag{6.78}
$$

These obey the complex conjugacy relations $G_+^* = G_-$ and $G_F^* = G_D$. The four Greens functions are not independent as they obey the identity

$$
G_F + G_D = G_+ + G_-. \tag{6.79}
$$

This means that we can choose a basis where we can choose one of the elements of the 2×2 is zero. One such basis is the Keldysh representation.

The momentum space representations of the four Greens functions can be computed from the mode functions $u(k, t)$ as follows. The Wightman functions are

$$
G_+(x - x') = \int \frac{d^3k}{(2\pi)^3} e^{i\mathbf{k}\cdot(\mathbf{x}-\mathbf{x}')} u(k, t)u^*(k, t')
$$
$$
= \int \frac{d^3k}{(2\pi)^3 2\omega_k} e^{i\mathbf{k}\cdot(\mathbf{x}-\mathbf{x}')} e^{-i\omega_k(t-t')} \tag{6.80}
$$

and

$$
G_-(x - x') = \int \frac{d^3k}{(2\pi)^3} e^{i\mathbf{k}\cdot(\mathbf{x}-\mathbf{x}')} u^*(k, t)u(k, t')
$$
$$
= \int \frac{d^3k}{(2\pi)^3 2\omega_k} e^{i\mathbf{k}\cdot(\mathbf{x}-\mathbf{x}')} e^{i\omega_k(t-t')}. \tag{6.81}
$$

The Feynman's propagator is

$$G_F(x, x') = \int \frac{d^3k}{(2\pi)^3} e^{i\mathbf{k}\cdot(\mathbf{x}-\mathbf{x}')}$$

$$\times \left(\theta(t - t') u(t, k) u^*(t', k) + \theta(t' - t) u^*(t, k) u(t', k) \right)$$

$$= \int \frac{d^3k}{(2\pi)^3} e^{i\mathbf{k}\cdot(\mathbf{x}-\mathbf{x}')} \left(\frac{\theta(t - t')}{2\omega_k} e^{-i\omega_k(t-t')} + \frac{\theta(t' - t)}{2\omega_k} e^{-i\omega_k(t'-t)} \right)$$

$$= \int \frac{d^3k}{(2\pi)^3} e^{i\mathbf{k}\cdot(\mathbf{x}-\mathbf{x}')} \int \frac{dk_0}{2\pi} \frac{i e^{-ik_0(t-t')}}{k_0^2 - \omega_k^2 + i\epsilon}. \tag{6.82}$$

The sign of the $i\epsilon$ term is important in giving the time ordering. This can be checked by carrying out the k_0 integral in the last line to obtain the line above. The k_0 integral can be carried out by contour-integration. The poles of the integrand are

$$k_0 = \pm(\omega_k^2 - i\epsilon)^{1/2} = \pm(\omega_k - i\epsilon). \tag{6.83}$$

The contour goes above the pole at $+\omega_k$ and for $t - t' > 0$ is closed from below for convergence. For the pole at $-\omega_k$ the contour goes below the pole and for $t - t' < 0$ is closed from above for convergence. The sign of the $i\epsilon$ term therefore decides the time ordering of the propagator.

From (6.82) we see that the Feynman propagator in momentum space is

$$G_F(k) = \frac{i}{k_0^2 - |\mathbf{k}|^2 - m^2 + i\epsilon}. \tag{6.84}$$

Similarly the Dyson propagator in the momentum space is

$$G_D(k) = \frac{i}{k_0^2 - |\mathbf{k}|^2 - m^2 - i\epsilon}. \tag{6.85}$$

6.7   The Keldysh Representation

In the Keldysh representation we take the following combinations of the fields and the corresponding sources [6]

$$\varphi_+ = (\varphi_1 + \varphi_2)/2, \quad \varphi_- = (\varphi_1 - \varphi_2),$$
$$J_+ = (J_1 + J_2)/2, \quad J_- = J_1 - J_2 \tag{6.86}$$

Classical fields are given by the average $\varphi_+ = (\varphi_1 + \varphi_2)/2$ and the quantum fluctuation is the difference $\varphi_- = (\varphi_1 - \varphi_2) = 0$.

The generating function in the (φ_1, φ_2) basis is

$$Z(J_+, J_-) = e^{iW(J_+, J_-)}$$

$$= \exp\left[\frac{i}{2}\int d^4x\, d^4y\, J_A(x) G^{AB}(x-y) J_B(y)\right], \quad A, B = 1, 2.$$

$$(6.87)$$

The different propagators in the $J_A = (J_+(x), J_-(y))$ basis are

$$G^{AB}(x-y) = \begin{pmatrix} 0 & G_{adv}(x-y) \\ G_{ret}(x-y) & \frac{i}{2}G_H(x-y) \end{pmatrix}, \quad (6.88)$$

where G_{adv} is the advanced, G_{ret} is the retarded and G_H is the Hadamard propagator. These are related to the Wightman functions as follows. The commutator (or causal) and anti-commutator (Hadamard) two point functions are

$$G_c(x-x') = \langle[\varphi(x), \varphi(x')]\rangle = G_+(x-x') - G_-(x-x'),$$
$$G_H(x-x') = \langle\{\varphi(x), \varphi(x')\}\rangle = G_+(x-x') + G_-(x-x'). \quad (6.89)$$

The retarded and advanced Greens functions are

$$G_{ret}(x-x') = \theta(t-t')G_c(x-x') = \theta(t-t')\left(G_+(x-x') - G_-(x-x')\right),$$
$$G_{adv}(x-x') = -\theta(t'-t)G_c(x-x') = \theta(t'-t)\left(G_-(x-x') - G_+(x-x')\right).$$

$$(6.90)$$

The momentum space representations of the causal Greens function are

$$G_c(x-x') = G_+(x-x') - G_-(x-x')$$

$$= \int \frac{d^3k}{(2\pi)^3 2\omega_k}\left[e^{-ik\cdot(x-x')} - e^{+ik\cdot(x-x')}\right]$$

$$= \int \frac{d^3k}{(2\pi)^3 2\omega_k} e^{i\mathbf{k}\cdot(\mathbf{x}-\mathbf{x}')}\left[e^{-i\omega_k(t-t')} - e^{+i\omega_k(t-t')}\right]$$

$$= i\int \frac{d^4k}{(2\pi)^3}\, \delta(k^2-m^2)\, \text{sgn}(k_0) e^{ik\cdot(x-x')}, \quad (6.91)$$

where $\text{sgn}(k_0)$ is the sign function $\text{sgn}(k_0) = \theta(k_0) - \theta(-k_0)$.

The momentum space representation of retarded and advanced Greens function can similarly be written in terms of the mode functions $u(k, t)$ of the homogenous equation. The retarded Greens function is

$$
\begin{aligned}
G_{\text{ret}}(x - x') &= \theta(t - t') \left[G_+(x - x') - G_-(x - x') \right] \\
&= \int \frac{d^3k}{(2\pi)^3 2\omega_k} \theta(t - t') \left[e^{-ik\cdot(x-x')} - e^{+ik\cdot(x-x')} \right] \\
&= \int \frac{d^3\mathbf{k}}{(2\pi)^3} e^{i\mathbf{k}\cdot(\mathbf{x}-\mathbf{x}')} \int \frac{dk_0}{2\pi} \frac{ie^{-ik_0(t-t')}}{(k_0 + i\epsilon)^2 - \omega_k^2}.
\end{aligned}
\tag{6.92}
$$

The poles are located at $k_0 = \pm\omega_k - i\epsilon$. The contour is chosen above both the poles and as $t - t' \geq 0$, closed from below.

Similarly the advanced propagator in the momentum representation is given by

$$
\begin{aligned}
G_{\text{adv}}(x - x') &= \int \frac{d^3k}{(2\pi)^3 2\omega_k} \theta(t' - t) \left[e^{-ik\cdot(x-x')} - e^{+ik\cdot(x-x')} \right] \\
&= \int \frac{d^3\mathbf{k}}{(2\pi)^3} e^{i\mathbf{k}\cdot(\mathbf{x}-\mathbf{x}')} \int \frac{dk_0}{2\pi} \frac{ie^{-ik_0(t-t')}}{(k_0 - i\epsilon)^2 - \omega_k^2}.
\end{aligned}
\tag{6.93}
$$

The poles are located at $k_0 = \pm\omega_k + i\epsilon$. The contour passes below both the poles and as $t - t' \leq 0$, is closed from above for convergence.

Finally the Hadamard Greens function is

$$
\begin{aligned}
\frac{i}{2} G_H(x - x') &= = \frac{i}{2} \left[G_+(x - x') + G_-(x - x') \right] \\
&= \frac{i}{2} \int \frac{d^3k}{(2\pi)^3 2\omega_k} e^{i\mathbf{k}\cdot(\mathbf{x}-\mathbf{x}')} \left[e^{-i\omega_k(t-t')} + e^{+i\omega_k(t-t')} \right] \\
&= \frac{-1}{2} \int \frac{d^3\mathbf{k}}{(2\pi)^3} e^{ik\cdot(x-x')} \delta(k^2 - m^2).
\end{aligned}
\tag{6.94}
$$

The Hadamard Greens function is therefore related to the imaginary part of the Feynman propagator as

$$
\frac{i}{2} G_H(x - x') = \operatorname{Im} G_F(x - x').
\tag{6.95}
$$

The classical source is $J_+ = J_c$ which is not set to zero after taking the derivatives for computing the correlation functions. In the presence of a classical source there is a non-zero classical field given by the one-point correlation function

$$
\begin{aligned}
\langle \varphi(x) \rangle_{\text{in-in}} &= \frac{-i\delta}{\delta J_-(x)} Z(J_+, J_-) \Big|_{J_-=0,\, J_+=J_c} \\
&= \int d^3x' G_{\text{ret}}(x - x') J_c(x').
\end{aligned}
\tag{6.96}
$$

The 'in-in' formalism gives the correct classical field for a given source in terms of the retarded propagator while in the 'in-out' case the one-point function is given in terms of the Feynman propagator (6.65),

$$\langle \varphi(x) \rangle_{\text{in-out}} = \int d^3x' G_F(x - x') J_c(x')$$

$$= \int d^3x' \frac{1}{2} \Big[G_{\text{ret}}(x - x') + G_{\text{adv}}(x - x') - i G_H(x - x') \Big] J_c(x')$$

$$(6.97)$$

which gives acausal solution for the classical field in terms of the source. The advanced Greens function $G_{\text{adv}}(x - x')$ is non-zero for $t - t' < 0$ while the Hadamard Greens function is non-zero even for spacelike separation $x - x' = |\mathbf{x} - \mathbf{x}'|$.

6.8   Gravitational Waves in Keldysh Formulation

The case of gravitational radiation which we will apply this formalism is basically going from $J_A(x) G^{AB}(x - y) J_B(y)$ for scalars to $J_A^{\mu\nu}(x) G^{AB}(x - y) T_B^{\alpha\beta} P_{\mu\nu\alpha\beta}$ for gravitons. The form of the functions $G^{AB}(x - y)$ will be identical for the two cases. The source $J^{\mu\nu}(x)$ for gravitons is the stress tensor $T^{\mu\nu}(x)$. We double the sources to $T_{1,2}^{\mu\nu}$. We then go to the Keldysh representation and express $T_{1,2}^{\mu\nu}(x)$ as $T_+^{\mu\nu}(x_\pm)$ and $T_-^{\mu\nu}(x_\pm)$.

We start with the generating function in the Keldysh representation

$$Z(J_\pm^{\mu\nu}) = \exp\left\{ \left(\frac{i}{2}\right) \int d^4x d^4x' J_A^{\mu\nu}(x) G_{\mu\nu\alpha\beta}^{AB}(x - x') J_B^{\alpha\beta}(x') \right\}, \qquad (6.98)$$

where

$$G_{\mu\nu\alpha\beta}^{AB}(x - x') = P_{\mu\nu\alpha\beta} \begin{pmatrix} 0 & G_{\text{adv}}(x - x') \\ G_{\text{ret}}(x - x') & \frac{1}{2} G_H(x - x') \end{pmatrix} \qquad (6.99)$$

The expectation value of the graviton field in the presence of a classical source $J_c^{\mu\nu}$ is given by

$$\langle h_{\mu\nu}(x) \rangle_{\text{in-in}} = \frac{-i\delta}{\delta J_-^{\mu\nu}(x)} Z(J_\pm^{\mu\nu}) \Big|_{J_-^{\rho\sigma} = 0, \ J_+^{\rho\sigma} = J_c^{\rho\sigma}}$$

$$= \int d^4x' P_{\mu\nu\alpha\beta} G_{\text{ret}}(x - x') J_c^{\alpha\beta}(x'). \qquad (6.100)$$

Gravity has a gauge symmetry $h_{\mu\nu} \to h_{\mu\nu} + \partial_{(\mu}\xi_{\nu)}$ which implies that the source stress tensor is $\partial_\mu J_c^{\mu\nu} = 0$. Making use of the gauge freedom to remove the redundant degrees of freedom, the free graviton is taken as the transverse–traceless projection of h_{ij}, $h_{ij}^{TT} = \Lambda_{ijkl}h^{kl}$. The expression for expectation value of h_{ij}^{TT} is then obtained from above by expressing the J_c^{0i} and J_c^{00} in terms of the spatial components J_c^{ij}. The expression for the TT graviton in terms of the source is then given by

$$\langle h_{ij}(x)^{TT}\rangle_{\text{in-in}} = \Lambda_{ijkl} \int d^4x' P_{klrs} G_{\text{ret}}(x - x') J_c^{rs}(x'). \qquad (6.101)$$

The retarded propagator is given by

$$G_{\text{ret}}(x - x') = \frac{1}{2\pi}\delta((t - t') - |\mathbf{x} - \mathbf{x}'|^2)$$

$$= \frac{1}{4\pi}\frac{\delta(t - t' - |\mathbf{x} - \mathbf{x}'|)}{|\mathbf{x} - \mathbf{x}'|}. \qquad (6.102)$$

The classical source which is the stress tensor $J_c^{ij}(x) = \frac{\kappa}{2}T^{ij}(x, t)$ (where $\kappa = \sqrt{32\pi G}$) can be written in terms of the mass quadrupole distribution Q_{ij} defined as

$$Q_{ij}(t) = \int d^3 T^{00}(\mathbf{x}, t)x_i x_j \qquad (6.103)$$

by making use of the conservation relations $\partial_\mu T^{\mu\nu} = 0$. These give two relations $\partial_i T^{i0} + \partial_0 T^{00} = 0$ and $\partial_i T^{ij} + \partial_0 T^{0j} = 0$. These can be used to relate write T_{ij} in terms of Q_{ij} as follows. Take a time derivative of (6.103) and use the conservation relations

$$\dot{Q}_{ij}(t) = \int d^3x\, \partial_0 T^{00}(\mathbf{x}, t)x_i x_j = -\int d^3x\, \partial_l T^{l0}(\mathbf{x}, t)x_i x_j$$

$$= 2\int d^3x\, T^{i0}(\mathbf{x}, t)x_j, \qquad (6.104)$$

where we dropped the velocity terms v_i and have integrated by parts. Take one more time derivative of (6.104) and again use the conservation equations,

$$\ddot{Q}_{ij}(t) = 2\int d^3x\, \partial_0 T^{i0}(\mathbf{x}, t)x_j = -2\int d^3 \partial_l T^{il}(\mathbf{x}, t)x_j$$

$$= 2\int d^3x\, T^{ij}(\mathbf{x}, t). \qquad (6.105)$$

Therefore to the order v^0 we have the relation

$$\frac{1}{2}\ddot{Q}_{ij}(t) = \int d^3x\, T^{ij}(\mathbf{x}, t).$$ (6.106)

Using (6.106) and (6.102) in (6.101) we obtain the result

$$\langle h_{ij}^{TT}(\mathbf{x}, t)\rangle_{\text{in-in}} = \frac{\kappa}{8\pi|x|} \int dt'\, \Lambda_{ijkl}\, P_{klrs}\frac{1}{2}\delta(t - t' - |x|)\ddot{Q}^{rs}(t'),$$ (6.107)

where we have taken the source to observer distance to be large compared to the size of the source $|\mathbf{x} - \mathbf{x}'| \simeq |\mathbf{x}|$.

The angular average of $\Lambda_{ijkl}(\hat{n})P_{klrs}(\hat{n})$ over the directions $\hat{n}$ of the observer w.r.t the source coordinates gives us

$$\int \frac{d\Omega_{\hat{n}}}{4\pi} \Lambda_{ijkl}(\hat{n})\, P_{klrs}(\hat{n}) = \frac{1}{5}\left(\delta_{ir}\delta_{js} + \delta_{is}\delta_{jr} - \frac{2}{3}\delta_{ij}\delta_{rs}\right).$$ (6.108)

Using this in (6.109) we obtain the expression for the graviton expectation value in terms of the source

$$\langle h_{ij}^{TT}(\mathbf{x}, t)\rangle_{\text{in-in}} = \frac{G}{5|\mathbf{x}|}\left(\ddot{Q}_{ij}(t') - \frac{1}{3}\delta_{ij}\ddot{Q}_l^l(t')\right)\Bigg|_{t'=t-|\mathbf{x}|}$$ (6.109)

which the classical quadrupole formula for gravitational wave signal from a slow moving source.

6.9 Dissipative Force from Backreaction of Gravitational Radiation

We now come to the main application of the non-equilibrium formulation which is to calculate the backreaction and the consequent dissipative force on the source due to the gravitational radiation [8–13]. We proceed by doubling the sources to $Q_{1,2}^{ij}$. We then go to the Keldysh representation and express $Q_{1,2}^{ij}$ as $Q^{ij}_+(x_\pm)$ and $Q^{ij}_-(x_\pm)$ by writing $\mathbf{x}_+ = (\mathbf{x}_1 + \mathbf{x}_2)/2$ and $\mathbf{x}_- = (\mathbf{x}_1 - \mathbf{x}_2)$. Consider the bodies as point particles with masses m_n ($n = 1, 2\ldots$) being the particle number. The quadrupole moments $Q_1(x_1)$ and $Q_2(x_2)$ are

$$Q_1^{ij}(t) = \sum_n m_n\left(x_{n1}^i x_{n1}^i - \frac{1}{3}x_{n1}^2\delta_{ij}\right), \quad Q_2^{ij}(t) = \sum_n m_n\left(x_{n2}^i x_{n2}^i - \frac{1}{3}x_{n2}^2\delta_{ij}\right).$$ (6.110)

The sources $Q_{\pm}^{ij}(\mathbf{x}_{\pm})$ are the given by

$$Q_{+}^{ij}(\mathbf{x}_{\pm}) = \frac{1}{2}\left(Q_1^{ij}(t) + Q_2^{ij}(t)\right) = \sum_n m_n \left(x_{n+}^i x_{n+}^j - \frac{1}{3}\delta^{ij}\mathbf{x}_{n+} \cdot \mathbf{x}_{n+}\right)$$

$$Q_{-}^{ij}(\mathbf{x}_{\pm}) = \left(Q_1^{ij}(t) - Q_2^{ij}(t)\right)$$

$$= \sum_n m_n \left(x_{n+}^i x_{n-}^j + x_{n-}^i x_{n+}^j - \frac{2}{3}\delta^{ij}\mathbf{x}_{n+} \cdot \mathbf{x}_{n-}\right) + O(x_-^3). \quad (6.111)$$

The effective action in the Keldysh variables is given by

$$iW(J_{\pm}) = \frac{i}{2}\frac{\kappa^2}{4}\int d^4x\, d^4x'\, \ddot{Q}_A^{ij}(t')\, P_{ijlm}\, \Lambda_{lmrs}\, \Lambda_{rspq}\, G^{AB}(x-x')\, \ddot{Q}_B^{pq}(t). $$
$$(6.112)$$

We have two TT projection operators as we are using the two point correlation of on-shell gravitons $\langle h_{ij}^{TT} h_{lm}^{TT}\rangle$ as we are calculating the radiation reaction of real gravitons. This is different from the propagators for virtual graviton line as used, for example, in calculating the Newtonian potential. In the virtual graviton case the TT projection operators are not required.

Dropping the space-time indices, the structure of the effective Lagrangian is

$$iW(J_{\pm}) = \frac{i}{2}\int d^4x\, d^4x'\, \ddot{Q}_A(t')\, G^{AB}(x-x')\, \ddot{Q}_B(t)$$

$$= \frac{i}{2}\int d^4x\, d^4x'\, \ddot{Q}_-(t')\, G^{-+}(x-x')\, \ddot{Q}_+(t) + \ddot{Q}_+(t')\, G^{+-}(x-x')\, \ddot{Q}_-(t)$$

$$+ \ddot{Q}_-(t')\, G^{--}(x-x')\, \ddot{Q}_-(t)$$

$$= i\int d^4x\, d^4x'\, \ddot{Q}_-(t')\, G_{\text{ret}}(x-x')\, \ddot{Q}_+(t), \quad (6.113)$$

where in the last line we used the relation $G^{-+}(x'-x) = G^{+-}(x-x')$, We also dropped the G^{--} term as it couples to terms quadratic in x_-^i which vanish when we take the classical limits of e.o.m $\mathbf{x}_+ = \mathbf{x}_{\text{cl}}$ and $\mathbf{x}_- = 0$.

Going back to (6.112) we do the angular average of $P_{ijlm}(\hat{n})\Lambda_{lmrs}(\hat{n})\Lambda_{rspq}(\hat{n})$ over the directions $\hat{n}$ of the observer w.r.t the source coordinates, to obtain

$$\int \frac{d\Omega_{\hat{n}}}{4\pi}\Lambda_{ijkl}(\hat{n})\, P_{klrs}(\hat{n}) = \frac{4}{5}\left(\delta_{ip}\delta_{jq} + \delta_{iq}\delta_{jp} - \frac{2}{3}\delta_{ij}\delta_{pq}\right) \quad (6.114)$$

and we have

$$\ddot{Q}_A^{ij}(t')\, P_{ijlm}\Lambda_{lmrs}\Lambda_{rspq}\,\ddot{Q}_B^{pq}(t) = \frac{8}{5}\left(\ddot{Q}_A^{ij}(t')\ddot{Q}_{ij\,B}(t) - \frac{1}{3}\ddot{Q}_{i\,A}^i(t')\,\ddot{Q}_{j\,B}^j(t)\right)$$

(6.115)

Putting (6.113) and (6.115) in (6.112) we have the expression for the effective action,

$$iW(J_\pm) = i8\pi G \int d^4x d^4x'$$

$$\times \frac{8}{5}\left(\ddot{Q}_-^{ij}(t')\ddot{Q}_{ij\,+}(t) - \frac{1}{3}\ddot{Q}_{i\,-}^i(t')\,\ddot{Q}_{j\,+}^j(t)\right) G_{\text{ret}}(x - x').$$

(6.116)

We can define the traceless quadrupole distribution as

$$I_{ij}(t) = \left(Q_{ij}(t) - \frac{1}{3}Q_i^i(t)\right).$$

(6.117)

The effective action (6.118) in terms of I_{ij} is

$$iW(J_\pm) = i8\pi G \int d^3x d^3x' dt dt' \frac{8}{5}\left(\ddot{I}_-^{ij}(t')\ddot{I}_{ij\,+}(t)\right) G_{\text{ret}}(x - x').$$

(6.118)

We can write in $G_{\text{ret}}(x - x')$ in frequency space

$$G_{\text{ret}}(x - x') = \int \frac{d^4k}{(2\pi)^4 2\omega_k} \frac{e^{-i(x-x')\cdot k}}{(\omega + i\epsilon)^2 - |\mathbf{k}|^2}$$

(6.119)

and obtain the effective action

$$iW(J_\pm) = i8\pi G \int \frac{d\omega}{2\pi} \frac{8}{5}\omega^4 \left(I_-^{ij}(\omega)I_{ij\,+}(\omega)G_{\text{ret}}(\omega)\right)$$

$$= i\frac{G}{5} \int \frac{d\omega}{2\pi}\omega^5 I_-^{ij}(\omega)I_{ij\,+}(\omega).$$

(6.120)

Transforming back to position space

$$iW(J_\pm) = i\frac{G}{5} \int dt\, I_-^{ij}(t)\frac{d^5}{dt^5}I_{ij\,+}(t).$$

(6.121)

The equation of motion for $\mathbf{x}_n$

$$\left.\frac{\delta W(J_\pm)}{\delta x_n}\right|_{\mathbf{x}_{n+}=\mathbf{x}_n,\mathbf{x}_{n-}=0} = 0$$

(6.122)

gives

$$\ddot{x}_n = -\frac{2}{5}\frac{d^5}{dt^5}I_{ij}(t)x^j. \tag{6.123}$$

This is the Burke-Thorne radiation reaction force.

6.10 Thermal Bath

In the thermal initial state, the state at $t = -\infty$ is not the zero-particle vacuum state but a statistical mixture of different occupation numbers given by

$$|0\rangle_\beta = \sum_{n_k} p_n(\omega_k)|n_k\rangle, \tag{6.124}$$

where $p_n(\omega_k)$ is the probability of the state with occupancy n_k which is given by the Boltzmann factor

$$p_n(\omega_k) = \frac{e^{-\beta\omega_k n_k}}{\sum_{n_k}e^{-\beta n_k \omega_k}}, \qquad \omega_k = (|\mathbf{k}|^2 + m^2)^{1/2}. \tag{6.125}$$

For fermions, n_k the occupancy of any state with momentum $\mathbf{k}$ can be 0 or 1. Therefore

$$\sum_{n_k}e^{-\beta n_k \omega_k} = 1 + e^{-\beta\omega_k} \qquad \text{(fermions)} \tag{6.126}$$

whereas for bosons the occupancy n_k can be any natural number therefore

$$\sum_{n_k}e^{-\beta n_k \omega_k} = \sum_{n=0}^{\infty}e^{-\beta n\omega_k} = \frac{1}{1 - e^{-\beta\omega_k}} \qquad \text{(bosons).} \tag{6.127}$$

The expectation value of any operator which is an eigenstate of a Fock state $|n_k\rangle$ w.r.t the thermal density matrix is then given by

$$\langle O\rangle_\beta = \frac{\sum_{n_k} p_n(\omega_k)\langle n_k|O|n_k\rangle}{\sum_{n_k}e^{-\beta n_k \omega_k}}. \tag{6.128}$$

The average occupation number of state with momenta $k = (\mathbf{k}, \omega_k)$ is

$$\bar{n}_k = \langle a_k^\dagger a_k\rangle_\beta = \frac{\sum_{n_k} p_n(\omega_k)n_k}{\sum_{n_k}e^{-\beta n_k \omega_k}}. \tag{6.129}$$

For fermions this is

$$\bar{n}_k = \frac{e^{-\beta\omega_k}}{1 + e^{-\beta\omega_k}} = \frac{1}{e^{\beta\omega_k} + 1} \tag{6.130}$$

which is the Fermi-Dirac distribution.

For bosons the average occupation number of a state with momentum k is

$$\bar{n}_k = \langle a_k^\dagger a_k \rangle_\beta = \frac{\sum_{n_k} P_n(\omega_k) n_k}{\sum_{n_k} e^{-\beta n_k \omega_k}}$$

$$= \left(1 - e^{-\beta\omega_k}\right) \sum_{n_k} n_k e^{-\beta n_k \omega_k} = \left(1 - e^{-\beta\omega_k}\right) \left(\frac{-d}{\omega_k d\beta}\right) \sum_{n_k} e^{-\beta n_k \omega_k}$$

$$= \left(1 - e^{-\beta\omega_k}\right) \left(\frac{-d}{\omega_k d\beta}\right) \frac{1}{\left(1 - e^{-\beta\omega_k}\right)}$$

$$= \frac{1}{e^{\beta\omega_k} - 1} \tag{6.131}$$

which is the Bose-Einstein distribution function.

The expectation value of the various combinations a_k, $a_k^\dagger$ operators in thermal initial state are

$$\langle a_k \rangle_\beta = 0, \quad \langle a_k^\dagger \rangle_\beta = 0, \quad \langle a_k^\dagger a_{k'}^\dagger \rangle_\beta = 0, \quad \langle a_k a_{k'}' \rangle_\beta = 0,$$

$$\langle a_{k'}^\dagger a_k \rangle_\beta = \bar{n}(\omega_k)\delta^3(\mathbf{k} - \mathbf{k'}),$$

$$\langle a_k a_{k'}^\dagger \rangle_\beta = (1 \pm \bar{n}(\omega_k))\delta^3(\mathbf{k} - \mathbf{k'}). \tag{6.132}$$

The mode function which satisfy the homogenous wave equation is

$$\hat{\phi}(x) = \int \frac{d^3k}{(2\pi)^3} \frac{1}{\sqrt{2\omega_k}} \left(a_k e^{-ik\cdot x} + a_k^\dagger e^{ik\cdot x}\right). \tag{6.133}$$

The Wightmans functions in the thermal background are then given by

$$G_+(x, x') = \langle \hat{\phi}(x)\hat{\phi}(x')^\dagger \rangle$$

$$= \int \frac{d^3k}{(2\pi)^3} \frac{1}{2\omega_k} e^{ik\cdot x} \left(\langle a_k a_k^\dagger \rangle e^{-i\omega_k(t-t')} + \langle a_k^\dagger a_k \rangle e^{i\omega_k(t-t')}\right)$$

$$= \int \frac{d^3k}{(2\pi)^3} \frac{1}{2\omega_k} e^{ik\cdot x} \left((1 \pm \bar{n}_k)e^{-i\omega_k(t-t')} + \bar{n}_k e^{i\omega_k(t-t')}\right) \tag{6.134}$$

(where the plus sign is for Bosons and $\bar{n}_k$ is the Bose-Einstein distribution while the minus sign is for fermions and the $\bar{n}_k$ for that case is the Fermi-Dirac distribution) and

$$G_-(x, x') = \langle \hat{\phi}(x)^\dagger \hat{\phi}(x') \rangle$$

$$= \int \frac{d^3k}{(2\pi)^3} \frac{1}{2\omega_k} e^{i\mathbf{k}\cdot\mathbf{x}} \left((1 \pm \bar{n}_k) e^{i\omega_k(t-t')} + \bar{n}_k e^{-i\omega_k(t-t')} \right), \quad (6.135)$$

where again the plus sign is for bosons and—sign for fermions.

Knowing the thermal form of the Wightmans functions we can evaluate the forms for all the Greens functions in a thermal bath. This is the form of the Greens functions when the particles which are propagating are in equilibrium with the heat bath. Next we will consider the non-equilibrium situation where the propagating particles are in a heat bath but not in thermal equilibrium with the heat bath.

6.11 Non-equilibrium Propagators in a Thermal Bath

A system initially is a thermal state is describe by the thermal density matrix is

$$\rho(t_0) = \frac{e^{-\beta H}}{\text{Tr}(e^{-\beta H})}, \quad (6.136)$$

where $\beta = 1/k_B T$ [14]. To describe the time evolution of operator, the thermal density matrix can be included in the path integral formalism by generalising the time coordinate to complex values $z = t - i\beta$. Formally we can write the expectation value of an Heisenberg operator at time z to be

$$\langle \hat{O}(z) \rangle = \frac{\text{Tr}\left\{ \hat{\mathcal{T}}_c\, e^{-i \int_\gamma dz' H(z')}\, \hat{O}(z) \right\}}{\text{Tr}\left\{ \hat{\mathcal{T}}_c\, e^{-i \int_\gamma dz' H(z')} \right\}}, \quad (6.137)$$

where the time evolution is along the contour γ shown in Fig. 6.3. Here the time ordering operator orders the operators following the contour which starts at $z' = 0^+$ and goes till $z' = t$ and then returns to $z' = 0^+$ and goes to $z' = -i\beta$.

When the system is not in equilibrium with the thermal bath at the initial time, then the time evolution of the system introduced into a thermal bath at some initial time is described by the non-equilibrium Greens functions which propagate the field φ_1 and φ_2 on the directed contour shown in Fig. 6.3. The combinations of two fields then give four types of Greens functions which we can write as

$$G_\beta^{ab}(t - t') \begin{pmatrix} \langle T \varphi_1(t)\varphi_1(t') \rangle & \eta\langle \varphi_1(t)\varphi_2(t') \rangle \\ \langle \varphi_2(t)\varphi_1(t') \rangle & \langle \bar{T} \varphi_2(t)\varphi_2(t') \rangle \end{pmatrix}, \quad (6.138)$$

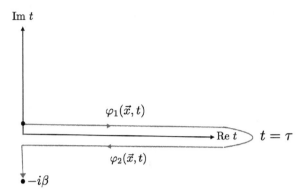

Fig. 6.3 Time evolution of φ_1 and φ_2 at finite temperature

where $\eta = 1$ for the commuting bosons and $\eta = -1$ for the anti-commuting fermions. The fields φ_1 are time ordered and φ_2 are anti-time ordered while no time ordering is needed for the off-diagonal terms as the ϕ_2 fields always appear 'later' along the contour w.r.t the φ_1 fields. Now we would like to see the explicit dependence on the thermal distributions functions. With this end take the initial vacuum to be the Fock state $|n(\omega)\rangle$ such that $a|n\rangle = \sqrt{(n)}|n-1\rangle$ and $a^\dagger|n(\omega)\rangle = \sqrt{(n+1)}|n+1\rangle$. This assumed for a and $a^\dagger$ from both fields φ_1 and φ_2. The occupation number as a function of energy is the statistical distribution

$$n(\omega) = \left[e^{\beta\omega} - \eta \right]^{-1}, \tag{6.139}$$

where $\eta = 1$ to represent Dirac the Bose-Einstein distribution of commuting boson fields and $\eta = -1$ for the Fermi-Dirac distribution of the anti-commuting fermion fields. Taking the initial vacuum as the thermal state we get expressions for the Greens functions in momentum space in the (φ_1, φ_2) basis, in terms of the statistical distribution function $n(\omega)$ given by

$$G_\beta^{ab}(\omega, \mathbf{k}) = \begin{pmatrix} \frac{i}{k^2 - m^2 - i\epsilon} & 2\pi\theta(-\omega)\delta(k^2 - m^2) \\ 2\pi\theta(\omega)\delta(k^2 - m^2) & \frac{i}{k^2 - m^2 + i\epsilon} \end{pmatrix}$$

$$+ 2\pi n(\omega)\delta(k^2 - m^2) \begin{pmatrix} 1 & 1 \\ 1 & 1 \end{pmatrix}. \tag{6.140}$$

In the finite temperature path integral the Greens functions in the Keldysh representation are generalised as follows. In the momentum representation the retarded and advanced Greens functions are at zero temperature

$$G^0_{ret}(\mathbf{p}, \omega) = \left[G^0_{adv}(\mathbf{p}, \omega)\right]^\dagger = \frac{1}{\omega - \epsilon_p - i\epsilon}, \qquad \epsilon_p = \sqrt{\mathbf{p}^2 + m^2}. \qquad (6.141)$$

At finite temperature the different Greens functions in the momentum representation are given by Keldysh [6]

$$\Delta^\beta_+(\mathbf{p}, \omega) = 2\pi i n_p \delta(\omega - \epsilon_p)$$

$$\Delta^\beta_-(\mathbf{p}, \omega) = -2\pi i (1 \pm n_p)\delta(\omega - \epsilon_p), \qquad +(\text{boson}), \quad -(\text{fermion})$$

$$\frac{1}{2}G^\beta_H(\mathbf{p}, \omega) = 2\pi i (2n_p \pm 1)\delta(\omega - \epsilon_p), \qquad +(\text{boson}), \quad -(\text{fermion})$$

$$(6.142)$$

where n_p is the distribution function

$$n_p = \left[exp(\beta(\epsilon_p - \mu)) \mp 1\right]^{-1}, \qquad -(\text{boson}), \quad +(\text{fermion}). \qquad (6.143)$$

The retarded Greens function at finite temperature is obtained from the combination of $\Delta^\beta_-(\mathbf{p}, \omega)$ and $\Delta^\beta_+(\mathbf{p}, \omega)$.

References

1. W.L. Burke, K.S. Thorne, Gravitational radiation damping, in *Relativity*, ed. by M. Carmeli, S.I. Fickler, L. Witten (Plenum, New York, 1970), pp. 209–228
2. B.R. Iyer, C.M. Will, Post-Newtonian gravitational radiation reaction for two-body systems. Phys. Rev. Lett., **70**, 113 (1993)
3. C.R. Galley, Classical mechanics of nonconservative systems. Phys. Rev. Lett. **110**(17), 174301 (2013). [arXiv:1210.2745 [gr-qc]]
4. C.R. Galley, M. Tiglio, Radiation reaction and gravitational waves in the effective field theory approach. Phys. Rev. D **79**, 124027 (2009). [arXiv:0903.1122 [gr-qc]]
5. J.S. Schwinger, Brownian motion of a quantum oscillator. J. Math. Phys. **2**, 407 (1961)
6. L.V. Keldysh, Diagram technique for nonequilibrium processes. Zh. Eksp. Teor. Fiz. **47**, 1515 (1964)
7. L.P. Kadanoff, G. Baym, *Quantum Statistical Mechanics* (Benjamin, New York, 1962)
8. C.R. Galley, A.K. Leibovich, Radiation reaction at 3.5 post-Newtonian order in effective field theory. Phys. Rev. D **86**, 044029 (2012). [arXiv:1205.3842 [gr-qc]]
9. C.R. Galley, A.K. Leibovich, R.A. Porto, A. Ross, Tail effect in gravitational radiation reaction: Time nonlocality and renormalization group evolution. Phys. Rev. D **93**, 124010 (2016). [arXiv:1511.07379 [gr-qc]]
10. J.F. Melo, The Propagator Matrix Reloaded (2021). [arXiv:2112.09119 [hep-th]]

11. W.D. Goldberger, J. Li, I.Z. Rothstein, Non-conservative effects on spinning black holes from world-line effective field theory. J. High Energy Phys. **06**, 053 (2021). [arXiv:2012.14869 [hep-th]]

12. G. Kälin, J. Neef, R.A. Porto, Radiation-Reaction in the Effective Field Theory Approach to Post-Minkowskian Dynamics (2022). [arXiv:2207.00580 [hep-th]]

13. C.R. Galley, B.L. Hu, Self-force on extreme mass ratio inspirals via curved spacetime effective field theory. Phys. Rev. D **79**, 064002 (2009). [arXiv:0801.0900 [gr-qc]]

14. U. Kraemmer, A. Rebhan, Advances in perturbative thermal field theory. Rept. Prog. Phys. **67**, 351 (2004). [arXiv:hep-ph/0310337 [hep-ph]]

Gravitational Waves from Blackhole Quasi-Normal Mode Oscillations

7

Abstract

When a binary pair of blackholes or neutron stars coalesces in a blackhole, the resulting blackhole will undergo oscillations in quasi-normal modes (QNM). The oscillations of the blackhole metric are accompanied by emission of gravitational waves hence the frequencies become 'quasi-normal' (complex), with the real part representing the actual frequency of the oscillation and the imaginary part representing the damping due to energy loss into gravitational waves. We study how observations of gravitational waves in the ring-down phase can test the no hair theorems of blackholes. We discuss how observations can distinguish blackhole mergers from bodies falling into other exotic objects like wormholes.

7.1   Introduction

When a binary pair of blackholes or neutron stars coalesces in a blackhole, the resulting blackhole will undergo oscillations in quasi-normal modes (QNM). The oscillations of the blackhole metric are accompanied by emission of gravitational waves hence the frequencies become 'quasi-normal' (complex), with the real part representing the actual frequency of the oscillation and the imaginary part representing the damping due to energy loss into gravitational waves. The existence of QNMs were first shown by Vishveshwara [1] in the example of the scattering of gravitational waves by a Schwarzschild blackhole. It was pointed out that when the blackhole is perturbed, it goes through a period where it oscillates as a damped oscillator with a single dominant frequency, and the oscillation frequency and the damping time depend on the intrinsic properties of blackholes like mass, charge and spin and are independent of the initial conditions of the perturbations.

A blackhole undergoing QNM oscillations will have the frequency and damping time characteristic of the blackhole [2–10]. The GW signals during QNM oscilla-

© The Author(s), under exclusive license to Springer Nature Switzerland AG 2023
S. Mohanty, *Gravitational Waves from a Quantum Field Theory Perspective*,
Lecture Notes in Physics 1013, https://doi.org/10.1007/978-3-031-23770-6_7

tions will be of the form [11–13]

$$h(t) = h_0 e^{-\pi f(t-T)/Q} \sin 2\pi f(t-T),\qquad (7.1)$$

where T is the time of merger and $t - T \geq 0$. Here Q is the quality factor which is related to the damping time τ of the oscillations and the frequency f as $Q = \pi f \tau$. The quality factor Q is the number of oscillations in one e-folding. For a Kerr blackhole characterised by mass M and rotation parameter $a = J/M$ the frequency of GW of the $(l, m, n) = (2, 2, 0)$ QNM mode is [11],

$$f \simeq \frac{1}{2\pi GM} \left[1 - \frac{63}{100}(1-a)^{3/10} \right]\qquad (7.2)$$

and the quality factor is [11],

$$Q = \pi f \tau = 2(1-a)^{-9/20}.\qquad (7.3)$$

Observations of the QNM gravitational waves would enable us to measure the mass and spin of the remnant Kerr BH (Fig. 7.1).

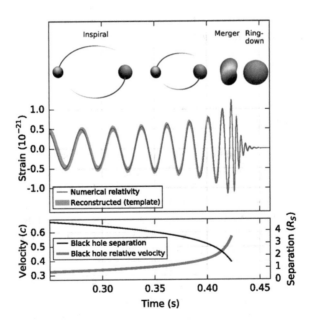

Fig. 7.1 Gravitational wave signal showing the post-merger ring-down phase. The complex QNM frequencies result in exponentially damped sinusoidal signal in the ring-down phase. Reprinted from [14]. Figure credit: ©The Author(s) 2016. Reproduced under CC-BY-3.0 license

7.2   Wave Equation in Spherically Symmetric Blackhole Background

Consider a spherically symmetric blackhole with a metric of the general form

$$ds^2 = -A(r)dt^2 + B(r)dr^2 + r^2 \left(d\theta^2 + \sin^2\theta d\phi^2 \right). \tag{7.4}$$

The equation of motion of a massless, minimally coupled scalar field $\Phi(t, r, \theta, \phi)$ in a general background metric is given by

$$\frac{1}{\sqrt{-g}} \partial_\mu \left(\sqrt{-g} g^{\mu\nu} \partial_\nu \Phi(t, r, \theta, \phi) \right) = 0. \tag{7.5}$$

We can separate the angular and radial equations by defining

$$\Phi(t, r, \theta, \phi) = \sum_{l,m} \frac{1}{r} \Psi_{l,m}(t, r) Y_{lm}(\theta, \phi), \tag{7.6}$$

where Y_{lm} are the eigenfunctions of the angular part of the wave operator

$$\Delta_{\Omega_2} Y_{lm} = -l(l+1) Y_{lm}, \tag{7.7}$$

where Δ_{Ω_2} is the Laplace-Beltrami operator,

$$\Delta_{\Omega_2} Y_{lm} \equiv \frac{1}{\sin\theta} \frac{\partial}{\partial\theta} \left(\sin\theta \frac{\partial Y_{lm}}{\partial\phi} \right) + \frac{1}{\sin^2\theta} \frac{\partial^2 Y_{lm}}{\partial\phi^2}. \tag{7.8}$$

Substituting (7.6) in (7.5) we have for the radial part of the wave equation

$$-\partial_t^2 \Psi_{lm}(t, r) + \frac{A}{B} \partial_r^2 \Psi_{lm}(t, r) + \frac{BA' - AB'}{2B^2} \partial_r \Psi_{lm}(t, r)$$
$$+ \frac{A(B' - 2l(l+1)B^2) - rBA'}{2r^2 B^2} \Psi_{lm}(t, r) = 0. \tag{7.9}$$

where $'$ denotes derivative w.r.t radial coordinate r. This equation may be written in the canonical wave equation form by going to tortoise coordinates defined by

$$dx = \left(\frac{B(r)}{A(r)} \right)^{1/2} dr. \tag{7.10}$$

In the tortoise coordinates, the wave equation (7.9) assumes the canonical form

$$- \partial_t^2 \Psi_{lm}(t, x) + \partial_x^2 \Psi_{lm}(t, x) - V(r) \Psi_{lm}(t, x) = 0, \tag{7.11}$$

where the potential

$$V(r) = A(r) \frac{l(l+1)}{r^2} + \frac{1}{2r} \frac{d}{dr} \frac{A(r)}{B(r)} . \tag{7.12}$$

For Schwarzschild blackholes

$$A(r) = B(r)^{-1} = \left(1 - \frac{2GM}{r} \right) \tag{7.13}$$

and the solution for the tortoise coordinate defined by (7.10) is

$$x = r + 2GM \log \left(\frac{r}{2GM} - 1 \right), \tag{7.14}$$

so that $x \to -\infty$ as $r \to 2GM$ (horizon) and $x \to \infty$ as $r \to \infty$ (spatial infinity). For the Schwarzschild metric the potential (7.12) takes the form

$$V(r) = \left(1 - \frac{2GM}{r} \right) \left[\frac{l(l+1)}{r^2} + \frac{2GM}{r^3} \right]. \tag{7.15}$$

We Fourier transform to frequency space

$$\Psi_{lm}(t, x) = \int dt \, e^{i\omega_{lm} t} \, \tilde{\Psi}_{lm}(\omega_{lm}, x) \tag{7.16}$$

and the wave equation for $\tilde{\Psi}_{lm}(\omega_{lm}, x)$ is given by

$$\left(\partial_x^2 + \omega_{lm}^2 - \left(1 - \frac{2GM}{r} \right) \left[\frac{l(l+1)}{r^2} + \frac{2GM}{r^3} \right] \right) \tilde{\Psi}_{lm}(\omega_{lm}, x) = 0. \tag{7.17}$$

This wave equation is generalised to massless bosonic fields of general spin s as

$$\left(\partial_x^2 + \omega_{lm}^2 - \left(1 - \frac{2GM}{r} \right) \left[\frac{l(l+1)}{r^2} + (1 - s^2) \frac{2GM}{r^3} \right] \right) \tilde{\Psi}_{lm}(\omega_{lm}, x) = 0, \tag{7.18}$$

where $s = 0$ for scalars, $s = 1$ for vector fields and $s = 2$ for axial gravitational perturbations [15].

7.3 Regge-Wheeler and Zerilli Equations

The axial gravitational perturbations are the ones which under parity transformation have the property $\tilde{\Psi}^-(\pi - \theta, \phi + \pi) \to (-1)^{(l+1)} \tilde{\Psi}^-(\theta, \phi)$. These are also referred to as 'odd' and 'vector' perturbations and obey the Regge-Wheeler equation [16]

$$\left(\partial_x^2 + \omega_{lm}^2 - V_l^-\right) \tilde{\Psi}_{lm}^-(\omega_{lm}, x) = 0 \,,$$

$$V_l^- = \left(1 - \frac{2GM}{r}\right)\left[\frac{l(l+1)}{r^2} - \frac{6GM}{r^3}\right]. \tag{7.19}$$

There are also an independent class of polar perturbations also called 'even' and 'scalar' perturbations which under parity have the property $\tilde{\Psi}^+(\pi - \theta, \phi + \pi) \to (-1)^l \tilde{\Psi}^+(\theta, \phi)$. These obey the Zerilli equation [17] given by

$$\left(\partial_x^2 + \omega_{lm}^2 - V_l^+\right) \tilde{\Psi}_{lm}^+(\omega_{lm}, x) = 0 \,,$$

$$V_l^+ = \left(1 - \frac{2GM}{r}\right)\left[\frac{2L^2(L+1)r^3 + 6GML^2r^2 + 18L(GM)^2r + 18(GM)^3}{r^3(Lr + 3GM)^2}\right],$$

$$L = \frac{1}{2}(l - 1)(l + 2). \tag{7.20}$$

7.4 Isospectral Property of Odd and Even Perturbation Equations

Consider a wave equation with the potential $V(x)$

$$-\frac{d^2\Psi(x)}{dx^2} + V(x, \omega)\Psi(x) = \omega^2\Psi(x) \tag{7.21}$$

which has an eigenfunction $\Psi(x)$ with eigenvalue ω^2. Now suppose any two potentials $V^\pm$ are related as follows:

$$V^\pm(x) = W^2(x) \pm \frac{dW(x)}{dx} + \beta, \tag{7.22}$$

where $W(x)$ is any function of the tortoise coordinate x and β is a constant.

Let $\Psi^+(x)$ be an eigenstate of (7.21) with potential $V^+(x)$ with eigenvalue ω^2. Then the eigenstate $\Psi^-(x)$ of (7.21) with potential $V^-(x)$ is given by

$$\Psi^-(x) \propto \left(W(x) - \frac{dW(x)}{dx}\right)\Psi^+(x) \tag{7.23}$$

with the same eigenvalue ω^2. The potentials $V\pm$ are isospectral as they give the same eigenvalue spectrum.

It was shown by Chandrasekhar [15] that the odd and even potentials given in (7.2) and (7.20), respectively, are isospectral. To see this take [18]

$$W(x) = \frac{2GM}{r^2} + \frac{3 + 2L}{3r} + \frac{3L + 2L^2}{3(2GM + Lr)} - \frac{L^2 + L}{3GM},$$

$$\beta = -\frac{L^2(L+1)^2}{9(MG)^2}, \quad L = \frac{1}{2}(l-1)(l+2). \tag{7.24}$$

Then from (7.22) the V^- and V^+ potentials are

$$V^-(x) = \left(1 - \frac{2GM}{r}\right)\left[\frac{l(l+1)}{r^2} - \frac{6GM}{r^3}\right],$$

$$V^+(x) = \left(1 - \frac{2GM}{r}\right)\left[\frac{2L^2(L+1)r^3 + 6GML^2r^2 + 18L(GM)^2r + 18(GM)^3}{r^3(Lr + 3GM)^2}\right]$$

$$\tag{7.25}$$

which are the Regge-Wheeler (7.2) and Zerilli (7.20) potentials, respectively. As a result of these relations between the odd and even perturbations the complex frequencies ω_{lm} of these two types of quasi-normal oscillations are identical [19].

7.5   Quasi-Normal Modes: Eikonal Calculation

Due to the isospectral relation between V^+ and V^- for determining the QNM frequencies it is sufficient to discuss the simpler Regge-Wheeler potential V^-. This potential is defined for the coordinate x in the range $x \in (-\infty, +\infty)$ and shown in Fig. 7.2 as a function of the radial coordinate r. The potential has a maxima at $r \simeq 3GM$ (depending on l), and goes to zero at the horizon $r = 2GM$ ($x = -\infty$) and at spatial infinity $r \to \infty$ ($x \to \infty$). High frequency waves follow the boundary conditions

$$\Psi(x \to \infty) \sim e^{i\omega x}, \tag{7.26}$$

and

$$\Psi(x \to -\infty) \sim e^{-i\omega x}, \tag{7.27}$$

which means that the waves are purely outgoing at spatial infinity $r \to \infty$ and are purely incoming at the horizon $r = 2GM$.

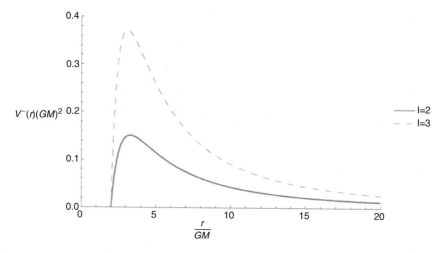

Fig. 7.2 Regge-Wheeler potential $V^-(r)$ plotted in the range of r from the blackhole horizon to spatial infinity for $l = 2$ and $l = 3$. The maxima is the light sphere radius which occurs at $r \sim 3GM$ (depending on l)

We can think of the blackhole perturbation as a scattering of an incident wave by the potential $V^-(x)$. The incident and scattered waves may be represented at high frequency as

$$\Psi(x) = A(x)e^{iS(x)/\epsilon}, \quad \epsilon \to 0, \tag{7.28}$$

where $A(x)$ is a slowly varying amplitude and $S(x)$ is a fast varying phase. We use the eikonal solution (7.28) in (7.2) and solve order by order in $1/\epsilon$ and l [20]. To keep consistent track of order of $1/\epsilon$ in this double expansion we replace l by l/ϵ. Substituting (7.28) in (7.2) we have

$$\partial_x^2 A(x) + \frac{i}{\epsilon}\left[2(\partial_x S)(\partial_x A) + A(x)\partial_x^2 S(x)\right]$$
$$+ \left[\omega^2 - \frac{1}{\epsilon^2}(\partial_x S)^2 - f(r)\left(\frac{\frac{l}{\epsilon}(\frac{l}{\epsilon}+1)}{r^2} - \frac{6GM}{r^3}\right)\right]A(r) = 0, \tag{7.29}$$

where for the Schwarzschild metric $f(r) = 1 - 2GM/r$. The frequency of the waves is of order $\omega \sim \partial_x S(x)/\epsilon$. Taking terms of order $1/\epsilon^2$ in (7.29) we obtain

$$\omega^2 - (\partial_x S)^2 - f(r)\frac{l^2}{r^2} = 0. \tag{7.30}$$

Define $U(r) \equiv f(r)/r^2$. Take derivative of (7.30) by x to get

$$2 \left(\partial_x S \right) \left(\partial_x^2 S \right) = -l^2 \frac{dr}{dx} \frac{dU}{dr} .$$ (7.31)

At $r = r_m$ the maxima of $U(r)$ we have $\frac{dU}{dr}|_{r_m} = 0$ and $(\partial_x S)|_{r_m} = 0$. This gives us $r_m = 3GM$, which is the radius of the light sphere for $l \gg 1$. At $r < r_m$ massless particles plunge into the blackhole. The quasi-normal modes are oscillations of the light sphere as first pointed out in [21].

Evaluate the Eq. (7.30) at $r = r_m$ where $(\partial_x S)|_{r_m} = 0$, and solve for ω. This gives us the real part of the QNM frequency to the leading order,

$$\omega_R^{(0)} = l \sqrt{U(r_m)} = \frac{l}{3\sqrt{3}GM} .$$ (7.32)

The leading order in ω goes as $\omega_R^{(0)} \propto l$ which is ϵ^{-1}. Writing the real and imaginary parts of QNM frequencies as a series in ϵ the next order terms $\omega_R^{(1)}$ and $\omega_I^{(1)}$ are of order ϵ^0. We write the complex QNM frequency as a series in ϵ as

$$\omega = \frac{1}{\epsilon} \omega_R^{(0)} + \omega_R^{(1)} + i \omega_I^{(1)} + O(\epsilon).$$ (7.33)

Substitute (7.33) in (7.29) and take terms of order $1/\epsilon$ to obtain,

$$2i \left(\partial_x S \right) \left(\partial_x A \right) + \left((\partial_x^2 S) + 2i \omega_R^{(0)} \omega_I^{(1)} + 2 \omega_R^{(0)} \omega_R^{(1)} - lU(r) \right) A(r) = 0.$$ (7.34)

In order to evaluate $(\partial_x^2 S)$ we start with (7.30), and Taylor expand $U(r)$ around $r = r_m$,

$$(\partial_x S)^2 = \omega_R^{(0)2} - l^2 \left(U(r_m) + U'(r_m) + \frac{1}{2} U''(r_m)(r - r_m)^2 \right)$$

$$= -\frac{1}{2} U''(r_m)(r - r_m)^2 .$$ (7.35)

Take the positive square root and differentiate w.r.t x to get

$$(\partial_x^2 S) = \frac{i}{\sqrt{2}} \frac{dr}{dx} \sqrt{U''(r_m)} .$$ (7.36)

Substitute (7.36) in (7.34) and evaluate the imaginary part at $r = r_m$ to obtain

$$\omega_I^{(1)} = -\frac{1}{2 \omega_R^{(0)}} (\partial_x^2 S) \Big|_{r=r_m} = -\frac{1}{6\sqrt{3}GM} .$$ (7.37)

Taking the real part of (7.34) at $r = r_m$ and solving for $\omega_R^{(1)}$ we obtain

$$\omega_R^{(1)} = \frac{1}{2\omega_R^{(0)}} l U(r_m) = \frac{1}{2}\sqrt{U(r_m)} = \frac{1}{6\sqrt{3}GM} . \tag{7.38}$$

We therefore have the real and imaginary parts of the QNM oscillation frequency in the large l and eikonal approximation given by Yagi [20],

$$\omega = \frac{1}{3\sqrt{3}GM}\left(l + \frac{1}{2}\right) - i\frac{1}{6\sqrt{3}GM} . \tag{7.39}$$

This corresponds to the frequency

$$f = \frac{\omega_R}{2\pi} = 1.2074\left(\frac{10M_\odot}{M}\right) \text{kHz} \tag{7.40}$$

and decay lifetime

$$\tau = \frac{1}{\omega_I} = 0.5537\left(\frac{M}{10M_\odot}\right) \text{ms} . \tag{7.41}$$

So the QNM signal is well within the range of observations of LIGO.

7.6   Gravitational Waves from Quasi-Normal Oscillations

The solutions $\tilde{\Psi}_{lm}^-(\omega_{lm}, x)$ of the Regge-Wheeler equation (7.2) and $\tilde{\Psi}_{lm}^+(\omega_{lm}, x)$ of the Zerilli equation (7.20) are related to the gravitational waves observed at a distance r as follows [22, 23],

$$h^+(t, r, \theta, \phi) = \frac{1}{2\pi r}\int_{-\infty}^{\infty} d\omega e^{-i\omega(t-x)}$$

$$\times \sum_{l,m}\left[\tilde{\Psi}_{lm}^+(\omega_{lm}, x)W^{lm}(\theta, \phi) - \frac{\tilde{\Psi}_{lm}^-(\omega_{lm}, x)}{i\omega}\frac{X^{lm}(\theta, \phi)}{\sin\theta}\right],$$

$$h^\times(t, r, \theta, \phi) = \frac{1}{2\pi r}\int_{-\infty}^{\infty} d\omega e^{-i\omega(t-x)}$$

$$\times \sum_{l,m}\left[\tilde{\Psi}_{lm}^+(\omega_{lm}, x)\frac{X^{lm}(\theta, \phi)}{\sin\theta} + \frac{\tilde{\Psi}_{lm}^-(\omega_{lm}, x)}{i\omega}W^{lm}(\theta, \phi)\right],$$

$$\tag{7.42}$$

where

$$W^{lm}(\theta, \phi) = \left(\partial_\theta^2 - \cot\theta\,\partial_\theta - \frac{1}{\sin^2\theta}\partial_\phi^2\right)Y^{lm}(\theta, \phi),$$

$$X^{lm}(\theta, \phi) = 2\left(\partial_{\theta\phi} - \cot\theta\,\partial_\phi\right)Y^{lm}(\theta, \phi), \tag{7.43}$$

where $Y^{lm}(\theta, \phi)$ are the scalar spherical harmonics.

7.7   Quasi-Normal Frequencies of Kerr Blackholes

The quasi-normal frequencies for the Kerr metric can be written in a analytical form for the slow rotation case where the rotation parameter $a = J/M \ll GM$. The perturbation equation for spin-2 in the slow rotation Kerr metric is [24],

$$\frac{d^2\Psi(x, \omega)}{dx^2} + \left(\omega^2 - V(r, \omega)\right)\Psi(x, \omega) = 0,$$

$$V(r, \omega) = \left(1 - \frac{2GM}{r}\frac{l(l+1)}{r^2} + 4\frac{am\omega GM}{r^3}\right), \tag{7.44}$$

where m is the azimuthal number. The quasi-normal frequencies using the Pöschl-Teller method are given by Ferrari and Mashhoon [24],

$$\omega = \frac{1}{3\sqrt{3}GM}\left[\pm\left(l + \frac{1}{2}\right) + \frac{2am}{3\sqrt{3}GM} - i\left(n + \frac{1}{2}\right)\right], \tag{7.45}$$

where the plus and minus signs are for co-rotating perturbations and minus sign is for counterrotating perturbations w.r.t the blackhole spin. Here $l \gg m \gg 1$ and $n = 0, 1, 2 \cdots \ll l$ is the overtone number.

7.8   Greens Functions and Late Time Tails

Although the primary component of the blackhole ring-down signal which behave as damped sinusoids, there are other components. For example, it was shown by Price [25] that at late times of the ring-down the gravitational wave signal has a polynomial damping of the form

$$|\Psi(t)| \sim t^{-(2l+3)} \tag{7.46}$$

for both Schwarzschild and Kerr blackholes [18].

To determine the complete gravitational wave signal expected at the detector from QNM oscillations we can employ Green's functions. If the perturbations at time $t = 0$ the perturbations have the initial data $\partial_t \Psi(x', t)|_{t=0}$ and $\Psi(x', t)|_{t=0}$ then this is propagated by the Greens function as

$$\Psi(x, t) = \int dx' G(x, x'; t) \, \partial_t \Psi(x', t)|_{t=0} + \int dx' \partial_t G(x, x'; t) \, \Psi(x', t)|_{t=0},$$

$$(7.47)$$

where $x = r + 2GM \log \left(\frac{r}{2GM} - 1 \right)$.

The Greens function is a solution of the wave equation with the delta function source

$$\left(-\frac{\partial}{\partial x^2} + \frac{\partial}{\partial t^2} + V(x) \right) G(x, x'; t) = \delta(t) \delta(x - x').$$

$$(7.48)$$

In Fourier space we can write this as

$$\left(-\frac{\partial}{\partial x^2} - \omega^2 + V(x) \right) \tilde{G}(x, x'; \omega) = \delta(x - x').$$

$$(7.49)$$

The boundary conditions on the perturbations are $\Psi(x, \omega) \sim e^{-i\omega x}$ at $x \to -\infty$ and $\Psi(x, \omega) \sim e^{i\omega x}$ at $x \to \infty$ which means that the waves are ingoing at the horizon and outgoing at spatial infinity. We consider mode functions $f(x, \omega)$ and $g(x, \omega)$ which satisfy the homogeneous wave equations, and f satisfies the b.c at $x \to -\infty$ while g satisfies the b.c at $x \to \infty$. The modes are normalised such that

$$f(\omega, x) = e^{-i\omega x}, \quad x \to -\infty,$$
$$g(\omega, x) = e^{i\omega x}, \quad x \to \infty.$$

$$(7.50)$$

Using the two modes we construct the Wronskian

$$W(\omega) = g \partial_x f - f \partial_x g.$$

$$(7.51)$$

The retarded Greens function can now be written in terms of the mode functions as

$$\tilde{G}(x, x'; \omega) = \frac{\theta(x' - x)}{W(\omega)} f(\omega, x) g(\omega, x') + \frac{\theta(x - x')}{W(\omega)} f(\omega, x') g(\omega, x)$$

$$(7.52)$$

The Greens function in time is then given by

$$G(x, x'; t) = \int d\omega_{-\infty}^{\infty} e^{i\omega t} \, \tilde{G}(x, x'; \omega).$$

$$(7.53)$$

The contributions to $G(x, x'; t)$ arise from the poles and the branch cuts of $\tilde{G}(x, x'; \omega)$ in the complex ω plane (Fig. 7.3).

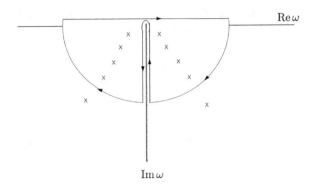

Fig. 7.3 Poles and the branch cut of the Greens function $\tilde{G}(x, x'; \omega)$ and the contour of integration is shown. The branch cut is $-Im\omega$ axis

$\tilde{G}(x, x'; \omega)$ has complex poles $\omega_R - i\omega_I$ which are the quasi-normal mode frequencies. These are responsible for the exponentially damped sinusoidal ringdown signals. The branch cut occurs because of the modes which initially fall inwards but are scattered in the outward direction by the curvature of the blackhole. These give rise to the polynomial suppression of the signal with time of the form $\Psi \sim t^{2l+3}$ at late stages of the ring-down and is called the tail effect [25–28].

7.9   Testing Blackhole No Hair Theorems

A static, axially symmetric and asymptotically flat blackhole spacetime cannot support a static configuration of scalar field outside the horizon. The proof of this no-scalar hair theorem was given by Hawking [29] and Bekenstein [30]. A proof given by Sotiriou and Faraoni [31] is as follows [32]:

Consider a scalar-tensor theory with the action

$$S = \int d^4x \sqrt{-g} \left(\frac{1}{16\pi G} R - \frac{1}{2}\partial_\mu\varphi\partial^\mu\varphi - V(\varphi) \right). \tag{7.54}$$

The scalar field obeys the Klein-Gordon equation

$$\Box\varphi - V'(\varphi) = 0, \tag{7.55}$$

where the prime denotes $\partial/\partial\varphi$. A static and axial symmetric spacetime has two Killing vectors ∂_t and ∂_ϕ associated with the conserved charges energy and angular momentum, respectively. Assuming that the scalar solution of the KG equation also has the same symmetries, we must have $\partial_t \varphi = \partial_\phi \varphi = 0$. Multiply the KG Eq. (7.55) by V' and integrate over spacetime,

$$\int d^4x \sqrt{-g} \left(V'(\varphi)\Box\varphi - (V'(\varphi))^2 \right) = 0. \tag{7.56}$$

Integrate the first term by parts. There are two surface terms, one at the horizon and one at $\to \infty$. Assuming that the scalar field drops-off as $\varphi \sim 1/r$ or faster, the surface term at spatial infinity is zero. We have after the integration by parts

$$\int d^4x \sqrt{-g} \left(-V''(\varphi)\partial_\mu\varphi\partial^\mu\varphi - (V'(\varphi))^2 \right) + \int_{\mathcal{H}} d\sigma\, V'(\varphi)n^\nu\partial_\mu\varphi = 0. \tag{7.57}$$

The event horizon of a static asymptotically flat spacetime is an Killing horizon. This means that the normal n^μ to the horizon is a linear combination of the Killing vectors ∂_t and ∂_ϕ. Therefore since $\partial_t \varphi = \partial_\varphi = 0$, we must have $n^\mu \partial_\mu \varphi = 0$ on the horizon and the surface term in (7.57) vanishes and we are left with

$$\int d^4x \sqrt{-g} \left(V''(\varphi)\partial_\mu\varphi\partial^\mu\varphi + (V'(\varphi))^2 \right) = 0. \tag{7.58}$$

If $V''(\varphi) < 0$ then there will be a tachyonic instability of scalar filed rolling down the potential. So we can assume that $V''(\varphi) \geq 0$. The static field configuration must have $\partial_a\varphi\partial_a\varphi \gtrless 0$. Therefore the only solutions that are allowed are the trivial solutions namely $V'(\varphi) = 0$ and $V''(\varphi) = 0$ or $V'(\varphi) = 0$ and $\partial_a\varphi = 0$.

7.10 Traversable Wormholes

The Morris-Thorne wormhole is a traversable wormhole whose metric can be written as [33]

$$ds^2 = -dt^2 + \frac{dr^2}{\left(1 - \frac{b(r)}{r}\right)} + r^2 \left(d\theta^2 + \sin^2\theta d\phi^2\right). \tag{7.59}$$

The 'throat' is located at the radial coordinate $r = b(r) = b_0$ where there is a coordinates singularity (Fig. 7.4). The proper radial distance from the throat is given by

$$l(r) = \pm \int_{b_0}^{r} \frac{dr'}{\left(1 - \frac{b(r')}{r'}\right)}. \tag{7.60}$$

The range of the proper distance from the throat $l(r) \in (0, \infty)$ in one direction and $l(r) \in (0, -\infty)$ in the opposite direction. The wormhole connects two distinct asymptotically flat regions at $r \to \infty$ through the throat at $r = b_0$.

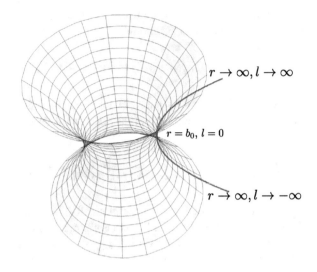

Fig. 7.4 Wormhole connecting two distinct asymptotic regions in the universe

$r \to \infty, l \to \infty$

$r = b_0, l = 0$

$r \to \infty, l \to -\infty$

The stress tensor which gives the wormhole metric (7.59) in a orthonormal reference frame given by Morris and Thorne [33], and Roman [34]

$$T_{\hat{t}\hat{t}} = \rho(r) = \frac{b'(r)}{8\pi r^2},$$

$$T_{\hat{r}\hat{r}} = -\tau(r) = -\frac{b(r)}{8\pi r^3},$$

$$T_{\hat{\theta}\hat{\theta}} = T_{\hat{\phi}\hat{\phi}} = p(r) = \frac{b(r) - b'(r)r}{8\pi r^3}, \tag{7.61}$$

where $\rho(r)$ is the energy density, $\tau(r)$ the radial tension and $p(r)$ is the pressure. In the orthonormal frame given by the tetrad

$$e_{\hat{t}} = e_t, \quad e_{\hat{r}} = (1 - b/r)^{1/2} e_r, \quad e_{\hat{\theta}} = r^{-1} e_\theta, \quad e_{\hat{\phi}} = (r \sin\theta)^{-1} e_\phi \tag{7.62}$$

the metric $g_{\hat{\mu}\hat{\nu}} = e_{\hat{\mu}} e_{\hat{\nu}} = \eta_{\hat{\mu}\hat{\nu}}$ but the Riemann and Ricci curvatures are not zero. The non-zero components of the Riemann tensor in the orthonormal frame are [33, 34],

$$R^{\hat{r}}_{\hat{\theta}\hat{r}\hat{\theta}} = \frac{b'r - b}{2r^3}, \quad R^{\hat{r}}_{\hat{\phi}\hat{r}\hat{\phi}} = \frac{b'r - b}{2r^3}, \quad R^{\hat{\theta}}_{\hat{\phi}\hat{\theta}\hat{\phi}} = \frac{b}{r^3} \tag{7.63}$$

and their permutations using the symmetry property of the Riemann tensor

$$R^{\hat{\mu}}_{\hat{\nu}\hat{\alpha}\hat{\beta}} = -R^{\hat{\mu}}_{\hat{\nu}\hat{\beta}\hat{\alpha}} = R^{\hat{\nu}}_{\hat{\mu}\hat{\beta}\hat{\alpha}} = R_{\hat{\alpha}\hat{\beta}}{}^{\hat{\nu}}{}_{\hat{\mu}}. \tag{7.64}$$

Another example of a wormhole that is traversable and where the tidal acceleration of the free fall observer can be tuned to be small has the metric [33]

$$ds^2 = -dt^2 + dl^2 + (l^2 + b_0^2)\left(d\theta^2 + \sin^2\theta \, d\phi^2\right). \tag{7.65}$$

The range of the proper distance from the throat is $l \in (0, \infty)$ in one direction
and $l \in (0, -\infty)$ in the opposite direction. The wormhole connects two distinct
asymptotically flat regions at $l \to \pm\infty$ through the throat at $l = 0$.

The corresponding stress tensor which generates the metric (7.65) is given by

$$T^{tt} = T^{ll} = -T^{\theta\theta} = -T^{\phi\phi} = -\frac{1}{8\pi G}\frac{b_0^2}{(b_0^2 + l^2)^2} \tag{7.66}$$

and the energy density $\rho = T^{tt} < 0$. The Riemann tensor from the metric (7.65)
have non-zero components given by

$$R_{\theta\phi\theta\phi} = -R_{l\theta l\theta} = -R_{l\phi l\phi} = \frac{b_0^2}{(b_0^2 + l^2)^2} \tag{7.67}$$

and other components related by the symmetries of the Riemann tensor.

The free fall observer has a world line $l = vt$, $\theta =$ constant and $\phi =$ constant
where v is a constant velocity. The local inertial frame of the free fall observer is
described by the tetrad pf basis vectors

$$e_{\hat{t}} = \gamma e_t + v\gamma e_l, \quad e_{\hat{l}} = \gamma e_l + v\gamma e_t, \quad e_{\hat{\theta}} = e_\theta, \quad e_{\hat{\phi}} = e_\phi, \tag{7.68}$$

where $\gamma = (1 - v^2)^{-1/2}$ is the Lorentz factor. The four-momentum of a free fall
particle of mass m in the local inertial frame $g_{\hat{\mu}\hat{v}} = e_{\hat{\mu}} \cdot e_{\hat{v}} = \eta_{\hat{\mu}\hat{v}}$ is $p^{\hat{t}} = \gamma m$,
$p^{\hat{l}} = \gamma m v$, $p^{\hat{\theta}} = p^{\hat{\phi}} = 0$. The tidal acceleration of the free fall observer is given by
the Riemann tensor in the local inertial frame which turns out to be

$$R_{\hat{\theta}\hat{t}\hat{\theta}\hat{t}} = R_{\hat{\phi}\hat{t}\hat{\phi}\hat{t}} = -v^2\gamma^2\frac{b_0^2}{(b_0^2 + l^2)^2}. \tag{7.69}$$

The tidal acceleration of the free fall observer depends upon the velocity and is small
for bodies with low free fall velocity.

7.11   Quasi-Normal Modes of Wormholes

Wormhole can mimic blackholes and have QNM oscillations of the metric when an object perturbs it by transversing it and going over to the other universe. The metric outside the throat at $r = r_0$ is the Schwarzschild metric given by

$$ds^2 = -F dt^2 + F^{-1} dr^2 + r^2 d\Omega^2, \quad r > r_0.$$

$$F = 1 - \frac{2GM}{r}. \tag{7.70}$$

To mimic the QNM's of blackholes the throat of the wormholes should be inside the light sphere $r_\gamma = 3GM$. Perturbations of the metric at the light sphere are observed as gravitational waves from the QNM oscillations. The wormhole can be created by a thin shell of exotic matter at the throat with the surface density and pressure given by Visser [35]

$$\sigma = \frac{1}{2\pi r_0} \sqrt{F(r_0)}, \qquad p = \frac{1}{4\pi r_0} \frac{(1 - \frac{GM}{r_0})}{\sqrt{F(r_0)}}. \tag{7.71}$$

For the wormhole to be traversable the weak energy condition has to be violated and $\sigma < 0$ [35]. The strong and null energy conditions are not violated as long as the throat is within the light sphere $r_\gamma < 3GM$. A body traversing the wormhole will source oscillations of the metric and obey the Zerilli equation [17]

$$[\omega^2 - V_l(r)]\Psi_{lm}(r) = S_l. \tag{7.72}$$

The solutions will generate gravitational waves which are then observed at a large distance from the wormhole. The amplitude of the gravitational waves is related to Ψ_l as [36],

$$h = h_+ - i h_\times = \frac{1}{r} \sum_{l,m} \sqrt{(l-1)l(l+1)(l+2)} \ \Psi_{lm} \ e^{i\omega_{nlm}(t-t_0)} \ _{-2}Y_{lm}. \tag{7.73}$$

The QNM's of the metric are obtained from the solutions of the homogenous equation with the boundary conditions the GW are outgoing at both asymptotic regions $r_* \to \pm\infty$,

$$\Psi_{lm} = e^{\pm i\omega r_*} \qquad \text{at} \qquad r_* \to \pm\infty, \tag{7.74}$$

where r_* is defined by

$$\frac{dr}{dr^*} = \pm F, \tag{7.75}$$

where $+$ sign is for our universe and $-$ sign is for the other universe (Fig. 7.5).

The wormholes perturbations are trapped between the two surfaces which are the light spheres of each universe. On the other hand the blackholes perturbations have

Fig. 7.5 Wormhole potential (top) compared to blackhole potential (bottom). Reprinted from [37]. Figure credit: ©2016 American Physical Society. Reproduced with permissions. All rights reserved

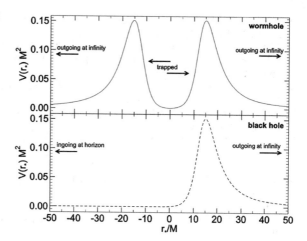

only one potential barrier at the light sphere. As a result the wormhole oscillations have a very slow decay time compared to blackhole perturbations. The QNM of blackholes are $\omega_{BH} GM = 0.3737 - 0.0890i$. The imaginary part of the wormhole frequencies becomes small as the throat moves closer and closer to the horizon. Taking $r_0 - 2GM = 10^{-5}$ the QNM frequencies of the wormholes is $\omega_{WH} GM = 0.07088 - 10^{-9} i$ [37]. For rotating wormholes the QNM frequencies can be close to those of Kerr blackholes in both real and imaginary part and it is not yet possible to distinguish between an Kerr BH and a rotating wormhole from LIGO observations [28].

References

1. C.V.Vishveshwara, Scattering of Gravitational Radiation by a Schwarzschild Black-hole. Nature **227**(5261), 936–938 (1970)
2. W.H. Press, Long wave trains of gravitational waves from a vibrating black hole. Astrophys. J. Lett. **170**, L105–L108 (1971)
3. K.D. Kokkotas, B.G. Schmidt, Quasinormal modes of stars and black holes. Living Rev. Rel. **2**, 2 (1999). https://doi.org/10.12942/lrr-1999-2. [arXiv:gr-qc/9909058 [gr-qc]]
4. V. Cardoso, J.P.S. Lemos, Gravitational radiation from the radial infall of highly relativistic point particles into Kerr black holes. Phys. Rev. D **67**, 084005 (2003). [arXiv:gr-qc/0211094 [gr-qc]]
5. V. Cardoso, Quasinormal modes and gravitational radiation in black hole spacetimes. [arXiv:gr-qc/0404093 [gr-qc]]
6. E. Berti, V. Cardoso, A.O. Starinets, Quasinormal modes of black holes and black branes. Class. Quant. Grav. **26**, 163001 (2009). [arXiv:0905.2975 [gr-qc]]
7. A. Nagar, L. Rezzolla, Gauge-invariant non-spherical metric perturbations of Schwarzschild black-hole spacetimes. Class. Quant. Grav. **22**, R167 (2005). [erratum: Class. Quant. Grav. **23**, 4297 (2006)]. [arXiv:gr-qc/0502064 [gr-qc]]
8. E. Berti, K. Yagi, H. Yang, N. Yunes, Extreme gravity tests with gravitational waves from compact binary coalescences: (II) Ringdown. Gen. Rel. Grav. **50**(5), 49 (2018). [arXiv:1801.03587 [gr-qc]]

9. V. Ferrari, B. Mashhoon, New approach to the quasinormal modes of a black hole. Phys. Rev. D **30**, 295–304 (1984)
10. G. Franciolini, L. Hui, R. Penco, L. Santoni, E. Trincherini, Effective field theory of black hole quasinormal modes in scalar-tensor theories. J. High Energy Phys. **02**, 127 (2019). [arXiv:1810.07706 [hep-th]]
11. F. Echeverria, Gravitational wave measurements of the mass and angular momentum of a black hole. Phys. Rev. D **40**, 3194–3203 (1989)
12. L.S. Finn, Detection, measurement and gravitational radiation. Phys. Rev. D **46**, 5236–5249 (1992). [arXiv:gr-qc/9209010 [gr-qc]]
13. E. Berti, V. Cardoso, C.M. Will, On gravitational-wave spectroscopy of massive black holes with the space interferometer LISA. Phys. Rev. D **73**, 064030 (2006). https://doi.org/10.1103/PhysRevD.73.064030. [arXiv:gr-qc/0512160 [gr-qc]]
14. B.P. Abbott et al., [LIGO Scientific and Virgo], Observation of Gravitational Waves from a Binary Black Hole Merger. Phys. Rev. Lett. **116**(6), 061102 (2016). [arXiv:1602.03837 [gr-qc]]
15. S. Chandrasekhar, *The Mathematical Theory of Black Holes* (Oxford University Press, Oxford, 1983)
16. T. Regge, J.A. Wheeler, Stability of a Schwarzschild Singularity. Phys. Rev. **108**, 1063 (1957)
17. F.J. Zerilli, Effective potential for even-parity Regge-Wheeler gravitational perturbation equations. Phys. Rev. Lett., **24**(13), 737 (1970)
18. R.A. Konoplya, A. Zhidenko, Quasinormal modes of black holes: From astrophysics to string theory. Rev. Mod. Phys. **83**, 793–836 (2011). [arXiv:1102.4014 [gr-qc]]
19. S. Chandrasekhar, S.L. Detweiler, The quasi-normal modes of the Schwarzschild black hole. Proc. Roy. Soc. Lond. A **344**, 441–452 (1975)
20. K. Yagi, Analytic estimates of quasi-normal mode frequencies for black holes in General Relativity and beyond. [arXiv:2201.06186 [gr-qc]].
21. C.J. Goebel, Comments on the "vibrations" of a Black Hole. Astrophys. J. Lett. **172**, L95 (1972)
22. L. Rezzolla, Gravitational waves from perturbed black holes and relativistic stars. ICTP Lect. Notes Ser. **14**, 255–316 (2003). [arXiv:gr-qc/0302025 [gr-qc]]
23. V. Ferrari, L. Gualtieri, Quasi-Normal Modes and Gravitational Wave Astronomy. Gen. Rel. Grav. **40**, 945–970 (2008). [arXiv:0709.0657 [gr-qc]]
24. V. Ferrari, B. Mashhoon, New approach to the quasinormal modes of a black hole. Phys. Rev. D **30**, 295–304 (1984). https://doi.org/10.1103/PhysRevD.30.295. N. Andersson, Evolving test fields in a black hole geometry. Phys. Rev. D **55**, 468–479 (1997). [arXiv:gr-qc/9607064 [gr-qc]]
25. R.H. Price, Nonspherical Perturbations of Relativistic Gravitational Collapse. II. Integer-Spin, Zero-Rest-Mass Fields. Phys. Rev. D **5**, 2439–2454 (1972)
26. E.W. Leaver, Spectral decomposition of the perturbation response of the Schwarzschild geometry. Phys. Rev. D **34**, 384–408 (1986)
27. N. Andersson, Evolving test fields in a black hole geometry. Phys. Rev. D **55**, 468–479 (1997). https://doi.org/10.1103/PhysRevD.55.468. [arXiv:gr-qc/9607064 [gr-qc]]
28. R.A. Konoplya, A. Zhidenko, Wormholes versus black holes: quasinormal ringing at early and late times. J. Cosmol. Astropart. Phys. **12**, 043 (2016). [arXiv:1606.00517 [gr-qc]]
29. S.W. Hawking, Black holes in the Brans-Dicke theory of gravitation. Commun. Math. Phys. **25**(2), 167–171 (1972)
30. J.D. Bekenstein, Nonexistence of baryon number for static black holes. Phys. Rev. D **5**, 1239–1246 (1972)
31. T.P. Sotiriou, V. Faraoni, Black holes in scalar-tensor gravity. Phys. Rev. Lett. **108**, 081103 (2012). [arXiv:1109.6324 [gr-qc]]
32. A. Kuntz, Testing gravity with the two-body problem. [arXiv:2010.05931 [gr-qc]]
33. M.S. Morris, K.S. Thorne, Wormholes in space-time and their use for interstellar travel: A tool for teaching general relativity. Am. J. Phys. **56**, 395–412 (1988)

34. T.A. Roman, Inflating Lorentzian wormholes. Phys. Rev. D **47**, 1370–1379 (1993). https://doi.org/10.1103/PhysRevD.47.1370. [arXiv:gr-qc/9211012 [gr-qc]]
35. M. Visser, *Lorentzian wormholes: From Einstein to Hawking* (AIP, Woodbury, 1996)
36. N. Sago, S. Isoyama, H. Nakano, Fundamental Tone and Overtones of Quasinormal Modes in Ringdown Gravitational Waves: A Detailed Study in Black Hole Perturbation. Universe **7**(10), 357 (2021). [arXiv:2108.13017 [gr-qc]]
37. V. Cardoso, E. Franzin, P. Pani, Is the gravitational-wave ringdown a probe of the event horizon?. Phys. Rev. Lett. **116**(17), 171101 (2016). [erratum: Phys. Rev. Lett. **117**(8), 089902 (2016)]. [arXiv:1602.07309 [gr-qc]]

Gravitational Radiation from Spin Dynamics in Binary Orbits

8

Abstract

We discuss the Lagrangian framework for studying spin dynamics of extended objects and we derive the Mathisson–Papapetrou–Dixon equation of spinning macroscopic bodies. We study the effect of spin and orbital angular momentum on the gravitational wave signals from spinning binaries.

8.1 Dynamics of Spinning Bodies in General Relativity

In this section we construct a Lagrangian to describe the spin of macroscopic bodies and we study the dynamics of spin from the equations of motion derived from this Lagrangian. Consider a spinning body along the world line $y^\alpha(\tau)$. Attach a local tetrad with the body centre $e^a_\mu(\tau)$ which moves with the world line $y^\alpha(\tau)$ of the particle.[1] The change in the local tetrad attached to a spin body along the trajectory of the body defines the antisymmetric Ricci rotation tensor $\omega^{\mu\nu}$,

$$\frac{De^\mu_a(\tau)}{D\tau} = u^\alpha \nabla_\alpha e^\mu_a(\tau) = -\omega^{\mu\nu} e_{va}, \tag{8.1}$$

---

[1] Latin indices denote the local inertial frame coordinates and Greek indices denote the general coordinates. The tetrads have the following orthogonality relations in the flat and curved space indices, $e^a_\mu e^{b\mu} = \eta^{ab}$ and $e^a_\mu e_{av} = g_{\mu\nu}$.

© The Author(s), under exclusive license to Springer Nature Switzerland AG 2023
S. Mohanty, *Gravitational Waves from a Quantum Field Theory Perspective*,
Lecture Notes in Physics 1013, https://doi.org/10.1007/978-3-031-23770-6_8

where $u^\alpha = dy^\alpha/d\tau$. Operating with $e^{\beta a}$ on both sides we have

$$e^{\beta a} u^\alpha \nabla_\alpha e^\mu_a(\tau) = \omega^{\beta\mu}. \tag{8.2}$$

The rotation tensor is antisymmetric $\omega^{\mu\nu} = -\omega^{\nu\mu}$ which can be seen from the fact that the metric is a covariant constant $\nabla_\nu g^{\mu\nu} = 0$,

$$0 = \frac{Dg^{\mu\nu}}{D\tau} = \frac{De_a{}^\mu}{D\tau} e^{a\nu} + e^\mu_a \frac{De^{a\nu}}{D\tau} = -(\omega^{\mu\nu} + \omega^{\nu\mu}). \tag{8.3}$$

Expanding the covariant derivative in (8.2) we have

$$\omega_{\alpha\beta} = \eta^{ab} e_{\alpha a} u^\rho \nabla_\rho e_{\beta b}$$

$$= \left(\frac{de_{\beta b}}{d\tau} - \Gamma^\rho_{\alpha\gamma} e_{\rho b} u^\gamma\right) \eta^{ab} e_{\alpha a}. \tag{8.4}$$

The variables u^α and $\omega^{\mu\nu}$ can be associated with conjugate momenta p_α and $S_{\alpha\beta}$, respectively. With these dynamical objects we construct the Lagrangian,

$$L = p_\alpha u^\alpha + \frac{1}{2} S_{\alpha\beta} \omega^{\alpha\beta}, \tag{8.5}$$

where the indices are contracted with the metric tensor $g_{\mu\nu}(x)$.

Varying the action w.r.t the rotation tensor $\omega^{\mu\nu}$ we get

$$\delta S = \int d\tau \left(\frac{DS^{\alpha\beta}}{D\tau} e_{a\alpha} - \frac{De_{a\alpha}}{D\tau} S^{\alpha\beta} + S^{\rho\nu} \frac{De_{b\nu}}{D\tau} e_{\rho a} e^{b\beta}\right) \delta e^a_\beta \tag{8.6}$$

which gives the equation of motion of the spin angular momentum tensor

$$\frac{DS^{\alpha\beta}}{D\tau} = S^{\alpha\mu} \omega_\mu{}^\beta - \omega^\alpha{}_\lambda S^{\lambda\beta}. \tag{8.7}$$

Varying the action w.r.t the four-velocity u^α we get the equation of motion of p_α given by

$$\frac{DS_{\alpha\beta}}{D\tau} = p_\alpha u_\beta - p_\beta u_\alpha, \tag{8.8}$$

where we made use of (8.4).

The spin tensor $S^{\mu\nu}$ is an antisymmetric tensor and in 4-dimension it has 6 degrees of freedom. The spin tensor must be further constrained by some spin supplementary conditions (SSC). We adopt the covariant SSC given by Tulczyjew [1],

$$S^{\mu\nu} p_\mu = 0. \tag{8.9}$$

Using the SSC (8.9) and the e.o.m (8.7) and (8.8) we see that

$$\frac{D\left(S_{\alpha\beta}S^{\alpha\beta}\right)}{D\tau} = 2S^{\alpha\beta}\left(p_\alpha u_\beta - p_\beta u_\alpha\right) = 0 \tag{8.10}$$

which means that the scalar $S^2 = \frac{1}{2}S_{\alpha\beta}S^{\alpha\beta}$ is a conserved quantity. In addition $S_{\mu\nu}S^{*\mu\nu}$ is a conserved, where $S^{*\mu\nu} = \frac{1}{2}\epsilon_{\mu\nu\alpha\beta}S^{\alpha\beta}$.

Using the equations of motion we can check that

$$\frac{DS_{ab}}{D\tau} = \frac{D\left(S_{\alpha\beta}e^\alpha{}_a e^\beta{}_b\right)}{D\tau} = 0 \tag{8.11}$$

and the spin in the local inertial frame is conserved.

In a spinning extended object the velocity and rotation coefficient between two near points in the body with a separation $\delta y^\alpha(\tau)$ will be different. We define the difference of the fields separated at two different points in the body by defining a covariant variation [2–4],

$$\Delta \equiv \delta y^\alpha \nabla_\alpha . \tag{8.12}$$

Since the metric is a covariant constant between the two points the covariant variation $\Delta g_{\mu\nu} = 0$ although $g_{\mu\nu}(y^\alpha) \neq g_{\mu\nu}(y^\alpha + \delta y^\alpha)$. The four-momentum p_μ and spin $S_{\alpha\beta}$ are attributes of the entire body therefore

$$\Delta p_\mu \equiv 0, \quad \Delta S_{\mu\nu} \equiv 0. \tag{8.13}$$

The variation of the velocity u^α and rotation tensor $\omega^{\mu\nu}$ between the two points in the body are then

$$\Delta u^\alpha = \frac{D\delta y^\alpha}{D\tau},$$

$$\Delta \omega^{\mu\nu} = \frac{D}{D\tau}\left(\Delta\Theta^{\mu\nu}\right) + \omega_\alpha{}^{[\mu}\Delta\Theta^{\nu]\alpha} + R^{\mu\nu}{}_{\alpha\beta}u^\alpha\delta y^\beta, \tag{8.14}$$

where

$$\Delta\Theta^{\mu\nu} \equiv e^{a[\mu}\Delta e_a{}^{\nu]}. \tag{8.15}$$

Varying the Lagrangian by δy^α and $\Delta\Theta^{\mu\nu}$ gives

$$\delta L(g_{\mu\nu}, y^\alpha, \omega^{\mu\nu}) = p_\alpha \frac{D\delta y^\alpha}{D\tau}$$

$$+ \frac{1}{2}S_{\mu\nu}\left(\frac{D}{D\tau}\left(\Delta\Theta^{\mu\nu}\right) + \omega_\alpha{}^{[\mu}\Delta\Theta^{\nu]\alpha} + R^{\mu\nu}{}_{\alpha\beta}u^\alpha\delta y^\beta\right).$$

$$\tag{8.16}$$

Variation of the action then gives us after integrating by parts,

$$\delta S = \int d\tau \left(-\frac{Dp_\beta}{D\tau} + R^{\mu\nu}{}_{\alpha\beta} S_{\mu\nu} \, u^\alpha \right) \delta y^\beta$$
$$+ \frac{1}{2} \left(-\frac{DS_{\mu\nu}}{D\tau} \left(\Delta\Theta^{\mu\nu} \right) + \omega_\alpha{}^{[\mu} \Delta\Theta^{\nu]\alpha} \right). \tag{8.17}$$

Minimising the action w.r.t δy^β gives us the equation of motion

$$\frac{Dp_\beta}{D\tau} = -\frac{1}{2} R^{\mu\nu}{}_{\beta\alpha} S_{\mu\nu} \, u^\alpha \tag{8.18}$$

which is the Mathisson–Papapetrou–Dixon equation of spinning macroscopic bodies [5–7].

Minimising the action w.r.t the variation $\Delta\Theta^{\mu\nu}$ gives us the equation of motion for $S_{\mu\nu}$

$$\frac{DS_{\mu\nu}}{D\tau} = S_{\alpha[\mu}\omega_{\nu]}{}^\alpha \tag{8.19}$$

which we have already obtained in (8.7).

We can define the kinematical mass of the of the body through the relation

$$m^2 = -g^{\alpha\beta} p_\alpha p_\beta. \tag{8.20}$$

From the SSC condition $S^{\mu\nu} p_\mu = 0$ we can write

$$\frac{DS^{\mu\nu}}{D\tau} p_\mu + S^{\mu\nu} \frac{Dp_\mu}{D\tau} = 0. \tag{8.21}$$

Using (8.8) and (8.18) we see that this may be written as

$$p_\mu (p \cdot u) + m^2 u_\mu = \frac{1}{2} u^\gamma R^\nu{}_{\gamma\alpha\beta} S^{\alpha\beta} S_{\mu\nu}. \tag{8.22}$$

Contracting with u^μ we obtain an expression for $p \cdot u$, for extended spinning bodies

$$(p \cdot u)^2 = m^2 + \frac{1}{2} u^\gamma u^\mu R^\nu{}_{\gamma\alpha\beta} S^{\alpha\beta} S_{\mu\nu}. \tag{8.23}$$

We define a spin four-vector S_μ by

$$S^{\mu\nu} = \frac{1}{m} \varepsilon^{\alpha\beta\mu\nu} p_\alpha S_\beta, \tag{8.24}$$

where $\varepsilon^{\alpha\beta\mu\nu}$ is the Levi-Civita tensor. We would like the spatial components of S_μ in the rest frame to be the spin vector. With that aim we impose the condition

$$S_\mu u^\mu = 0. \tag{8.25}$$

The spin vector in the local inertial frame is $S_a = e_a^\mu S_\mu$. By choosing $e_a^\mu = u^\mu$ we ensure due to the orthogonality condition (8.25) that $S_a = (0, \mathbf{S})$ has only spatial components. From the conservation of $S^\alpha S_\alpha$ we see that $S^\alpha S_\alpha = S^a S_a = |\mathbf{S}|^2$ is conserved for a rigid body barring other interactions.

From (8.11) we obtain the e.o.m for the spin vector in the local Lorentz frame

$$\frac{dS^a}{d\tau} = \omega^{ab} S_b. \tag{8.26}$$

Defining the rotation tensor in the local frame as $\Omega^{ab} = \frac{d\tau}{dt}\omega^{ab}$ we have equation

$$\frac{dS^a}{dt} = \Omega^{ab} S_b. \tag{8.27}$$

Defining the rotation vector as $\Omega^a = \frac{-1}{2}\varepsilon^{abc}\Omega_{bc}$ we get the e.o.m for the spin vector $\mathbf{S}$ given by

$$\frac{d\mathbf{S}}{dt} = \mathbf{\Omega} \times \mathbf{S}. \tag{8.28}$$

8.2   Binary System of Spinning Black Holes

In a binary system consisting of black holes of masses m_1 and m_2 with spins $\mathbf{S}_1$ and $\mathbf{S}_2$, respectively, orbiting each other with an orbital angular momentum L the orbital plane of the binary will precess around the total angular momentum $\mathbf{J} = \mathbf{S}_1 + \mathbf{S}_2 + \mathbf{L}$ as long as the total spin is not aligned with the angular momentum $(\mathbf{S}_1 + \mathbf{S}_2) \times \mathbf{L} \neq 0$ [8–17] (Fig. 8.1).

The angular momenta change due to post-Newtonian order spin-orbital and spin-spin couplings. The change in angular momenta due to these effects are

$$\frac{d\mathbf{L}}{dt} = \frac{1}{r^3}\left[\frac{4m_1 + 3m_2}{2m_1}\mathbf{S}_1 + \frac{4m_2 + 3m_1}{2m_2}\mathbf{S}_1\right] \times \mathbf{L}$$

$$- \frac{3}{2r^3}\left[(\mathbf{S}_2 \cdot \hat{\mathbf{L}})\mathbf{S}_1 + (\mathbf{S}_1 \cdot \hat{\mathbf{L}})\mathbf{S}_2\right] \times \hat{\mathbf{L}} - \frac{32}{5}\frac{\mu^2}{r}\left(\frac{M}{r}\right)^{5/2}\hat{\mathbf{L}}$$

$$\frac{d\mathbf{S}_1}{dt} = \frac{1}{r^3}\left[\frac{4m_1 + 3m_2}{2m_1}\mathbf{L}\right] \times \mathbf{S}_1 + \frac{1}{r^3}\left[\frac{1}{2}\mathbf{S}_2 - \frac{3}{2}(\mathbf{S}_1 \cdot \hat{\mathbf{L}})\hat{\mathbf{L}}\right] \times \mathbf{S}_1$$

$$\frac{d\mathbf{S}_2}{dt} = \frac{1}{r^3}\left[\frac{4m_2 + 3m_1}{2m_2}\mathbf{L}\right] \times \mathbf{S}_2 + \frac{1}{r^3}\left[\frac{1}{2}\mathbf{S}_1 - \frac{3}{2}(\mathbf{S}_2 \cdot \hat{\mathbf{L}})\hat{\mathbf{L}}\right] \times \mathbf{S}_2. \tag{8.29}$$

Fig. 8.1 The orbital angular
momentum vector **L**
precesses around the total
angular momentum **J**. The
direction of the observer is **n**

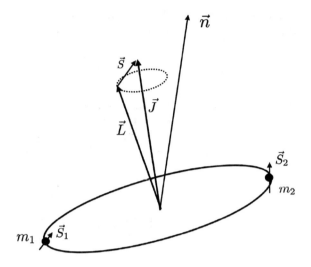

Fig. 8.1 The orbital angular momentum vector **L** precesses around the total angular momentum **J**. The direction of the observer is **n**

The last term in the equation for $d\mathbf{L}/dt$ is the change in **L** due to loss of angular momentum by gravitational wave radiation. The rate of energy loss in binaries due to gravitational radiation is

$$\frac{dE}{dt} = -\frac{32}{5}\frac{\mu^2 M^3}{r^5}. \tag{8.30}$$

Due to this energy loss there is a change in the separation distance between the black holes $dr/dt = (dE/dt)(dr/dE)$ which using $E = -(1/2)\mu M/r$ is

$$\frac{dr}{dt} = -\frac{64}{5}\frac{\mu M^2}{r^3}. \tag{8.31}$$

Solving this equation gives us the time dependence of the separation distance $r(t)$

$$r(t) = \left(\frac{256}{5}\mu M^2\right)^{1/4}(t_c - t)^{1/4}, \tag{8.32}$$

where t_c is the time of coalescence of the black holes.

The orbital angular momentum in the Newtonian orbit is

$$\mathbf{L} = \mu M^{1/2} r^{1/2}\hat{\mathbf{L}} \tag{8.33}$$

from which we have $d|\mathbf{L}|/dt = (d|\mathbf{L}|/dr)(dr/dt)$ and the rate of change of orbital angular momentum due to GW radiation is given by

$$\left.\frac{d\mathbf{L}}{dt}\right|_{GW} = -\frac{32}{5}\frac{\mu^2}{r}\left(\frac{M}{r}\right)^{5/2}\hat{\mathbf{L}}. \tag{8.34}$$

8.3   Signature of Black Hole Spins in Gravitational Wave Events

In the Newtonian orbit approximation the gravitational wave amplitudes of the two polarisations are

$$h_+ = -\frac{2\mu M}{rd}\left[1 + (\hat{L} \cdot \hat{n})^2\right]\cos(2\Phi(t))$$

$$h_\times = \frac{2\mu M}{rd}\left[2(\hat{L} \cdot \hat{n})\right]\sin(2\Phi(t)), \qquad (8.35)$$

where $r(t)$ is the separation between the two masses in retarded time.

The strain at the detector will depend on the direction $\hat{n}(\theta, \phi)$ and the plane of polarisation ψ,

$$h(t) = F_+(\theta, \phi, \psi)h_+(t) + F_\times(\theta, \phi, \psi)h_\times(t) \qquad (8.36)$$

where

$$F_+(\theta, \phi, \psi) = \frac{1}{2}\left(1 + \cos^2\theta\right)\cos 2\phi \cos 2\psi - \cos\theta \cos 2\phi \sin 2\psi ,$$

$$F_\times(\theta, \phi, \psi) = \frac{1}{2}\left(1 + \cos^2\theta\right)\cos 2\phi \sin 2\psi + \cos\theta \sin 2\phi \cos 2\psi . \qquad (8.37)$$

References

1. W. Tulczyjew, Motion of multipole particles in general relativity theory. Acta Phys. Pol. **18**, 37 (1959)
2. J. Steinhoff, Canonical formulation of spin in general relativity. Annalen Phys. **523**, 296–353 (2011). [arXiv:1106.4203 [gr-qc]]
3. D. Knickmann, General relativistic dynamics of spinning particles. PhD, Thesis (Universit¨at Bremen, Bremen, 2015). https://d-nb.info/1098374932/34
4. J. Vines, D. Kunst, J. Steinhoff, T. Hinderer, Canonical Hamiltonian for an extended test body in curved spacetime: To quadratic order in spin. Phys. Rev. D **93**(10), 103008 (2016). [erratum: Phys. Rev. D **104**, no.2, 029902 (2021)]. [arXiv:1601.07529 [gr-qc]]
5. M. Mathisson, Neue Mechanik materieller Systeme. Acta Phys. Polon. **6**, 163–2900 (1937)
6. A. Papapetrou, Spinning test particles in general relativity. 1. Proc. Roy. Soc. Lond. **A 209**, 248–258 (1951)
7. W.G. Dixon, Dynamics of extended bodies in general relativity. I. Momentum and angular momentum. Proc. Roy. Soc. Lond. **A 314**, 499–527 (1970)
8. T.A. Apostolatos, C. Cutler, G.J. Sussman, K.S. Thorne, Spin-induced orbital precession and its modulation of the gravitational waveforms from merging binaries. Phys. Rev. D **49**(12), 6274–6297 (1994)
9. L.E. Kidder, C.M. Will, A.G. Wiseman, Spin effects in the inspiral of coalescing compact binaries. Phys. Rev. D **47**(10), R4183–R4187 (1993)
10. S. Miller, T.A. Callister, W. Farr, The low effective spin of binary black holes and implications for individual gravitational-wave events. Astrophys. J. **895**(2), 128 (2020). [arXiv:2001.06051[astro-ph.HE]]

11. C.S. Reynolds, Observational Constraints on Black Hole Spin. Ann. Rev. Astron. Astrophys. **59**, 117–154 (2021). https://doi.org/10.1146/annurev-astro-112420-035022. [arXiv:2011.08948 [astro-ph.HE]]
12. R.A. Porto, Post-Newtonian corrections to the motion of spinning bodies in NRGR. Phys. Rev. **D 73**, 104031 (2006). https://doi.org/10.1103/PhysRevD.73.104031. [arXiv:gr-qc/0511061 [gr-qc]]
13. R.A. Porto, Next to leading order spin-orbit effects in the motion of inspiralling compact binaries. Class. Quant. Grav. **27**, 205001 (2010). [arXiv:1005.5730 [gr-qc]]
14. M. Levi, Next to Leading Order gravitational Spin-Orbit coupling in an Effective Field Theory approach. Phys. Rev. **D 82**, 104004 (2010). [arXiv:1006.4139 [gr-qc]]
15. M. Levi, J. Steinhoff, Spinning gravitating objects in the effective field theory in the post-Newtonian scheme. J. High Energy Phys. **09**, 219 (2015). [arXiv:1501.04956 [gr-qc]]
16. Z. Bern, A. Luna, R. Roiban, C.H. Shen, M. Zeng, Spinning black hole binary dynamics, scattering amplitudes, and effective field theory. Phys. Rev. **D 104**(6), 065014 (2021). [arXiv:2005.03071 [hep-th]]
17. C.F. Cho, N.D. Hari Dass, Equivalence principle, stress tensor, and long-range behavior of gravitational interactions. Phys. Rev. **D 14**, 2511 (1976)

Refractive Index and Damping of Gravitational Waves in a Medium

9

Abstract

Speed of gravitational waves remains c when it propagates through medium with stress tensor of a perfect fluid. Gravitational waves are absorbed in matter with non-zero shear viscosity. The speed of gravitational waves can differ from c when propagating through in-homogenous scalar field configurations and in modified theories of gravity. Observationally, the damping of gravitational waves and the change in speed of gravity can be probed using multi-messenger signals (γ rays and gravitational waves) from neutron-star mergers.

9.1 Introduction

The corrections to the propagation equation in theories beyond Einstein's gravity or in propagation through a medium can be described by additional terms in to the wave equation

$$\ddot{h}_{ij} + (3+v)H\dot{h}_{ij} + c_g^2 \frac{\mathbf{k}^2}{a^2} h_{ij} + M_{ijkl}h^{kl} = 0, \tag{9.1}$$

where c_g^2 is the speed of gravitational waves in the medium and v is the dissipative term which changes the amplitude of the gravitational waves. For propagation over cosmological distances it is convenient to switch to conformal time $\tau = \int dt/a$ then the wave equation (9.1) can be written as

$$h_{ij}'' + (2+v)\mathcal{H}h_{ij}' + c_g^2 \mathbf{k}^2 h_{ij} = 0, \tag{9.2}$$

© The Author(s), under exclusive license to Springer Nature Switzerland AG 2023
S. Mohanty, *Gravitational Waves from a Quantum Field Theory Perspective*,
Lecture Notes in Physics 1013, https://doi.org/10.1007/978-3-031-23770-6_9

where $v = \frac{2}{3}\alpha_M$ and primes denote derivative w.r.t conformal time τ and $\mathcal{H} = a'/a$. The gravitational wave signal is modified relative to its vacuum-GR form as follows [1,2],

$$h = h_{GR}\ \underbrace{e^{-\frac{1}{2}\int v\mathcal{H}d\tau}}_{\text{amplitude}}\ \underbrace{e^{i|\mathbf{k}|\int(c_g^2-1)d\tau}}_{\text{phase}}. \tag{9.3}$$

The damping of the amplitude changes the effective luminosity distance of gravitational waves. The standard luminosity distance is a flat universe is

$$d_L = (1+z)\int_0^z \frac{c_g}{H(z')}dz',$$

$$H(z) = H_0\left(\Omega_m(1+z)^3 + \Omega_r(1+z)^4 + \Omega_\Lambda\right)^{1/2}. \tag{9.4}$$

In case of gravitational wave damping and/or change in the speed of gravity the effective luminosity distance of gravitational waves will be

$$d_L^{gw} = (1+z)\frac{c_g(z)}{c_g(0)}\exp\left[\int_0^z \frac{v}{1+z'}dz'\right]\int_0^z \frac{c_g(z')}{H(z')}. \tag{9.5}$$

The luminosity distance is related to the amplitude of the $+$ and $\times$ polarization gravitational waves as (2.94),

$$h_+ \propto \frac{(1+\cos^2 i)}{2d_L^{gw}} \quad\text{and}\quad h_\times \propto \frac{\cos i}{d_L^{gw}}, \tag{9.6}$$

where i is the inclination angle of the binary orbit plane w.r.t. the direction of propagation. By measuring both polarizations in a GRW signal, which is possible with 3 detectors it may be possible to determine the inclination angle i and luminosity distance d_L^{gw} [3]. To measure the damping or deviation of c_g from unity there must be determination of the source red-shift z. This can be possible with simultaneous observations of gamma rays and GW waves from neutron star mergers [4].

9.2   Graviton Propagation in a Medium

We start with Einstein's equation

$$R^{\mu\nu} - \frac{1}{2}g^{\mu\nu}R = \frac{\kappa^2}{4}T^{\mu\nu} \tag{9.7}$$

and we expand the metric $g_{\mu\nu} = \bar{g}_{\mu\nu} + h_{\mu\nu}$ where $\bar{g}_{\mu\nu}$ is the background metric and $h_{\mu\nu}$ the gravitational wave in propagating in the $\bar{g}_{\mu\nu}$ background. The stress tensor of a general matter field is defined as

$$T^{\mu\nu} = \frac{-2}{\sqrt{-g}} \frac{\delta\left(\sqrt{-g}\mathcal{L}_m\right)}{\delta g_{\mu\nu}}. \tag{9.8}$$

In order to obtain the equations of motion of $h_{\mu\nu}$ we expand $G^{\mu\nu} \equiv R^{\mu\nu} - \frac{1}{2}g^{\mu\nu}R = \frac{\kappa^2}{4}T^{\mu\nu}$ to the linear order in $h_{\mu\nu}$ around the background metric $\bar{g}_{\mu\nu}$.

The zeroth order equation in $h_{\mu\nu}$ is the background equation

$$\bar{R}^{\mu\nu} - \frac{1}{2}\bar{g}^{\mu\nu}\bar{R} = \frac{\kappa^2}{4}\bar{T}^{\mu\nu}. \tag{9.9}$$

To obtain the first order corrections in $h_{\mu\nu}$ of (9.7) is

$$\delta^{(1)}R^{\mu\nu} - \frac{1}{2}\delta^{(1)}g^{\mu\nu}\bar{R} - \frac{1}{2}\bar{g}^{\mu\nu}\delta^{(1)}R = \frac{\kappa^2}{4}\delta^{(1)}T^{\mu\nu}. \tag{9.10}$$

The first order correction to the Christoffel connection is

$$\delta^{(1)}\Gamma^{\mu}_{\nu\alpha} = -\frac{1}{2}h^{\mu\beta}\left(\partial_{\nu}\bar{g}_{\alpha\beta} + \partial_{\alpha}\bar{g}_{\nu\beta} - \partial_{\beta}\bar{g}_{\nu\alpha}\right) + \frac{1}{2}\bar{g}^{\mu\beta}\left(\partial_{\nu}h_{\alpha\beta} + \partial_{\alpha}h_{\nu\beta} - \partial_{\beta}\bar{h}_{\nu\alpha}\right)$$

$$= \frac{1}{2}\bar{g}^{\mu\beta}\left(\nabla_{\nu}h_{\alpha\beta} + \nabla_{\alpha}h_{\nu\beta} - \nabla_{\beta}\bar{h}_{\nu\alpha}\right), \tag{9.11}$$

where the covariant derivatives ∇_{μ} are calculated w.r.t the background metric $\bar{g}_{\mu\nu}$.

The first order corrections to the Ricci tensor can be obtained from the Palatini identity

$$\delta^{(1)}R_{\mu\nu} = \nabla_{\alpha}\left(\delta^{(1)}\Gamma^{\alpha}_{\mu\nu}\right) - \nabla_{\mu}\left(\delta^{(1)}\Gamma^{\alpha}_{\alpha\nu}\right)$$

$$= \frac{1}{2}\left(\nabla_{\alpha}\nabla_{\mu}h_{\nu}{}^{\alpha} + \nabla_{\alpha}\nabla_{\nu}h_{\mu}{}^{\alpha} - \Box h_{\mu\nu} - \nabla_{\mu}\nabla_{\nu}h\right), \tag{9.12}$$

where $\Box = \bar{g}^{\mu\nu}\nabla_{\mu}\nabla_{\nu}$ and $h = \bar{g}^{\mu\nu}h_{\mu\nu}$.

The first order correction to the Ricci scalar is

$$\delta^{(1)}R = \delta^{(1)}\left(g^{\alpha\beta}R_{\alpha\beta}\right)$$

$$= -h^{\alpha\beta}\bar{R}_{\alpha\beta} + \bar{g}^{\alpha\beta}\delta^{(1)}R_{\alpha\beta}. \tag{9.13}$$

We can use (9.12) to evaluate the last term in (9.13),

$$\bar{g}^{\alpha\beta}\delta^{(1)}R_{\alpha\beta} = \nabla_\alpha\nabla_\beta h^{\alpha\beta} - \Box h, \tag{9.14}$$

where we have used the property $\nabla_\alpha\bar{g}^{\mu\nu} = 0$.

Using these relations we see that Eq. (9.10) can be written as

$$-\nabla_\alpha\nabla_\mu h_\nu{}^\alpha - \nabla_\alpha\nabla_\nu h_\mu{}^\alpha + \Box h_{\mu\nu} + \nabla_\mu\nabla_\nu h + \bar{g}_{\mu\nu}\left(\nabla_\alpha\nabla_\beta h^{\alpha\beta} - \Box h\right)$$
$$+ \bar{R}h_{\mu\nu} - \bar{g}_{\mu\nu}h^{\alpha\beta}\bar{R}_{\alpha\beta} = -\frac{\kappa^2}{2}\delta^{(1)}T^{\mu\nu}. \tag{9.15}$$

We can simplify the equation by making the TT gauge choice $\nabla_\mu h^{\mu\nu} = 0$ and $h = 0$. The first term in (9.15) can be written using the commutator identity

$$\nabla_\alpha\nabla_\mu h_\nu{}^\alpha = \nabla_\mu\nabla_\alpha h_\nu{}^\alpha + h_\nu{}^\alpha\bar{R}_{\alpha\mu} - \bar{R}_{\alpha\mu\nu\beta}h^{\alpha\beta}$$
$$= h_\nu{}^\alpha\bar{R}_{\alpha\mu} - \bar{R}_{\alpha\mu\nu\beta}h^{\alpha\beta}, \tag{9.16}$$

where the first term in the RHS is zero due to the gauge condition. Similarly the second term in (9.15) can be written as

$$\nabla_\alpha\nabla_\nu h_\mu{}^\alpha = h_\mu{}^\alpha\bar{R}_{\alpha\nu} - \bar{R}_{\alpha\nu\mu\beta}h^{\alpha\beta}. \tag{9.17}$$

With these simplifications the wave equation for $h_{\mu\nu}$ in the TT gauge reduces to

$$\Box h_{\mu\nu} + 2\bar{R}_{\alpha\mu\nu\beta}h^{\alpha\beta} + \bar{R}h_{\mu\nu} - \bar{g}_{\mu\nu}h^{\alpha\beta}\bar{R}_{\alpha\beta} - h_\mu{}^\beta\bar{R}_{\beta\nu} - h_\nu{}^\beta\bar{R}_{\beta\mu} = -\frac{\kappa^2}{2}\delta^{(1)}T^{\mu\nu}. \tag{9.18}$$

Taking the trace we get the relation

$$\bar{R}_{\mu\nu}h^{\mu\nu} = \frac{\kappa^2}{8}\bar{g}^{\mu\nu}\delta^{(1)}T_{\mu\nu}. \tag{9.19}$$

This can be used to replace the Ricci terms in (9.18) to obtain the wave equation for the tensor perturbations $h_{\mu\nu}$ in the TT gauge [5],

$$\Box h_{\mu\nu} + 2\bar{R}_{\alpha\mu\nu\beta}h^{\alpha\beta} = \frac{\kappa^2}{4}\left(h_\mu{}^\alpha\bar{T}_{\alpha\nu} + h_\nu{}^\alpha\bar{T}_{\alpha\mu} + \frac{1}{2}\bar{g}_{\mu\nu}\bar{g}^{\alpha\beta}\delta^{(1)}T_{\alpha\beta} - 2\delta^{(1)}T_{\mu\nu}\right). \tag{9.20}$$

We will use the master equation (9.20) to calculate the effect of different types of matter on the propagation of gravitational waves.

9.3   Propagation of Gravitational Waves Through Perfect Fluid Medium

We consider gravitational waves in a general background $g_{\mu\nu} = \bar{g}_{\mu\nu} + h_{\mu\nu}$. For cosmological applications $\bar{g}_{\mu\nu}$ is taken to be the flat FLRW metric $\bar{g}_{00} = 1$, $\bar{g}_{ij} = -a^2\delta_{ij}$. The fluid through which the GR waves propagate is taken to have a stress tensor of the perfect fluid form

$$T_{\alpha\beta} = (\rho + p)u_\alpha u_\beta - pg_{\alpha\beta}. \tag{9.21}$$

The first order perturbation of $T_{\alpha\beta}$ is given by

$$\delta^{(1)}T_{\alpha\beta} = -ph_{\alpha\beta} + \frac{1}{2}(\rho + p)\bar{g}_{\alpha\beta}h_{\mu\nu}u^\mu u^\nu + (\rho + p)\left(u_\alpha u^\nu h_{\nu\beta} + u_\beta u^\nu h_{\nu\alpha}\right). \tag{9.22}$$

Taking the trace we have the relation

$$\bar{g}^{\alpha\beta}\delta^{(1)}T_{\alpha\beta} = 4(\rho + p)h_{\mu\nu}u^\mu u^\nu. \tag{9.23}$$

Using (9.22) in the Ricci tensor relation (9.19) we obtain

$$\bar{R}_{\alpha\beta}h^{\alpha\beta} = \frac{\kappa^2}{4}(\rho + p)h_{\mu\nu}u^\mu u^\nu \tag{9.24}$$

From these two equations we have the consistency relations

$$(\rho + p)h_{\mu\nu}u^\mu u^\nu = 0. \tag{9.25}$$

Using these relations in (9.20) we obtain the wave equation of graviton propagating through a perfect fluid medium is

$$\Box h_{\alpha\beta} + 2\bar{R}_{\mu\alpha\beta\nu}h^{\mu\nu} + \frac{\kappa^2}{4}(\rho + p)\left(u_\alpha u^\nu h_{\nu\beta} + u_\beta u^\nu h_{\nu\alpha}\right) = 0, \tag{9.26}$$

where the D'Alembertian $\Box$ and the Riemann tensor $\bar{R}_{\mu\alpha\beta\nu}$ are defined w.r.t the background metric $\bar{g}_{\mu\nu}$. In the case of a comoving fluid $u^0 = 1$, $u^i = 0$ and for the TT gauge $h_{\mu0} = 0$ the terms $u^\nu h_{\nu\beta} = 0$ and the wave equation reduces to

$$\Box h_{\alpha\beta} + 2\bar{R}_{\mu\alpha\beta\nu}h^{\mu\nu} = 0. \tag{9.27}$$

For the FRLW metric $\bar{g}_{00} = 1$, $\bar{g}_{ij} = -a^2\delta_{ij}$ we have

$$\bar{\Gamma}^i_{0j} = H\delta^i_j, \quad \bar{\Gamma}^0_{ij} = a^2H\delta_{ij}, \quad \bar{R}_{kijl}h^{kl} = H^2h_{ij}, \tag{9.28}$$

where $H = \dot{a}/a$, and the gravitational wave equation in the FLRW universe is

$$\ddot{h}^i{}_j + 3H\dot{h}^i{}_j - \frac{\partial^2}{a^2}h^i{}_j = 0, \tag{9.29}$$

where $h^i{}_j = \bar{g}^{ik}h_{kj}$ and where $\partial^2 = \delta^{ij}\partial_i\partial_j$. The speed of gravitational waves c_g is the coefficient of the ∂^2/a^2 term and in this case $c_g = 1$. We see that despite the non-zero energy density and pressure of the fluid medium gravitational waves propagate through the medium with the speed of light.

9.4   Propagation Through Dissipative Medium

We now consider a more general case of a fluid with shear and bulk viscosities. We will see that a non-zero shear viscosity dampens the amplitude of gravitational waves. The stress tensor is given by

$$\tilde{T}_{\alpha\beta} = (\rho + \tilde{p})\,u_\alpha u_\beta - \tilde{p}g_{\alpha\beta} - \eta\left[u_\alpha u^\rho\nabla_\rho u_\beta + u_\beta u^\rho\nabla_\rho u_\alpha - \nabla_\alpha u_\beta - \nabla_\beta u_\alpha\right],$$

$$\tilde{p} \equiv p - \left(\xi - \frac{2}{3}\eta\right)\nabla_\lambda u^\lambda, \tag{9.30}$$

where ρ is the energy density, p the pressure, $\tilde{p}$ the total effective pressure with contributions from the bulk viscosity ξ and shear viscosity η. The expression for stress tensor of dissipative fluids (9.30) can also be written in the more compact form (3.32).

The first order correction to $\tilde{T}_{\alpha\beta}$ is

$$\delta^{(1)}\tilde{T}_{\alpha\beta} = \delta^{(1)}T_{\alpha\beta}(\tilde{p}) - \eta[h_{\alpha\rho}u^\rho u^\lambda\nabla_\lambda u_\beta - h_{\beta\sigma}u_\alpha u^\lambda\nabla_\lambda u^\sigma$$

$$-h_{\alpha\sigma}\nabla_\beta u^\sigma + u_\alpha u^\lambda u^\rho(\lambda_\lambda h_{\beta\sigma} - \frac{1}{2}\lambda_\beta h_{\lambda\rho}) + (\alpha \leftrightarrow \beta)]$$

$$+\eta u^\lambda\nabla_\lambda h_{\alpha\beta}, \tag{9.31}$$

where $\delta^{(1)}T_{\alpha\beta}(\tilde{p})$ is the perfect fluid expression (9.22) with p replaced with $\tilde{p}$. The Ricci condition (9.19) gives us the consistency condition

$$2(\rho + \tilde{p})h_{\mu\nu}u^\mu u^\nu = \eta\left[2h^{\mu\nu}\nabla_\mu u_\nu + u^\beta u^\lambda u^\rho\nabla_\lambda h_{\beta\rho}\right]. \tag{9.32}$$

We consider a comoving medium with fluid velocity $u_0 = 1$ and $u_i = 0$. In the TT gauge $h_{0\mu} = 0$. In this case the only non-zero contribution comes from the last term of (9.31) and we have

$$\delta^{(1)}\tilde{T}_{ij} = \eta u^\lambda\nabla_\lambda h_{ij} = \eta\left(\dot{h}_{ij} - 2Hh_{ij}\right). \tag{9.33}$$

Substituting (9.33) in (9.20) we see that the wave equation for the TT tensor modes h_{ij} gets a contribution from the shear viscosity and is given by

$$\Box h_{ij} + 2R_{kijl}h^{kl} + \eta\frac{\kappa^2}{2}\left(\dot{h}_{ij} - 2Hh_{ij}\right) = 0. \tag{9.34}$$

In terms of $h^i{}_j = \bar{g}^{ik}h_{kj}$, with the TT conditions $\partial^j h^i{}_j = 0$ and $h^i{}_i = 0$, the wave equation is

$$\ddot{h}^i{}_j + \left(3H + \eta\frac{\kappa^2}{2}\right)\dot{h}^i{}_j - \frac{\partial^2}{a^2}h^i{}_j = 0. \tag{9.35}$$

Switching to conformal time $d\tau = dt/a$ the wave equation in the viscous medium is given by Hawking [6], Prasanna [7], and Madore [8],

$$h^i{}''_j + 2\left(\mathcal{H} + \eta\frac{\kappa^2}{4}a\right)h^i{}'_j - \partial^2 h^i{}_j = 0. \tag{9.36}$$

The gravitational waves are damped compared to propagation in a non-dissipative medium by a damping factor given by

$$h_{GW} = h^{(0)}\,e^{-\eta\frac{\kappa^2}{4}\int\frac{a}{\mathcal{H}}d\tau}. \tag{9.37}$$

This damping if observed in signals of inspiralling binary NS or BH at LIGO and can give constraints on shear viscosity of dark matter [9–13].

9.5 Propagation Through Scalar Field Dark Matter

Ultralight scalar fields are candidates for being dark matter [14, 15]. Axion-like particles with coupling $f \simeq 10^{16} - 10^{18}$ GeV and mass in the range $m_a \sim 10^{-22} - 10^{-21}$ eV can solve the core-cusp problem of galactic dark matter and at the same time provide the closure density of the universe to be the dark matter at cosmological scales. In this section we study the propagation of gravitational waves through the scalar dark matter at cosmological scales and through the galaxy.

The Lagrangian of a minimally coupled scalar field is

$$\mathcal{L} = \sqrt{-g}\left[\frac{1}{2}g^{\rho\sigma}\partial_\rho\phi\partial_\sigma\phi - \frac{1}{2}m^2\phi^2\right]. \tag{9.38}$$

The corresponding stress tensor of the scalar field is

$$T_{\alpha\beta} = \partial_\alpha\phi\partial_\beta\phi - g_{\alpha\beta}\left(\frac{1}{2}g^{\rho\sigma}\partial_\rho\phi\partial_\sigma\phi - \frac{1}{2}m^2\phi^2\right). \tag{9.39}$$

We write the metric $g_{\mu\nu} = \bar{g}_{\mu\nu} + h_{\mu\nu}$ as a background metric $\bar{g}_{\mu\nu}$ and the first order tensor perturbation $h_{\mu\nu}$. We can expand the stress tensor into a background part and a first order correction in $h_{\mu\nu}$

$$T_{\mu\nu} = \bar{T}_{\mu\nu} + \delta^{(1)} T_{\mu\nu}, \tag{9.40}$$

where $\bar{T}_{\mu\nu}$ is of the same form as (9.39) with $g_{\mu\nu}$ replaced by $\bar{g}_{\mu\nu}$ and the first order correction in $h_{\mu\nu}$ of (9.39) is

$$\delta^{(1)} T_{\mu\nu} = -h_{\alpha\beta} \left(\frac{1}{2} \bar{g}^{\rho\sigma} \partial_\rho \phi \partial_\sigma \phi - \frac{1}{2} m^2 \phi^2 \right) + \frac{1}{2} \bar{g}_{\alpha\beta} h^{\mu\nu} \partial_\mu \phi \partial_\nu \phi. \tag{9.41}$$

The wave equation for gravitational waves $h_{\mu\nu}$ propagating through the scalar field dark matter is

$$\Box h_{\alpha\beta} + 2R_{\mu\alpha\beta\nu} h^{\mu\nu} = \frac{\kappa^2}{4} \left(h_\alpha{}^\nu \bar{T}_{\nu\beta} + h_\beta{}^\nu \bar{T}_{\nu\alpha} + \frac{1}{2} \bar{g}_{\alpha\beta} \bar{g}^{\mu\nu} \delta^{(1)} T_{\mu\nu} - 2\delta^{(1)} T_{\alpha\beta} \right)$$

$$= \frac{\kappa^2}{4} \left(h_\alpha{}^\nu \partial_\beta \phi + h_\beta{}^\nu \partial_\alpha \phi \right) \partial_\nu \phi. \tag{9.42}$$

For the TT modes $h^i{}_j = \bar{g}^{ik} h_{kj}$, with $\partial^j h^i{}_j = 0$ and $h^i{}_i = 0$, the wave equation is

$$\ddot{h}^i{}_j + 3H\dot{h}^i{}_j - \frac{\partial^2}{a^2} h^i{}_j = \frac{\kappa^2}{4} \left(h_i{}^k \partial_j \phi + h_j{}^k \partial_i \phi \right) \partial_k \phi \tag{9.43}$$

To deviate the gravitational waves from its trajectory the scalar fields should have spatial inhomogeneity $\partial_i \phi \neq 0$. The RHS can give rise to birefringence, i.e. polarization dependent speed of the gravitational waves in the in-homogenous scalar field medium.

9.6   Modified Gravity Theories

In multi-messenger observations as observation of GW170817 by Ligo and Virgo [16] and GRB 170817A by Fermi [17], the gamma rays arrived after 2 s delay from the arrival of the gravitational waves. The delay of photons is due to the delay of the photons in the dense envelop of matter blown off from the merged neutron stars. This timing of the gravitational waves provides an opportunity for testing theories of GR like Born-Infeld gravity where the velocity of gravitational waves differs from velocity of light [18].

The action for Born-Infeld gravity is [19]

$$S_{BI} = \frac{c^3}{8\pi G\kappa'} \int d^4x \left[\sqrt{-|g_{\mu\nu} + \kappa' R_{\mu\nu}(\Gamma)|} - \lambda\sqrt{-g} \right] + S_M(g_{\mu\nu}), \tag{9.44}$$

where $\lambda = \kappa'\Lambda + 1$, with Λ being the cosmological constant. Here κ' is the parameter of the theory with dimensions of [length]^2 and, for sufficiently small κ', the action reduces to the Einstein–Hilbert action.

The gravitational wave propagation equation for the Born-Infeld gravity is [20, 21],

$$\ddot{h}_{ij} + \left(3H + \frac{\dot{\alpha}}{\alpha}\right)\dot{h}_{ij} - c^2\beta^2\frac{\nabla^2}{a^2}h_{ij} = 0, \tag{9.45}$$

where

$$\alpha = \lambda + \frac{8\pi G\kappa'\rho}{c^2}, \qquad \beta^2 = \left|\frac{\lambda - \frac{8\pi G\kappa' p}{c^4}}{\lambda + \frac{8\pi G\kappa'\rho}{c^2}}\right|,$$

and ρ and p are the energy density and pressure of the matter in the Universe, $H = \dot{a}/a$ is the Hubble parameter, and a is the scale factor in the FRW spacetime. From the coefficient of $\dot{h}_{ij}$ in Eq. (9.45), we see that in addition to the cosmological damping proportional to $3H$, there is also an extra damping factor due to in Born-Infeld gravity proportional to $\frac{\dot{\alpha}}{\alpha}$ [20].

The speed of gravitational waves can be given as

$$v_{gw} = c\beta = c\left|\frac{\lambda - \frac{8\pi G\kappa' p}{c^4}}{\lambda + \frac{8\pi G\kappa'\rho}{c^2}}\right|^{1/2}. \tag{9.46}$$

The difference between the speed of gravity c_{gw} and the speed of light in vacuum c becomes

$$\frac{v_{gw} - c}{c} = \left|\frac{\lambda - \frac{8\pi G\kappa' p}{c^4}}{\lambda + \frac{8\pi G\kappa'\rho}{c^2}}\right|^{1/2} - 1. \tag{9.47}$$

From the bound on the speed of gravity derived from the time delay of GW signals at different Earth-based detectors, we obtained $|\kappa'| \lesssim 10^{21}\, m^2$ [18]. In the same way, from the time delay in the signals from GW170817 and GRB 170817A events, in the background of a FRW Universe, we obtained the bound $|\kappa'| \lesssim 10^{37}\, m^2$ [18].

9.7 Čerenkov Radiation of Gravitational Waves

In some theories of gravity like the Born-Infeld gravity [20, 21] and Horndeski gravity [22] the dispersion relations for gravitons gets modified and gravitons velocity is slower than the velocity of light (9.46). The refractive index is defined as $n = \frac{c}{c_{gw}} = c\frac{|\mathbf{k}|}{\omega}$. When the refractive exceeds unity then it becomes kinematically possible for particles travelling with velocities $v > c_{gw}$ to radiate gravitational waves by a process analogous to the Čerenkov radiation of photons [23–28].

$$(9.48)$$

Consider the graviton emission from a particle by the Čerenkov process $f(p) \rightarrow f(p') + h(k)$. The conservation of energy momentum $p'^{\mu} = p^{\mu} - k^{\mu}$ implies that

$$g_{\hat{\mu}\hat{v}} p'^{\hat{\mu}} p'^{\hat{v}} = g_{\hat{\mu}\hat{v}} p^{\hat{\mu}} p^{\hat{v}} + g_{\hat{\mu}\hat{v}} k^{\hat{\mu}} k^{\hat{v}} - 2 g_{\hat{\mu}\hat{v}} p^{\hat{\mu}} k^{\hat{v}}$$

$$\Rightarrow g_{\hat{\mu}\hat{v}} k^{\hat{\mu}} k^{\hat{v}} - 2 \left(p_{\hat{0}} k_{\hat{0}} - \mathbf{p} \cdot \mathbf{k} \right) = 0, \qquad (9.49)$$

where we used on-shell relation for the particle $g_{\hat{\mu}\hat{v}} p'^{\hat{\mu}} p'^{\hat{v}} = g_{\hat{\mu}\hat{v}} p^{\hat{\mu}} p^{\hat{v}} = m^2$, with m the mass of the particle. We shall take reference frame to be the local inertial frame of the particles and $g_{\hat{\mu}\hat{v}} = \eta_{\hat{\mu}\hat{v}} = $ diagonal $(1, -1, -1, -1)$. From this relation we find that angle between $\mathbf{p}$ and $\mathbf{k}$ vectors which is the angle of the Čerenkov cone is given by

$$\cos\theta \equiv \frac{\mathbf{p} \cdot \mathbf{k}}{|\mathbf{k}||\mathbf{p}|} = \frac{p_{\hat{0}} k_{\hat{0}}}{|\mathbf{p}||\mathbf{k}|} \left(1 - \frac{\eta_{\hat{\mu}\hat{v}} k^{\hat{\mu}} k^{\hat{\mu}}}{2 p_{\hat{0}} k_{\hat{0}}} \right). \qquad (9.50)$$

Writing the dispersion relation of gravitons of four-momentum $k_{\hat{\mu}} = (k_{\hat{0}}, \mathbf{k})$ in terms of the refractive index $n = |\mathbf{k}|/k_{\hat{0}}$ and the units $c = 1$,

$$\eta_{\hat{\mu}\hat{v}} k^{\hat{\mu}} k^{\hat{v}} = k_{\hat{0}}^2 - |\mathbf{k}|^2 = -k_{\hat{0}}^2 (n^2 - 1). \qquad (9.51)$$

The Čerenkov angle (9.50) in terms of the refractive index is

$$\cos\theta = \frac{p_{\hat{0}}}{n|\mathbf{p}|} \left(1 + \frac{(n^2 - 1)k_{\hat{0}}}{2 p_{\hat{0}}} \right). \qquad (9.52)$$

The necessary condition for the Čerenkov process can be written as $|\cos\theta| \leq 1$ which gives the relation

$$-1 \leq \frac{p_{\hat{0}}}{n|\mathbf{p}|} \left[1 + (n^2 - 1) \frac{k_{\hat{0}}}{2 p_{\hat{0}}} \right] \leq 1. \qquad (9.53)$$

The rate of energy radiated in the Čerenkov radiation process $f(p) \rightarrow f(p') + h(k)$ in the local inertial of the particle is given by

$$\frac{dE}{dt} = \frac{1}{2p_{\hat{0}}} \int \frac{d^3\mathbf{k}\, dk_{\hat{0}}}{(2\pi)^4} 2\pi \delta \left(n^2 k_{\hat{0}}^2 - |\mathbf{k}|^2 \right)$$

$$\times 2\pi \delta \left((p_{\hat{0}} - k_{\hat{0}})^2 - |\mathbf{p} - \mathbf{k}|^2 - m^2 \right) k_{\hat{0}} |\mathcal{M}|^2. \tag{9.54}$$

In (9.54) we can write the delta function for the mass-shell condition of the outgoing particle as

$$\delta \left((p_{\hat{0}} - k_{\hat{0}})^2 - |\mathbf{p} - \mathbf{k}|^2 - m^2 \right) = \frac{1}{2|\mathbf{p}||\mathbf{k}|} \delta \left(\frac{2p_{\hat{0}}k_{\hat{0}} - k_{\hat{\mu}}k^{\hat{\mu}}}{2|\mathbf{p}||\mathbf{k}|} - \cos\theta \right). \tag{9.55}$$

This is the same relation for the Čerenkov angle as (9.52). We can perform the integration over $d|\mathbf{k}|$ making use of the first delta function in (9.54) which simplifies to

$$\frac{dE}{dt} = \frac{1}{8\pi p_{\hat{0}}} \int d(\cos\theta)\, dk_{\hat{0}} \frac{1}{2|\mathbf{p}|} \delta \left(\frac{2p_{\hat{0}}k_{\hat{0}} - k_{\hat{\mu}}k^{\hat{\mu}}}{2n|\mathbf{p}|k_{\hat{0}}} - \cos\theta \right) k_{\hat{0}} |\mathcal{M}|^2. \tag{9.56}$$

The amplitude for a graviton emission from a scalar is

$$\mathcal{M} = \frac{i\kappa}{2} \epsilon_\lambda^{*\hat{\mu}\hat{\nu}}(k) T_{\hat{\mu}\hat{\nu}}(p, p'), \tag{9.57}$$

where the stress tensor for the scalar field with incoming momentum $p_{\hat{\mu}}$ and outgoing momentum $p'_{\hat{\mu}} = p_{\hat{\mu}} - k_{\hat{\mu}}$ is given by

$$T_{\hat{\mu}\hat{\nu}} = \left[p_{\hat{\mu}} p'_{\hat{\nu}} + p_{\hat{\nu}} p'_{\hat{\mu}} - \eta_{\hat{\mu}\hat{\nu}}(p \cdot p' - m^2) \right]. \tag{9.58}$$

The amplitude squares is

$$|\mathcal{M}|^2 = \frac{\kappa^2}{4} T_{\hat{\mu}\hat{\nu}}(p, p') T_{\hat{\alpha}\hat{\beta}}^*(p, p') \sum_\lambda \epsilon_\lambda^{*\hat{\mu}\hat{\nu}}(k) \epsilon_\lambda^{\hat{\alpha}\hat{\beta}}(k)$$

$$= \frac{\kappa^2}{4} \left(T_{\hat{\mu}\hat{\nu}}(p, p') T^{*\hat{\mu}\hat{\nu}}(p, p') - \frac{1}{2} |T_{\hat{\mu}}^{\ \hat{\mu}}(p, p')|^2 \right). \tag{9.59}$$

Using (9.58) and the conservation of stress tensor relations $k^{\hat{\mu}} T_{\hat{\mu}\hat{\nu}} = k^{\hat{\nu}} T_{\hat{\mu}\hat{\nu}} = 0$ we obtain

$$
\begin{aligned}
|M|^2 &= \frac{\kappa^2}{4}\left(4m^4 + 2(p_{\hat{\mu}}k^{\hat{\mu}})\right)^2 \\
&= \frac{\kappa^2}{4}\left(4m^4 + p_{\hat{0}}k_{\hat{0}} - |\mathbf{p}||\mathbf{k}|\cos\theta\right)^2 \\
&= \frac{\kappa^2}{8}(n^2 - 1)^2 k_{\hat{0}}^4.
\end{aligned}
\tag{9.60}
$$

For the case of high energy particles we can drop m^4 term compared to $(p \cdot k)^2$ and we have used the delta function in (9.56) to write $\cos\theta$ in terms of $p_{\hat{\mu}}$ and $k_{\hat{\mu}}$. Using (9.60) in the expression (9.56) for the rate of Čerenkov radiation we have,

$$
\frac{dE}{dt} = \frac{G}{4}\frac{1}{p_{\hat{0}}|\mathbf{p}|}\int_0^{k_{max}} dk_{\hat{0}}(n^2 - 1)^2 k_{\hat{0}}^5.
\tag{9.61}
$$

The upper limit of $k_{\hat{0}}$ comes from the (9.53) and is given by

$$
k_{\hat{0}} \leq 2p_{\hat{0}}\frac{(\beta' n - 1)}{n^2 - 1},
\tag{9.62}
$$

where $\beta' = |\mathbf{p}|/p_{\hat{0}}$. From (9.62) we see that the necessary conditions for Čerenokov radiation to take place are $n > 1$ and $\beta' n > 1$. In general the refractive index will be a function of $k_{\hat{0}}$ and the explicit form of $n(k_{\hat{0}})$ is required for performing the integration in (9.61).

For n independent of frequency, the rate of energy radiated by Čerenkov radiation is from (9.61) given by

$$
\begin{aligned}
\frac{dE}{dt} &= \frac{G}{24}\frac{1}{p_{\hat{0}}|\mathbf{p}|}(n^2 - 1)^2 k_{max}^6 \\
&= \frac{G}{24}\frac{1}{p_{\hat{0}}|\mathbf{p}|}(n^2 - 1)^2\left(2p_{\hat{0}}\frac{(\beta' n - 1)}{n^2 - 1}\right)^6.
\end{aligned}
\tag{9.63}
$$

In the case of Born-Infeld gravity, the refractive index is independent of frequency [20, 21] and is given by (9.46),

$$
n = \frac{c}{v_{gw}} = \left|\frac{\lambda + \frac{8\pi G \kappa' p}{c^4}}{\lambda - \frac{8\pi G \kappa' \rho}{c^2}}\right|^{1/2}.
\tag{9.64}
$$

9.8 Gravitational Radiation in a Background of Gravitons: Stimulated Emission

One of the important questions in physics is to determine whether Einstein's General Relativity is a theory of geometry or whether it can be studied as a quantum field theory like electromagnetism, weak and strong interactions. In particular, what is the experimental signature to prove that gravitons are quantised. The analogy of the experimental probes of gravitational waves with the history of our understanding of electromagnetic waves can be instructive. Maxwell in 1865 predicted that oscillations of electric and magnetic fields would propagate as electromagnetic at the speed of light. Heinrich Hertz was the first, in 1887, to generate and detect the waves from oscillating electromagnetic fields in an inductance-capacitance circuit. In 1900 Planck gave the thermal radiation distribution and introduced the h as a new fundamental constant. To derive the thermal distribution for photons Planck had to assume that the radiation energy is absorbed and radiated in units of h. In 1917 Einstein demonstrated that atoms and radiation can be in equilibrium resulting in the Planck law for radiation, if atoms had a spontaneous emission in addition to the stimulated emission and absorption. The relation between Einstein's A and B coefficients which he postulated was derived from quantum mechanics in 1927 by Dirac. The absorption rate of photons of frequency v is proportional to $n(v)$ the photon occupancy of the heat bath, while the emission of photons is proportional to $n(v) + 1$. This follows from the Bose statistics of the photon field and indeed the Planck distribution law itself was derived by Bose on the basis is commutation relations of the electromagnetic fields. In this chapter we study the absorption and emission of photons from sources like binaries if they are immersed in a graviton heat bath.

The rate of atomic transitions with photon emission is enhanced if the atom is in the background of other photons. This property called stimulated emission follows from the commutation properties of the creation and annihilation operators for bosons.

Consider the background into which the binary stars radiate as a Fock state with non-zero occupation number $|n_k\rangle$ instead of the zero graviton vacuum state. The Fock states have the properties

$$a_k|n_k\rangle = \sqrt{n_k}|n_k - 1\rangle, \quad a_k^\dagger|n_k\rangle = \sqrt{n_k + 1}|n_k + 1\rangle, \quad a_k^\dagger a_k|n_k\rangle = n_k|n_k\rangle.$$

$$(9.65)$$

Using these we see that

$$\langle n_k + 1|a_k^\dagger|n_k\rangle = \sqrt{n_k + 1} \quad \text{and} \quad \langle n_k - 1|a_k|n_k\rangle = \sqrt{n_k}.$$ $$(9.66)$$

This implies that the amplitude of emission of a graviton into a state with occupation number n_k is enhanced by a factor $\sqrt{n_k + 1}$ and the probability of graviton emission or the rate of graviton emission will be enhanced by a factor $n_k + 1$. The amplitude of absorption of a graviton from a state with occupation number n_k is proportional to $\sqrt{n_k}$ and the probability of graviton absorption is proportional to the occupation number of gravitons n_k in the initial state.

Consider the radiation of gravitons by binary pulsar when the initial state is a Fock state with non-zero occupation at different momenta $|n_k\rangle$. The matrix element of the graviton emission by a stress tensor $T_{\mu\nu}$ is given by

$$
\begin{aligned}
S_{if}^e &= -i\langle n_k + 1| \int d^4x \, \mathcal{L}_{int}(x)|n_k\rangle \\
&= -i\frac{\kappa}{2}\int d^4x \, \langle n_k + 1|\sqrt{2\omega_k}\epsilon_\lambda^{\mu\nu}(\mathbf{k})T_{\mu\nu}(x)\hat{h}^{\mu\nu}(x)|n_k\rangle \\
&= \sqrt{1+n_k}\, S_{if}^0,
\end{aligned} \tag{9.67}
$$

where S_{fi}^0 is the matrix element of graviton emission from a zero graviton vacuum state. Similarly the absorption rate from a thermal bath will be

$$
\begin{aligned}
S_{if}^a &= -i\langle n_k - 1| \int d^4x \, \mathcal{L}_{int}(x)|n_k\rangle \\
&= -i\frac{\kappa}{2}\int d^4x \, \langle n_k - 1|\sqrt{2\omega_k}\epsilon_\lambda^{\mu\nu}(\mathbf{k})T_{\mu\nu}(x)\hat{h}^{\mu\nu}(x)|n_k\rangle \\
&= \sqrt{n_k}\, S_{if}^0.
\end{aligned} \tag{9.68}
$$

9.9   The Thermal Background of Gravitons

If the emission and absorption of gravitons is taking place not in a vacuum state but a thermal background of gravitons, then the initial state is the thermal 'vacuum' given by

$$
|0\rangle_\beta = \sum_{n_k} p_n(\omega_k)|n_k\rangle, \tag{9.69}
$$

where $p_n(\omega_k)$ is the probability of the state with occupancy n_k which is given by the Boltzmann factor

$$
p_n(\omega_k) = \frac{e^{-\beta\omega_k n_k}}{\sum_{n_k} e^{-\beta n_k \omega_k}}, \qquad \omega_k = (|\mathbf{k}|^2 + m^2)^{1/2}. \tag{9.70}
$$

For bosons the occupancy n_k can be any natural number therefore

$$
\sum_{n_k} e^{-\beta n_k \omega_k} = \sum_{n=0}^{\infty} e^{-\beta n \omega_k} = \frac{1}{1 - e^{-\beta\omega_k}} \qquad \text{(bosons)} \tag{9.71}
$$

The expectation value of any operator which is an eigenstate of a Fock state $|n_k\rangle$ w.r.t the thermal density matrix is then given by

$$\langle O \rangle_\beta = \frac{\sum_{n_k} P_n(\omega_k) \langle n_k | O | n_k \rangle}{\sum_{n_k} e^{-\beta n_k \omega_k}}. \tag{9.72}$$

The average occupation number of state with momenta $k = (\mathbf{k}, \omega_k)$ is

$$\bar{n}_k = \langle a_k^\dagger a_k \rangle_\beta = \frac{\sum_{n_k} P_n(\omega_k) n_k}{\sum_{n_k} e^{-\beta n_k \omega_k}}. \tag{9.73}$$

For bosons the average occupation number of a state with momentum k is the Bose-Einstein distribution function.

$$\bar{n}_k = \langle a_k^\dagger a_k \rangle_\beta = \frac{\sum_{n_k} P_n(\omega_k) n_k}{\sum_{n_k} e^{-\beta n_k \omega_k}}$$

$$= \frac{1}{e^{\beta \omega_k} - 1}. \tag{9.74}$$

The expectation value of the various combinations a_k, $a_k^\dagger$ operators in thermal graviton initial state are

$$\langle a_k \rangle_\beta = 0, \quad \langle a_k^\dagger \rangle_\beta = 0, \quad \langle a_k{}^\dagger a_{k'}^\dagger \rangle_\beta = 0, \quad \langle a_k a_{k'}' \rangle_\beta = 0,$$

$$\langle a_{k'}^\dagger a_k \rangle_\beta = n(\omega_k) \delta^3(\mathbf{k} - \mathbf{k}'),$$

$$\langle a_k a_{k'}^\dagger \rangle_\beta = (1 + n(\omega_k)) \delta^3(\mathbf{k} - \mathbf{k}'). \tag{9.75}$$

9.10   Ensemble of Binaries in Thermal Graviton Background

Consider an ensemble of binary stars in a heat bath of thermal gravitons at temperature T. Consider the emission from a state of energy E_m to the state E_n by radiating a graviton of frequency $\omega = (E_m - E_n)$. The emission can be spontaneous or it can be stimulated by the gravitons in the heat bath. On the other hand the transition from the lower energy state to higher energy state can only take place by the stimulated absorption of gravitons from the heat bath. The stimulated emission or absorption processes will be proportional to the graviton intensity $u(\omega)$ define such that the graviton number density is

$$\rho = \int d\omega \, u(\omega). \tag{9.76}$$

The number of binaries in the energy state n is thus governed by the rate equation

$$\frac{dN_n}{dt} = A_{m \to n} N_m + B_{m \to n} N_m u(\omega) - B_{n \to m} N_n u(\omega). \tag{9.77}$$

If the binaries are in thermal equilibrium with the graviton bath at temperature T then the ration of their number at difference energy level will be given by the Boltzmann factor

$$\frac{N_n}{N_m} = e^{-(E_n - E_m)/T} = e^{-\omega/T}. \tag{9.78}$$

At thermal equilibrium $dN_n/dt = 0$ and we obtain from (9.77) and (9.78),

$$A_{m \to n} = B_{m \to n} \left(e^{\omega/T - 1} - 1 \right) u(\omega), \tag{9.79}$$

where we have assumed time reversal symmetry of the microscopic quantum theory that causes the transitions and assume $B_{m \to n} = B_{n \to m}$. From the first principle calculations of the transition rates we can relate the absorption and emission coefficients (as we will show explicitly by calculating the A and B coefficients the binary stars from quantum field theory)

$$A_{m \to n} = \frac{\omega^3}{\pi^2} B_{m \to n}. \tag{9.80}$$

From (9.79) and (9.80) we find that

$$u(\omega) = \frac{\omega^3}{\pi^2} \frac{1}{e^{\omega/T} - 1} \tag{9.81}$$

which is the Planck distribution of the graviton number density. This shows that gravitons in thermal equilibrium with an ensemble of absorbers/scatterers like binary stars will follow the Planck distribution.

The rate of absorption of gravitons from the heat bath is

$$\Gamma^a_{n \to m} = B_{m \to n} u(\omega)$$

$$= B_{m \to n} \frac{\omega^3}{\pi^2} n(\omega), \tag{9.82}$$

where $n(\omega)$ is the Bose-Einstein distribution

$$n(\omega) = \frac{1}{e^{\omega/T} - 1}. \tag{9.83}$$

The total rate of emission in a heat bath is the sum of spontaneous and stimulated emissions and is given by

$$\Gamma^e_{m \to n} = A_{m \to n} + B_{m \to n} u(\omega)$$

$$= \frac{\omega^3}{\pi^2} B_{m \to n} \left(1 + n(\omega)\right). \tag{9.84}$$

Therefore there is an enhancement by a factor of $(1 + n(\omega))$ of graviton emission in a thermal graviton background compared to the emission in vacuum. This is the phenomenon of stimulated emission expected from the gravitons being bosons.

At low frequencies, $\omega \ll T$ the enhancement factor due to stimulated emission is

$$1 + n(\omega) = \frac{1}{2} + \frac{T}{\omega} + \frac{1}{12} \frac{\omega}{T} - \frac{1}{720} \left(\frac{\omega}{T}\right)^3 + \cdots \quad . \tag{9.85}$$

9.11   Einstein's A and B Coefficients for an Ensemble of Binary Stars in Thermal Graviton Background

The rate of graviton absorption by a binary is from the thermal initial state is given using (9.68) by Fermi's Golden Rule

$$\Gamma^a_{n \to m} = \int n(\omega) \left|M(\omega')\right|^2 2\pi \delta(\omega' - \omega) g(\omega) \frac{d\omega}{2\omega}, \tag{9.86}$$

where $\omega' = E_m - E_n = 2\Omega$ and $g(\omega) = \omega^2/(2\pi^2)$ is the density of state of free gravitons. Here amplitude squared is

$$\left|M(\omega')\right|^2 = \frac{\kappa^2}{4} \frac{2}{5} \left(T_{ij}(\omega') T^*_{ji}(\omega') - \frac{1}{3} |T^i{}_i(\omega')|^2\right)$$

$$= \frac{16\pi G}{5} \mu^2 a^4 \Omega^4, \tag{9.87}$$

where we used (4.14) averaged over the 4π directions. The rate of graviton spontaneous and stimulated emission by the graviton heat bath is using (9.67) given by

$$\Gamma^e_{m \to n} = \int (n(\omega') + 1) \left|M(\omega')\right|^2 2\pi \delta(\omega' - \omega) \frac{4\pi \omega^2 d\omega}{(2\pi)^3 2\omega}. \tag{9.88}$$

Here we integrate over the final state graviton phase space and assume isotropic emission. The emitted graviton has an angular frequency $\omega = \omega' = 2\Omega$ but can be emitted in all directions. Here the factor $(n(\omega') + 1)$ represents the sum of stimulated emission which is the expression in (9.88) with the factor $n(\omega')$ and spontaneous emission the terms without the factor $n(\omega')$. Recall that Einstein's A and B coefficients are defined through the emission rate $\Gamma^e = A + u(\omega)B$ and absorption rates $\Gamma^a = u(\omega)B$ where $u(\omega)$ is the graviton number density $u(\omega) = (\omega^3/\pi^2)n(\omega)$. Using these in the explicit expressions for the absorption and emission rates (9.86) and (9.88), respectively, we can see that Einstein's A and B coefficients are related as $A = (\omega^3/\pi^2)B$. This relation was assumed by Einstein to prove that an ensemble of atoms in thermal equilibrium will give rise to a Planck distribution of photons in the heat bath as we have seen in (9.80) and (9.81).

Consider an ensemble of binaries in circular orbit. Using (4.10) and (4.14) in (9.86) and (9.88) we can calculate the absorption and emission rates for binaries in a graviton heat bath. We obtain for the stimulated absorption rate

$$\Gamma^a = \frac{16}{5}G\mu^2 a^4 \Omega^5 n(2\Omega), \tag{9.89}$$

where $n(2\Omega) = (e^{2\Omega/T}-1)^{-1}$, while the total stimulated plus spontaneous emission rate is

$$\Gamma^e = \frac{16}{5}G\mu^2 a^4 \Omega^5 \left(1 + n(2\Omega)\right). \tag{9.90}$$

For an ensemble of binary stars of orbital frequency Ω the Einstein's emission and absorption coefficients are, respectively,

$$A = \frac{16}{5}G\mu^2 a^4 \Omega^5, \quad B = \frac{126}{5}G\pi^2 \mu^2 a^4 \Omega^3. \tag{9.91}$$

In a heat bath the binaries absorb energy from the bath at the rate

$$\frac{dE^a}{dt} = \frac{32}{5}G\mu^2 a^4 \Omega^6 n(2\Omega) \tag{9.92}$$

and the rate of energy emission in stimulated and spontaneous emission is

$$\frac{dE^e}{dt} = \frac{32}{5}G\mu^2 a^4 \Omega^6 \left(1 + n(2\Omega)\right). \tag{9.93}$$

The net energy loss $\Delta E = E^e - E^a$ will be the same as the energy radiated in vacuum,

$$\frac{d\Delta E}{dt} = \frac{32}{5}G\mu^2 a^4 \Omega^6 \tag{9.94}$$

so at this order there is no effect of the graviton heat bath on the time period loss of binary pulsars. Hence it will be difficult to observe the effect of a possible graviton radiation background from the time period observations of binary pulsars. This is also seen in the propagation of gravitational waves through a thermal gas of hot electrons in the galaxy [29]. The absorption of GW by the hot gas is compensated by stimulated emission and any net absorption rate is suppressed by a factor $1 - N_m/N_n \sim 1 - e^{-\Omega/T}$ and is unobservably small.

9.12 Absorption of Gravitational Waves by Bound State Systems

Bound state objects like binaries and macroscopic scale gravitational atoms [30]. Gravitational waves radiated during coalescence events of blackholes or compact stars can be absorbed by these objects and scattered to a higher energy levels. Gravitational signals passing through an ensemble of such bound state objects will suffer an attenuation. We calculate the absorption cross section of compact binaries and gravitational atoms in the following sections.

9.12.1 Absorption of Gravitational Waves by Compact Binaries

In the absence of a thermal bath of gravitons, gravitons emitted from other sources can be absorbed by a compact stars binary. The absorption rate of gravitational waves in a binary is

$$
\begin{aligned}
\Gamma^a &= \int \left|M(\omega')\right|^2 2\pi \delta(\omega' - \omega) \left(\frac{\omega^2}{2\pi^2}\right) \frac{d\omega}{2\omega} \\
&= \frac{16}{5} G\mu^2 a^4 \Omega^5 .
\end{aligned}
\tag{9.95}
$$

The total cross section of gravitational wave scattering from a binary is given by the Breit-Wigner form

$$
\sigma_T = \frac{4\pi(2l+1)}{\omega^2} \frac{\Gamma^{a2}}{(\omega - \omega_0)^2 + \Gamma^{a2}},
\tag{9.96}
$$

where ω_0 is the resonant frequency of absorption and emission in circular orbits binaries is related to the angular frequency of the binary as $\omega_0 = 2\Omega$. The cross section of absorption of binaries for gravitational waves is given by

$$
\begin{aligned}
\sigma_a &= Im\left(\frac{1}{(\omega - \omega_0)^2 - i\Gamma\omega}\right) \\
&\simeq \frac{\Gamma^a}{\omega^3} = \frac{16}{5} \frac{G\mu^2 a^4 \Omega^5}{\omega^3}
\end{aligned}
\tag{9.97}
$$

at gravitational wave frequencies away from the resonant frequency $\omega \gg \omega_0$.

9.12.2 Absorption of Gravitational Waves by Scalar Cloud Around Blackholes

Ultralight scalars in the mass range $m_a < 10^{-19}$eV constitute a class of dark matter called 'fuzzy dark matter' [31, 32]. These long wavelength scalars can form clouds of bound states around blackholes and around binaries as giant gravitational atoms [30]. Gravitational waves can be absorbed by transitions between two energy levels of the gravitationally bound atoms if the transition between the two eigenstates Ψ_{lmn} and $\Psi_{l'm'n'}$ results in a change in the quadrupole moment of the atom [33]. Ultralight scalars with Compton wavelength larger than Schwarzschild radius of blackholes, $\lambda = 2\pi/m_a >> r_s$ ($r_s = 2GM$), can form a bound state with the blackhole resulting in a macroscopic sized atom. The bound state will be formed by the gravitational potential $V = \alpha/r$ where $\alpha = GMm_a$. The quantised orbits of the scalar particles will have energy levels [30]

$$E_{nlm} = m_a \left(1 - \frac{\alpha^2}{2n^2} - \frac{\alpha^4}{8n^4} + \cdots \right). \tag{9.98}$$

Consider a gravitational wave incident on a blackhole or a compact binary with a scalar cloud described by the wavefunction Ψ_{lmn} which have quantised energy levels. The rate of absorption by the atoms is by the Fermi Golden Rule

$$\Gamma^a = \int \left| \langle \Psi_{n'l'm'} | \frac{\kappa}{2} T^{\mu\nu} h_{\mu\nu} | \Psi_{nlm}, h_{\alpha\beta}(\mathbf{k}', \omega') \rangle \right|^2 2\pi \delta(\omega' - \omega) \left(\frac{\omega^2}{2\pi^2} \right) \frac{d\omega}{2\omega}$$

$$= \frac{\kappa^2}{4} \frac{2}{5} \left(T_{ij}(\omega') T_{ji}^*(\omega') - \frac{1}{3} |T^i{}_i(\omega')|^2 \right) \frac{\omega'}{2\pi}. \tag{9.99}$$

The quantum mechanical expectation value of T_{ij} are related to the wavefunctions as

$$T_{ij}(\omega') = m_a \langle \dot{x}_i \dot{x}_j \rangle = -m_a \omega'^2 \langle x_i x_j \rangle$$

$$= -m_a \omega'^2 \int d^3x \; \Psi_{nlm}^*(x) \, \Psi_{n'l'm'}(x) \; x_i x_j. \tag{9.100}$$

Here $\omega' = E_{nlm} - E_{n'l'm'}$. There is a change in the quadrupole moment for the transitions between the states,

$$\Psi_{100} = \frac{1}{\sqrt{\pi} r_B^{3/2}} e^{-r/r_B}, \quad \Psi_{322} = \frac{1}{162\sqrt{\pi} r_B^{3/2}} \left(\frac{r^2}{r_B^2} \right) e^{-r/3r_B} \sin^2 \theta e^{2i\phi},$$

$$\tag{9.101}$$

where r_B is the Bohr radius given by $r_B = (\mu\alpha)^{-1} = (GM\mu^2)^{-1}$.

The non-zero components of the change in the stress tensor (9.100) resulting from a transition between the Ψ_{100} and Ψ_{322} are given by [33],

$$T_{xx} = -T_{yy} = iT_{xy} = -\frac{3^4}{2^8}m_a r_B^2 \omega'^2 . \tag{9.102}$$

Using this in (9.99) we find that the rate of absorption of gravitational atoms for incident gravitational waves of frequency ω' is given by

$$\Gamma^a = \left(\frac{3^8}{2^{11}}\right)\frac{G}{5}m_a^2 r_B^4 \omega'^5 . \tag{9.103}$$

For the transitions between Ψ_{100} and Ψ_{322} states $\omega' = E_{322} - E_{100} = (4/9)m_a\alpha^2$.

The absorption cross section for gravitational waves at frequencies ω away from the resonant frequency ω' is $\sigma_a = \Gamma_a/\omega^3$.

References

1. J.M. Ezquiaga, M. Zumalacárregui, Dark Energy in light of Multi-Messenger Gravitational-Wave astronomy. Front. Astron. Space Sci. **5**, 44 (2018). [arXiv:1807.09241 [astro-ph.CO]]
2. J.B. Jiménez, J.M. Ezquiaga, L. Heisenberg, Probing cosmological fields with gravitational wave oscillations. J. Cosmol. Astropart. Phys. **04**, 027 (2020). [arXiv:1912.06104 [astro-ph.CO]]
3. B.P. Abbott et al. [LIGO Scientific and Virgo Collaborations], GW170814: A Three-Detector Observation of Gravitational Waves from a Binary Black Hole Coalescence. Phys. Rev. Lett. **119**(14), 141101 (2017). [arXiv:1709.09660 [gr-qc]]
4. B.P. Abbott et al., Multi-messenger Observations of a Binary Neutron Star Merger. Astrophys. J. Lett. **848**(2), L12 (2017). [arXiv:1710.05833 [astro-ph.HE]]
5. G. Fanizza, M. Gasperini, E. Pavone, L. Tedesco, Linearized propagation equations for metric fluctuations in a general (non-vacuum) background geometry. J. Cosmol. Astropart. Phys. **07**, 021 (2021). https://doi.org/10.1088/1475-7516/2021/07/021. [arXiv:2105.13041 [gr-qc]]
6. S.W. Hawking, Perturbations of an expanding universe. Astrophys. J. **145**, 544–554 (1966)
7. A.R. Prasanna, Propagation of gravitational waves through a dispersive medium. Phys. Lett. A **257**, 120–122 (1999)
8. J. Madore, The dispersion of gravitational waves. Commun. Math. Phys. **27**, 291–302 (1972)
9. G. Goswami, G.K. Chakravarty, S. Mohanty, A.R. Prasanna, Constraints on cosmological viscosity and self interacting dark matter from gravitational wave observations. Phys. Rev. D **95**(10), 103509 (2017). [arXiv:1603.02635 [hep-ph]]
10. G. Baym, S.P. Patil, C.J. Pethick, Phys. Rev. D **96**(8), 084033 (2017). https://doi.org/10.1103/PhysRevD.96.084033. [arXiv:1707.05192 [gr-qc]]
11. R. Flauger, S. Weinberg, Phys. Rev. D **97**(12), 123506 (2018). [arXiv:1801.00386 [astro-ph.CO]]
12. I. Brevik, S. Nojiri, Gravitational waves in the presence of viscosity. Int. J. Mod. Phys. D **28**(10), 1950133 (2019). [arXiv:1901.00767 [gr-qc]].
13. C. Ganguly, J. Quintin, Microphysical manifestations of viscosity and consequences for anisotropies in the very early universe. [arXiv:2109.11701 [gr-qc]]
14. W. Hu, R. Barkana, A. Gruzinov, Cold and fuzzy dark matter. Phys. Rev. Lett. **85**, 1158–1161 (2000). [arXiv:astro-ph/0003365 [astro-ph]]

15. L. Hui, J.P. Ostriker, S. Tremaine, E. Witten, Ultralight scalars as cosmological dark matter. Phys. Rev. D **95**(4), 043541 (2017). [arXiv:1610.08297 [astro-ph.CO]]
16. B. Abbott et al. [LIGO Scientific and Virgo], GW170817: Observation of gravitational waves from a binary neutron star inspiral. Phys. Rev. Lett. **119**(16), 161101 (2017)
17. G.P. Lamb, S. Kobayashi, GRB 170817A as a jet counterpart to gravitational wave triggerGW 170817. Mon. Not. Roy. Astron. Soc. **478**(1), 733–740 (2018)
18. S. Jana, G.K. Chakravarty, S. Mohanty, Constraints on Born-Infeld gravity from the speed of gravitational waves after GW170817 and GRB 170817A. Phys. Rev. D **97**(8), 084011 (2018). [arXiv:1711.04137 [gr-qc]]
19. M. Banados, P.G. Ferreira, Eddington's theory of gravity and its progeny. Phys. Rev. Lett. **105**, 011101 (2010). [arXiv:1006.1769 [astro-ph.CO]]
20. C. Escamilla-Rivera, M. Banados, P.G. Ferreira, A tensor instability in the Eddington inspired Born-Infeld Theory of Gravity. Phys. Rev. D **85**, 087302 (2012). https://doi.org/10.1103/PhysRevD.85.087302. [arXiv:1204.1691 [gr-qc]]
21. J. Beltran Jimenez, L. Heisenberg, G.J. Olmo, D. Rubiera-Garcia, On gravitational waves in Born-Infeld inspired non-singular cosmologies. J. Cosmol. Astropart. Phys. **10**, 029 (2017). [erratum: JCAP **08**, E01 (2018)]. [arXiv:1707.08953 [hep-th]]
22. T. Kobayashi, Horndeski theory and beyond: a review. Rep. Prog. Phys. **82**(8), 086901 (2019). https://doi.org/10.1088/1361-6633/ab2429. [arXiv:1901.07183 [gr-qc]]
23. M. Pardy, The gravitational Cerenkov radiation with radiative corrections. Phys. Lett. **B336**, 362–367 (1994)
24. G.D. Moore, A.E. Nelson, Lower bound on the propagation speed of gravity from gravitational Cherenkov radiation. J. High Energy Phys. **09**, 023 (2001). [arXiv:hep-ph/0106220 [hep-ph]]
25. J.W. Elliott, G.D. Moore, H. Stoica, Constraining the new Aether: Gravitational Cherenkov radiation. J. High Energy Phys. **08**, 066 (2005). [arXiv:hep-ph/0505211 [hep-ph]]
26. R. Kimura, K. Yamamoto, Constraints on general second-order scalar-tensor models from gravitational Cherenkov radiation. J. Cosmol. Astropart. Phys. **07**, 050 (2012). [arXiv:1112.4284 [astro-ph.CO]]
27. M. De Laurentis, S. Capozziello, G. Basini, Gravitational Cherenkov Radiation from Extended Theories of Gravity. Mod. Phys. Lett. A **27**, 1250136 (2012). [arXiv:1206.6681 [gr-qc]]
28. V.A. Kostelecký, J.D. Tasson, Constraints on Lorentz violation from gravitational Čerenkov radiation. Phys. Lett. B **749**, 551–559 (2015). [arXiv:1508.07007 [gr-qc]]
29. R. Flauger, S. Weinberg, Absorption of gravitational waves from distant sources. Phys. Rev. D **99**(12), 123030 (2019). [arXiv:1906.04853 [hep-th]]
30. D. Baumann, H.S. Chia, J. Stout, L. ter Haar, The spectra of gravitational atoms. J. Cosmol. Astropart. Phys. **12**, 006 (2019). https://doi.org/10.1088/1475-7516/2019/12/006. [arXiv:1908.10370 [gr-qc]]
31. W. Hu, R. Barkana, A. Gruzinov, Cold and fuzzy dark matter. Phys. Rev. Lett. **85**, 1158 (2000). [astro-ph/0003365]
32. L. Hui, J.P. Ostriker, S. Tremaine, E. Witten, Phys. Rev. D **95**(4), 043541 (2017). [arXiv:1610.08297 [astro-ph.CO]]
33. A. Palessandro, M.S. Sloth, Gravitational absorption lines. Phys. Rev. D **101**(4), 043504 (2020). [arXiv:1910.01657 [hep-th]]

Stochastic Gravitational Waves

10

Abstract

In particle physics models of Grand Unified Theories (GUTs), we have a series of symmetry breaking in the early universe. For example we may have the symmetry breaking chain $SO(10) \rightarrow SU(4)_c \times SU(2)_L \times SU(2)_R \rightarrow SU(3) \times SU(2) \times U(1)_Y$. There are two sources of gravitational waves. First, there is energy released in First Order Phase Transitions in the form of gravitational waves, whose characteristic frequency in the present universe depends on the temperature of the phase transition. Second, the symmetry breakings produce topological defects like monopoles, cosmic strings and domain walls which can produce gravitational waves. We study these sources of stochastic gravitational waves with the aim of predicting signals of this phenomenon for PTA, LISA and LIGO/VIRGO/KAGRA observations.

10.1 Introduction

The gravitational waves produced from inflation are vacuum fluctuations of the graviton which are amplified during the exponential expansion of the universe. There are other sources of stochastic background gravitational waves namely scalar field fluctuations following inflation in a period called pre-heating. Gravitational waves are also produced from the oscillation or decay of topological defects like cosmic strings and domain walls and during phase transitions. In this section we derive the general expression for the stochastic GW energy density of the universe as a function of the source stress tensor [1–5]. We then apply this in the following sections to derive the GW density from cosmic strings, domain walls and phase transitions.

© The Author(s), under exclusive license to Springer Nature Switzerland AG 2023
S. Mohanty, *Gravitational Waves from a Quantum Field Theory Perspective*,
Lecture Notes in Physics 1013, https://doi.org/10.1007/978-3-031-23770-6_10

10.2   Gravitational Waves from Extended Sources

In this section we compute gravitational waves from sources which are extended in space, i.e. with size $d \geq \lambda_{gw}$ and which last for a time $\delta t \leq f_{gw}$, where λ_{gw} and f_{gw} are the wavelengths and frequencies of gravitational waves at the time of emission. If the duration of the sources is comparable to the Hubble expansion rate, then one must take into account the expansion of the universe in the emission process. We therefore consider gravitons h_{ij} as tensor perturbations of the FRW metric. We compute the waveform h_{ij} for different sources using Green's function method. The energy density of stochastic GW is then computed from the two point correlations of h_{ij}.

The metric for the tensor perturbations in an expanding universe can be written as

$$ds^2 = -dt^2 + a(t)^2 \left(\delta_{ij} + h_{ij} \right) dx^i dx^j, \tag{10.1}$$

where we choose a gauge where h_{ij} is transverse and traceless, i.e. $\partial_i h_{ij} = 0$ and $h_{ii} = 0$.

The e.o.m. of $h_{ij}(t, \mathbf{x})$ with in the presence of sources is

$$\ddot{h}_{ij}(t, \mathbf{x}) + 3 \frac{\dot{a}}{a} \dot{h}_{ij}(t, \mathbf{x}) - \frac{\nabla^2}{a^2} h_{ij}(t, \mathbf{x}) = \frac{16\pi G}{a^2} \Pi_{ij}^{TT}(t, \mathbf{x}). \tag{10.2}$$

Here Π_{ij} is the anisotropic stress tensor given by

$$a^2 \Pi_{ij} = T_{ij} - \bar{p} g_{ij}, \tag{10.3}$$

where $\bar{p}$ is the homogenous background pressure and Π_{ij}^{TT} is the transverse–traceless projection of Π_{ij} which we shall define after transforming to momentum space. We change the time coordinate to conformal time $d\tau = dt/(a(t))$ and write the e.o.m. (10.2) for $h_{ij}(\tau, \mathbf{x})$ as

$$h''(\tau, \mathbf{x}) + 2 \frac{a'}{a} h'_{ij}(\tau, \mathbf{x}) - \nabla^2 h_{ij}(\tau, \mathbf{x}) = 16\pi G a^2 \Pi_{ij}^{TT}(\tau, \mathbf{x}), \tag{10.4}$$

where $'$ denotes derivative w.r.t. conformal time τ. We transform the equation to momentum space by taking

$$h_{ij}(\tau, \mathbf{x}) = \int \frac{d^3 k}{(2\pi)^{3/2}} h_{ij}(\tau, \mathbf{k}) e^{-\mathbf{k} \cdot \mathbf{x}} \tag{10.5}$$

and defining $\tilde{h}_{ij}(\tau, \mathbf{k}) = a h_{ij}(\tau, \mathbf{k})$. The e.o.m. for $\tilde{h}_{ij}$ is then given by

$$\tilde{h}''(\tau, \mathbf{k}) + \left(|\mathbf{k}|^2 - \frac{a''}{a} \right) \tilde{h}_{ij}(\tau, \mathbf{k}) = 16\pi G a^3 \Pi_{ij}^{TT}(\tau, \mathbf{k}). \tag{10.6}$$

The traceless–transverse projection of Π_{ij} can be obtained by the operating with projection operator w.r.t. the direction $\hat{\mathbf{k}}$ which projects out components of $\Pi_{ij}(\tau, \mathbf{k})$ which are orthogonal to $\hat{\mathbf{k}} = \mathbf{k}/|\mathbf{k}|$. This projection operator is written as

$$\Pi_{ij}^{TT}(\tau, \mathbf{k}) = \Lambda_{ij,lm}(\hat{\mathbf{k}})\Pi_{ij}(\tau, \mathbf{k}), \tag{10.7}$$

where we can define $\Lambda_{ij,lm}$ as

$$\Lambda_{ij,lm}(\hat{\mathbf{k}}) = P_{il}(\hat{\mathbf{k}})P_{jm}(\hat{\mathbf{k}}) - \frac{1}{2}P_{ij}(\hat{\mathbf{k}})P_{lm}(\hat{\mathbf{k}}) \tag{10.8}$$

and where

$$P_{ij}(\hat{\mathbf{k}}) = \delta_{ij} - \hat{k}_i\hat{k}_j . \tag{10.9}$$

The projection operator $\Lambda_{ij,lm}(\hat{\mathbf{k}})$ has the properties

$$\Lambda_{ij,lm}(\hat{\mathbf{k}}) = \Lambda_{lm,ij}(\hat{\mathbf{k}}), \quad \Lambda_{ij,lm}(\hat{\mathbf{k}}) = \Lambda_{ji,lm}(\hat{\mathbf{k}}), \quad k_i\Lambda_{ij,lm}(\hat{\mathbf{k}}) = 0, \quad \Lambda_{ii,lm}(\hat{\mathbf{k}}) = 0.$$

With these properties of the projection operator, we have $k_i\Pi_{ij}^{TT}(\tau, \mathbf{k}) = \Pi_{ii}^{TT}(\tau, \mathbf{k}) = 0$. The traceless–transverse anisotropic stress tensor is also given by

$$a^2\Pi_{ij}^{TT}(\hat{\mathbf{k}}) = T_{lm}^{TT}(\hat{\mathbf{k}}) = \Lambda_{ij,lm}T_{lm}(\hat{\mathbf{k}}), \tag{10.10}$$

and we can express (10.11) as

$$\tilde{h}''(\tau, \mathbf{k}) + \left(k^2 - \frac{a''}{a}\right)\tilde{h}_{ij}(\tau, \mathbf{k}) = 16\pi G a\, T_{ij}^{TT}(\tau, \mathbf{k}), \tag{10.11}$$

where $k = |\mathbf{k}|$. We consider gravitational waves with wavelengths smaller than the horizon scale. We can then drop the a''/a term compared to the $|\mathbf{k}|^2$ term as for the sub-horizon gravitational waves $a''/a \sim a^2H^2 \ll k^2$. For GW produced in the radiation era, the a''/a term is zero as $a \propto \tau$ in the radiation dominated universe. So finally the wave equation for sub-horizon gravitational waves generated by source with stress tensor T_{ij} is given by

$$\tilde{h}''(\tau, \mathbf{k}) + k^2\tilde{h}_{ij}(\tau, \mathbf{k}) = 16\pi G a\, T_{ij}^{TT}(\tau, \mathbf{k}). \tag{10.12}$$

Green's function $G(\tau - \tau', \mathbf{k})$ is the solution of (10.12) with a delta function source

$$G''(\tau - \tau', k) + k^2 G(\tau - \tau', k) = \delta(\tau - \tau'). \tag{10.13}$$

The sources are transient over the cosmological times. Suppose the source exists in the time interval $\tau_i < \tau < \tau_f$. For times $\tau \leq \tau_i$, there are no gravitational waves, so the appropriate boundary conditions are $\tilde{h}_{ij}(\tau = \tau_i, \mathbf{k}) = \tilde{h}'_{ij}(\tau = \tau_i, \mathbf{k}) = 0$. Green's function solution of (10.13) with these boundary conditions is

$$G(\tau - \tau', k) = \frac{1}{k} \sin k(\tau - \tau'). \tag{10.14}$$

The solution of (10.12) can therefore be written as a convolution of Green's function (10.14) with the source term

$$\tilde{h}_{ij}(\tau, \mathbf{k}) = \frac{16\pi G}{k} \int_{\tau_i}^{\tau} d\tau' \sin k(\tau - \tau') a(\tau') T_{ij}^{\mathrm{TT}}(\tau', \mathbf{k}), \quad \tau_i \leq \tau \leq \tau_f, \tag{10.15}$$

and its derivative w.r.t. time τ is

$$\tilde{h}'_{ij}(\tau, \mathbf{k}) = 16\pi G \int_{\tau_i}^{\tau} d\tau' \cos k(\tau - \tau') a(\tau') T_{ij}^{\mathrm{TT}}(\tau', \mathbf{k}), \quad \tau_i \leq \tau \leq \tau_f. \tag{10.16}$$

At times $\tau > \tau_f$, the source ceases to exist. The gravitational waves produced by the source then are solutions of the homogenous equation (10.12) with the general solution

$$\tilde{h}_{ij}(\tau, \mathbf{k}) = A_{ij}(\mathbf{k}) \sin k(\tau - \tau_f) + B_{ij}(\mathbf{k}) \cos k(\tau - \tau_f), \quad \tau > \tau_f. \tag{10.17}$$

The coefficients A_{ij} and B_{ij} can be evaluated by matching the solutions for h_{ij} and h'_{ij} with (10.15) and (11.135) at $\tau = \tau_f$. This gives us the coefficients

$$A_{ij}(\mathbf{k}) = \frac{16\pi G}{k} \int_{\tau_i}^{\tau_f} d\tau' \cos k(\tau_f - \tau') a(\tau') T_{ij}^{\mathrm{TT}}(\tau', \mathbf{k}),$$

$$B_{ij}(\mathbf{k}) = \frac{16\pi G}{k} \int_{\tau_i}^{\tau_f} d\tau' \sin k(\tau_f - \tau') a(\tau') T_{ij}^{\mathrm{TT}}(\tau', \mathbf{k}). \tag{10.18}$$

10.3   Energy Density of Gravitational Waves from Extended Sources

In this section we compute the energy density of GW with waveform given in (10.17) and (10.18). The stress tensor of gravitational waves is derived by taking perturbations of Einstein's equation up to second order in $h_{\mu\nu}$. The Landau–Lifshitz stress tensor or the gravitational waves is defined as

$$\tau_{\mu\nu} = \frac{1}{8\pi G} \langle R_{\mu\nu}^{(2)} - \frac{1}{2} \bar{g}_{\mu\nu} R^{(2)} \rangle, \tag{10.19}$$

where the angular brackets indicate time-averaging over the oscillatory terms. The Ricci tensor at the second order in $h_{\mu\nu}$ in the TT gauge $\partial^\mu h_{\mu\nu} = 0$, $h^\alpha{}_\alpha = 0$ and $\Box h_{\alpha\beta} = 0$ is

$$R^{(2)}_{\mu\nu} = \frac{1}{4} \partial_\mu h_{\alpha\beta} \partial_\nu h^{\alpha\beta}, \tag{10.20}$$

and the Landau–Lifshitz stress tensor of the gravitational perturbations is

$$\tau_{\mu\nu} = \frac{1}{32\pi G} \langle \partial_\mu h_{\alpha\beta} \partial_\nu h^{\alpha\beta} \rangle. \tag{10.21}$$

The energy density of gravitational waves h^{TT}_{ij} is therefore

$$\rho_{gw}(t) = \frac{1}{32\pi G} \langle \dot{h}_{ij}(t, \mathbf{x}) \dot{h}_{ij}(t, \mathbf{x}) \rangle_V, \tag{10.22}$$

where the angular brackets denote averaging over volume $V \gg \lambda^3_{GW}$. The volume average can be written as

$$\rho_{gw} = \frac{1}{32\pi G V} \int d^3\mathbf{x}\, \dot{h}_{ij}(t, \mathbf{x})\, \dot{h}_{ij}(t, \mathbf{x})$$

$$= \frac{1}{32\pi G V a(\tau)^3} \int d^3\mathbf{x}\, \tilde{h}'_{ij}(\tau, \mathbf{x})\, \tilde{h}'_{ij}(\tau, \mathbf{x}). \tag{10.23}$$

Transforming to Fourier space, we get

$$\rho_{gw}(\tau) = \frac{1}{32\pi G V a(\tau)^3} \int d^3\mathbf{k}\, \tilde{h}'_{ij}(\tau, \mathbf{k})\, \tilde{h}'^*_{ij}(\tau, \mathbf{k}). \tag{10.24}$$

In going from (10.23) to (10.24), we used the relation

$$\int d^3\mathbf{x}\, e^{-i(\mathbf{k}+\mathbf{k}')\cdot\mathbf{x}} = (2\pi)^3 \delta^3(\mathbf{k}+\mathbf{k}') \tag{10.25}$$

and the property of the Fourier transform of real fields $\tilde{h}(\tau, -\mathbf{k}) = \tilde{h}^*(\tau, \mathbf{k})$.

We can use the sinusoidal solutions (10.17) in (10.24) to compute the energy density of GW. At time scales of observation larger than the time period of the waves $T = 2\pi/\omega$, we can time average the oscillations using the results

$$\langle \sin^2 k\tau \rangle_T = \langle \cos^2 k\tau \rangle_T = \frac{1}{2}, \quad \langle \sin k\tau \cos k\tau \rangle_T = 0. \tag{10.26}$$

Using these time averages, we write the expression for the energy density of GW

$$\rho_{gw} = \frac{1}{32\pi G a(\tau)^4} \int d^3\mathbf{k}\, \frac{k^2}{2} \left(|A_{ij}(\mathbf{k})|^2 + |B_{ij}(\mathbf{k})|^2 \right). \tag{10.27}$$

The expression can be further simplified using the explicit expressions for $A_{ij}(k)$ and $B_{ij}(k)$ from (10.18). Expanding $\sin k(\tau_f - \tau')$ and $\cos k(\tau_f - \tau')$ and simplifying, we get

$$
\rho_{gw} = \frac{4\pi G}{a(\tau)^4} \int d^3 \mathbf{k} \left\{ \left| \int_{\tau_i}^{\tau_f} d\tau' \cos(k\tau') a(\tau') T_{ij}^{\mathrm{TT}}(\tau', \mathbf{k}) \right|^2 \right.
$$

$$
\left. + \left| \int_{\tau_i}^{\tau_f} d\tau' \sin(k\tau') a(\tau') T_{ij}^{\mathrm{TT}}(\tau', \mathbf{k}) \right|^2 \right\}. \tag{10.28}
$$

We can further simplify this expression by taking $\tau_i = 0$ and $\tau_f = \infty$ and writing $\cos k\tau$ and $\sin k\tau$ in terms of $e^{ik\tau}$ and $e^{-k\tau}$, write $k\tau = \omega t$, where ω is the physical frequency which is related to the comoving wavenumber as $\omega = k/a$ and do the time integral by taking $dt = a d\tau$. With these substitutions, we can write (10.28) in a form given in [6]

$$
\rho_{gw} = \frac{2G}{a(t)^4} \int \Lambda_{ij,lm}(\hat{\mathbf{k}}) \, d\Omega \int dk k^2 T_{ij}(\omega, \mathbf{k}) T_{lm}^*(\omega, \mathbf{k}), \tag{10.29}
$$

where

$$
T_{ij}(\omega, \mathbf{k}) = \int_0^\infty \frac{dt}{2\pi} e^{i\omega t} \int d^3 \mathbf{x} \, e^{-i\mathbf{k}\cdot\mathbf{x}} T_{ij}(t, \mathbf{x}). \tag{10.30}
$$

We have used the identities of the projection operators to write

$$
T_{ij}^{\mathrm{TT}}(\omega, \mathbf{k}) T_{ij}^{*\,\mathrm{TT}}(\omega, \mathbf{k}) = \Lambda_{ij,rs} \Lambda_{rs,lm} T_{ij} T_{lm}^* = \Lambda_{ij,lm} T_{ij} T_{lm}^*. \tag{10.31}
$$

10.4   Symmetry Breaking and Topological Defects

It is believed that the standard model gauge group arises from some larger Grand Unification group like $SO(10)$ via a series of spontaneous symmetry breakings $G_1 \to H_1 \to H_2 \to \cdots G_{SM}$, where $G_{SM} = SU(3)_c \times SU(2)_L \times U(1)_Y$ is the standard model gauge group. G_{SM} is broken by the electroweak Higgs when the temperature of universe falls below ~ 100 GeV to $SU(3)_c \times U(1)_{em}$ [7,8]. Some examples of intermediate symmetry breaking routes are

$$
SO(10) \to \left\{ \begin{array}{c} SU(3)_c \times SU(2)_L \times SU(2)_R \times U(1)_{B-L} \\ SU(4)_c \times SU(2)_L \times SU(2)_R \\ SU(5) \times U(1) \end{array} \right\} \to G_{SM} \times \mathbb{Z}_2.
$$

$$
\tag{10.32}
$$

Here $\mathbb{Z}_2$ is the R-parity of supersymmetry which is unbroken. The R-parity operator for particles is $P_R = (-1)^{3(B-L)+2s}$ which is $+1$ for the standard model particles and

−1 for their supersymmetric partners. Conservation of R-parity ensures stability of supersymmetric dark matter and prevents rapid proton decay.

Another important symmetry is the D-parity [9], which is a symmetry of $SU(2)_L \times SU(2)_R$ multiplets. The D-parity needs to be broken in order that the left and the right gauge couplings be different which is useful in order to generate baryon asymmetry.

$$SO(10) \rightarrow SU(3)_c \times SU(2)_L \times SU(2)_R \times U(1)_{B-L} \times D$$
$$\rightarrow SU(3)_c \times SU(2)_L \times SU(2)_R \times U(1)_{B-L} \rightarrow G_{SM} \times \mathbb{Z}_2. \quad (10.33)$$

The presence of the D-parity discrete symmetry determines whether there are domain walls after symmetry breaking.

When a symmetry group G is broken spontaneously to a subgroup H, the symmetry group of the lowest energy state, i.e. the vacuum state is the manifold $M = G/H$ the left coset set. This can be seen as follows. In the unbroken theory, the vacuum state $|0\rangle$ is invariant under G, so $g|0\rangle = |0\rangle$. Suppose the state $|\psi\rangle$ is a vacuum state that breaks G but obeys H symmetry. This means that $\exists g \in G$ such that $g|\psi\rangle \neq |\psi\rangle$ but $h|\psi\rangle = |\psi\rangle$, $\forall h \in H$. Two inequivalent vacuum states of the broken symmetric theories will be states like $g_1|\psi\rangle$ and $g_2|\psi\rangle$ if $g_1 \neq g_2h$ $\forall h \in H$. The manifold of the inequivalent vacuum states of the broken symmetric theory is the coset set $M = G/H$.

The type of defect formed depends on the homotopy class of the vacuum manifold M. Homotopy is the equivalence class of mapping from an n-sphere to M denoted by $\pi_n(M)$. Topological defects that cannot be undone without overcoming large energy barriers are formed when the homotopy class of M is non-trivial, i.e. $\pi_n(M) \neq I$. In the d-dimensional space, a p-dimensional topological defect will be formed if $\pi_{d-p-1}(G/H) \neq I$.

In 3-space dimensions, the defects formed from the spontaneous symmetry breaking $G \rightarrow H$ are of the following types:

$$\pi_0(G/H) \neq I \quad \Rightarrow \quad \text{Domain walls},$$
$$\pi_1(G/H) \neq I \quad \Rightarrow \quad \text{Cosmic strings},$$
$$\pi_2(G/H) \neq I \quad \Rightarrow \quad \text{Monopoles},$$
$$\pi_3(G/H) \neq I \quad \Rightarrow \quad \text{Textures}. \quad (10.34)$$

Of these defects, textures are not topologically stable and can decay quickly.

10.5   Domain Walls

The simplest realisation of a domain wall is from a real scalar field ϕ with a potential with a Z_2 symmetry $\phi \to \pm\phi$.

$$V(\phi) = \frac{\lambda}{4}\left(\phi^2 - v^2\right)^2.$$

(10.35)

This potential has two minima $\langle\phi\rangle = \pm v$. At the minima of the potential, Z_2 is spontaneously broken. The potential has a static solution of the equation of the e.o.m.

$$\frac{d\phi^2}{dz^2} - \frac{\partial V}{\partial \phi} = 0,$$

(10.36)

where ϕ is assumed to vary only along the z-axis. This equation has a solution

$$\phi(z) = v \tanh\left(\sqrt{\frac{\lambda}{2}}vz\right).$$

(10.37)

The solution goes to different vacua $\phi_{wall}(z \pm \infty) = \pm v$ on the two sides of the wall.
The stress tensor for this solution is given by

$$T_{\mu v} = \left(\frac{1}{2}\left(\frac{d\phi_{wall}}{dz}\right)^2 + V(\phi_{wall})\right) \text{diagonal}(1, -1, -1, 0)$$

$$= \frac{\lambda}{2}v^4 \, \mathrm{sech}^2\left(\sqrt{\frac{\lambda}{2}}vz\right) \text{diagonal}(1, -1, -1, 0).$$

(10.38)

From (10.38), we see that the energy density in the field configuration (10.37) is

$$\rho = T_{00} = \frac{\lambda}{2}v^4 \, \mathrm{sech}^4\left(\sqrt{\frac{\lambda}{2}}vz\right)$$

(10.39)

is maximum at $z = 0$ and goes to zero at $z \pm \infty$. This represents a domain wall between two different vacua of the potential. The thickness of the wall is

$$\delta = m_\phi^{-1} = \frac{1}{\sqrt{2\lambda}v}.$$

(10.40)

The energy density per area or the surface tension of the wall is

$$\sigma = \int_{-\infty}^{+\infty} dz \, \rho = \frac{2\sqrt{2}}{3} \sqrt{\lambda} v^3 . \tag{10.41}$$

Stable domain walls will dominate the density of the universe at a later time, as we discuss below. To make the domain walls unstable, we need a potential for ϕ which breaks the Z_2 symmetry explicitly. An example of a Z_2 breaking potential is

$$V_{Z_2}(\phi) = \frac{\mu}{2} \left(\phi^3 - v^3 \right) . \tag{10.42}$$

This breaks the degeneracy of the minima at $\langle \phi \rangle \pm v$ in the total potential $V_{tot} = V(\phi) + V_{Z_2}$. The difference in potential between the two sides of the domain wall

$$\Delta V = V_{tot}(\phi = v) - V_{tot}(\phi = -v) = \mu v^3 \tag{10.43}$$

acts as a volume pressure on the wall. Domain walls will annihilate when the volume pressure ΔV equals the surface tension ρ.

Other examples of domain walls are the pseudo-Nambu–Goldstone–Boson walls. When the global $U(1)$ symmetry breaks to a Z_N discrete symmetry we have massless Nambu–Goldstone–Bosons. Explicit $U(1)$ symmetry breaking terms will give mass to the 'axions' while retaining the shift symmetry $a \to a + \frac{2\pi}{N} f$, where f is the symmetry breaking scale. The $U(1)$ breaking and Z_2 conserving potential for the PNGB a is of generic form

$$V_a = \frac{m_a^2 f^2}{N^2} \left(1 - \cos \frac{Na}{f} \right) . \tag{10.44}$$

The leading order expansion of V_a in Na/v gives the mass m_a of the PNGB. The potential has N distinct minima at $a/f = 2\pi k/N$, $k = 0, 1, \cdots, N-1$.

The Lagrangian for the 'axion' with the symmetry breaking terms is

$$\mathcal{L} = \frac{1}{2}(\partial_\mu a)^2 + \frac{m_a^2 f^2}{N^2} \left(1 - \cos \frac{Na}{f} \right) . \tag{10.45}$$

There exists a static solution which varying in one spatial direction is given by Vilenkin and Shellard [10]

$$a(z) = \frac{2\pi f k}{N} + \frac{4f}{N} \tan^{-1} \exp(mz), \quad k = 0, 1, \cdots N-1. \tag{10.46}$$

This solution represents a domain wall in the x–y plane. This solution connects the k vacuum on one side of the wall with the $k+1$ vacuum on the other side.

Fig. 10.1 Axion field
configuration corresponding
to a domain wall

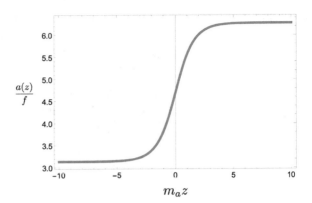

For $N = 2$ and $k = 1$, this solution represents a domain wall between two asymptotic vacua $a(z \to -\infty) = \pi f$ and $a(z \to \infty) = 2\pi f$, where $V_a = 0$. The two asymptotic regions $z \to \pm\infty$ are the two distinct spontaneously broken Z_2 vacua which are separated by the domain wall. The field value $a(z)$ is plotted in Fig. 10.1.

It can be seen that solution (10.46) represents a domain wall by plotting the energy density of the field configuration as a function of distance from the wall. The energy density is

$$\rho(z) = \frac{1}{2}\left|\frac{\partial a(z)}{\partial z}\right|^2 + V_a$$

$$= \frac{2f^2 m_a^2}{N^2}\,\mathrm{sech}^2(m_a z) + \frac{m_a^2 f^2}{N^2}\left(1 - \cos\frac{Nz(z)}{f}\right) \tag{10.47}$$

For the $N = 2$ case, the energy density as a function of z is shown in Fig. 10.2 from which it is clear that the static solution (10.46) represents a domain wall.

The thickness of the wall is $\delta = m_a^{-1}$, while the mass density or the surface tension is defined as

$$\sigma = \int_{-\infty}^{\infty} dz\; \rho(z) = \frac{8m_a f^2}{N^2}. \tag{10.48}$$

For the $N = 1$ case, the domain walls are unstable and disappear quickly and decay into relativistic axions. For the $N \geq 2$ case, the domain walls are stable at cosmological timescales and can over-close the universe. This is the domain wall problem in the PNGB models [11, 12].

One way out of the domain wall problem in PNGB models with an unbroken discrete symmetry is to make the vacua on the two sides of the domain wall non-degenerate. In this case, there will be a transition from the higher vacuum to the lower vacuum and energy radiated in the form of gravitational waves.

Fig. 10.2 The energy density as a function of distance for $N = 2$

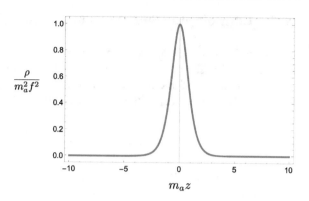

Fig. 10.3 The total potential V_{tot} for $N = 2$ with degeneracy of the vacua on the two sides of the domain wall lifted due to the explicit Z_2 breaking V_{Z_2} term (10.49)

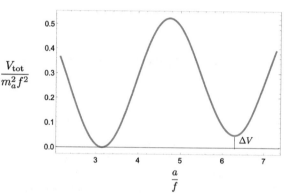

This can be achieved by adding to the Lagrangian an explicit Z_2 breaking term for example of the type

$$V_{\not{Z}_2} = \frac{1}{2}\mu^4\left(e^{ia/f} + hc\right) + \frac{1}{2}\mu^4. \tag{10.49}$$

If we take $\mu^4 \ll m_a^2 f^2/N^2$, then the shift in the location of the minima is small. This potential breaks the degeneracy of the minima on the two sides of the domain wall. The plot of the potential $V_{\text{tot}} = V_a + V_{\not{Z}_2}$ is shown in Fig. 10.3.

The energy difference between the two minima is

$$\Delta V = V_{\text{tot}}(a = 2\pi f) - V_{tot}(a = \pi f) \simeq \mu^4. \tag{10.50}$$

We will now study the decay of the biased domain wall and the production of the gravitational waves in this model.

10.6   Cosmological Evolution of Domain Walls

Domain walls will be formed when the temperature of the universe goes below the symmetry scale. The number density is one domain wall per causal horizon [13] and the planar dimensions of the domain walls are the horizon size, $R \sim H^{-1}$. The energy density in the walls is therefore of the order

$$\rho_{\text{wall}} \sim \frac{R^2 \sigma}{R^3} \sim \frac{\sigma}{H^{-1}} \sim \frac{\sigma}{t}, \tag{10.51}$$

as in the radiation era $H^{-1} = 2t$ and in the matter era $H^{-1} = (3/2)t$. This relation can made more exact by defining the constant $\mathcal{A}$ as

$$\rho_{\text{wall}} \equiv \mathcal{A} \frac{\sigma}{t}, \tag{10.52}$$

and numerical simulations [14] give the value

$$\mathcal{A} = 0.8 \pm 0.1 \tag{10.53}$$

for Z_2 domain walls.

The domain wall density falls off with scale factor as $\rho_{\text{wall}} \sim a^{-2}$ in the radiation era and as $\rho_{\text{wall}} \sim a^{-3/2}$, which means that it can dominate the density of the universe at a time when $\rho_{\text{wall}} = \rho_c$, where $\rho_c = 3M_p^2/4t^2$ is the critical density of the universe. The time for wall domination is therefore given by

$$t_{\text{dom}} = \frac{3M_P^2}{4\mathcal{A}\sigma} = 2.93 \times 10^3 \sec \left(\frac{\text{TeV}^3}{\sigma} \right). \tag{10.54}$$

In order that domain wall dominance does not occur before the BBN time of ~ 1 s, we must have $t_{\text{dom}} > 1$ s, which implies that $\sigma < 10^3$ TeV.

The most stringent bound comes from CMB, as domain walls would cause a density perturbation in the present universe of

$$\frac{\delta\rho}{\rho} = \frac{\rho_{\text{wall}}}{\rho_c} = \frac{4\mathcal{A}}{3M_P^2} \sigma t_0 = 10^{12} \left(\frac{\sigma}{\text{TeV}^3} \right), \tag{10.55}$$

and since CMB constraints $\delta\rho/\rho < 10^{-5}$, we have the bound [15]

$$\sigma < \text{MeV}^3. \tag{10.56}$$

To evade the problem of wall dominance, a Z_2 degeneracy breaking potential of the form Fig. 10.3 is introduced which gives rise to an energy density difference ΔV given in (10.50) between the two sides of the domain wall. This will cause an annihilation of the wall into PNGB and a part of the energy of the wall is released in the form of gravitational waves. The annihilation occurs when the pressure in the wall tension ρ_{wall} falls below the volume pressure given by ΔV. The time of annihilation is given by

$$t_a = C_a \frac{\mathcal{A}\sigma}{\Delta V}$$

$$= 6.58 \times 10^{-4} \sec C_a \mathcal{A}\left(\frac{\sigma}{\text{TeV}^3}\right)\left(\frac{\text{MeV}^4}{\Delta V}\right), \tag{10.57}$$

where from simulations $C_a \sim 5 - 10$ [16]. In the radiation era, this corresponds to an annihilation temperature

$$T_a = 3.41 \times 10^{-2} \, \text{GeV} \, C_a^{-1/2} \, \mathcal{A}^{-1/2} \left(\frac{10}{g_*}\right)^{1/4}\left(\frac{\text{TeV}^3}{\sigma}\right)^{1/2}\left(\frac{\Delta V}{\text{MeV}^4}\right)^{1/2}. \tag{10.58}$$

10.7 Gravitational Waves from Domain Wall Annihilation

The frequency of gravitational waves produced when the wall annihilates [17–20] will be $f \simeq H(t_a)$, which in the present epoch will red-shift to the value

$$f_{\text{peak}} = H(t_a)\frac{a(t_a)}{a(t_0)} = 1.1 \times 10^{-9} \, \text{Hz} \left(\frac{g_*(T_{\text{ann}})}{10}\right)^{1/6}\left(\frac{T_{\text{ann}}}{10^{-2}\text{GeV}}\right). \tag{10.59}$$

The gravitational wave energy spectrum from the annihilation of domain wall is

$$\left(\frac{d\rho_{gw}}{d\ln f}\right)_{\text{peak}} = \epsilon_{gw} G \mathcal{A}^2 \sigma^2, \tag{10.60}$$

where $\epsilon_{gw} = 0.7 \pm 0.4$ from numerical estimates [14]. The fractional energy in gravitational waves at the time of annihilation is

$$\Omega_{gw}(t_a, f_{\text{peak}}) \equiv \frac{1}{\rho_c(t_a)}\left(\frac{d\rho_{gw}}{d\ln f}\right)_{\text{peak}} = \frac{8\pi G}{3}\frac{\epsilon_{gw} G^2 \mathcal{A}^2 \sigma^2}{H(t_a)}. \tag{10.61}$$

The energy density of gravitational wave dilutes as $\rho_{gw} \propto a^{-4}$ with expansion and the fractional density of gravitational waves in the present epoch is

$$\Omega_{gw}(t_0, f)h^2 = 7.2 \times 10^{-18} \, \epsilon_{gw} \mathcal{A}^2 \left(\frac{g_*(T_a)}{10}\right)^{-4/3}$$

$$\times \left(\frac{\sigma}{\text{TeV}^3}\right)^2 \left(\frac{T_a}{10^{-2}\text{GeV}}\right)^{-4}\left(\frac{f}{f_{\text{peak}}}\right)^{n_{gw}}. \tag{10.62}$$

The spectral shape is a broken power law with $n_{gw} = 3$ for $f \leq f_{peak}$ and $n_{gw} = -1$ for $f > f_{peak}$.

10.8 Gravitational Radiation from Cosmic Strings

For the calculation of gravitational radiation, we can treat cosmic strings as one-dimensional objects characterised by the mass per unit length μ [21–23]. The location of any point of the string as a function of time is then a function (σ, t), where σ characterises the distance along the string length. A moving string is governed by the action

$$S = -\mu \int d\sigma \, dt \sqrt{-det(g_{\mu\nu}x^{\mu}_{,a}x^{\nu}_{,b})}, \tag{10.63}$$

where $g_{\mu\nu}$ is the spacetime metric and $x^{\mu}_{,a} = \{\partial x^{\mu}/\partial t, \partial x^{\mu}/\partial \sigma\}$. From string action (10.63), we obtain the stress tensor

$$T^{\mu\nu}(\mathbf{x}, t) = \mu \int d\sigma \, (\dot{x}^{\mu}\dot{x}^{\nu} - x^{\mu\prime}x^{\nu\prime})\delta^{3}(\mathbf{x} - \mathbf{x}(\sigma, t)), \tag{10.64}$$

where dots denote d/dt and primes denote $d/d\sigma$. The mass of the string is

$$M = \mu \int d\sigma . \tag{10.65}$$

Equations of motion for a string are

$$\ddot{\mathbf{x}} - \mathbf{x}'' = 0, \quad \dot{\mathbf{x}} \cdot \mathbf{x}' = 0, \quad \dot{\mathbf{x}}^2 - \mathbf{x}'^2 = 1. \tag{10.66}$$

Solutions of the e.o.m. can be written as

$$\mathbf{x}(\sigma, t) = \frac{1}{2}[\mathbf{a}(\sigma - t) + \mathbf{b}(\sigma + t)] . \tag{10.67}$$

We will consider gravitational radiation from closed strings of length L which have the boundary condition $\mathbf{x}(\sigma + L, t) = \mathbf{x}(\sigma, t)$. From the solution (10.67), it is clear that the boundary condition for closed strings can also be written as

$$\mathbf{x}\left(\sigma + \frac{L}{2}, t + \frac{L}{2}\right) = \mathbf{x}(\sigma, t). \tag{10.68}$$

The time period is therefore $L/2$ and the oscillations of the strings will be multiples of the fundamental frequency

$$\omega = \frac{4\pi}{L} . \tag{10.69}$$

The gravitational radiation rate per solid angle is given by

$$\frac{dE}{dt d\Omega} = \sum_{n=1}^{\infty} \frac{G\omega_n^2}{\pi} \left[T_{\mu\nu}^*(\omega_n, \mathbf{k}) T^{\mu\nu}(\omega_n, \mathbf{k}) - \frac{1}{2} |T_{\mu}^{\mu}(\omega_n, \mathbf{k}| \right], \qquad (10.70)$$

where $\omega_n = \omega n$ and $|\mathbf{k}| = \omega_n$ and $T^{\mu\nu}(\omega_n, \mathbf{k})$ is the Fourier transform of the stress tensor

$$T^{\mu\nu}(\omega_n, \mathbf{k}) = \frac{2}{L} \int_0^{L/2} dt\, e^{i\omega_n t} \int d^3 x\, e^{-\mathbf{k}\cdot\mathbf{x}} T_{\mu\nu}(\mathbf{x}, t). \qquad (10.71)$$

The specific solution of the string e.o.m. (10.66) is

$$\mathbf{x}(\sigma, t) = \frac{L}{2\pi} \left[\hat{e}_1 \sin\frac{2\pi\sigma}{L} \cos\frac{2\pi t}{L} + \hat{e}_2 \cos\frac{\phi}{2} \cos\frac{2\pi\sigma}{L} \cos\frac{2\pi t}{L} \right.$$

$$\left. + \hat{e}_3 \sin\frac{\phi}{2} \sin\frac{2\pi\sigma}{L} \sin\frac{2\pi t}{L} \right], \qquad (10.72)$$

where ϕ is a constant parameter.

Using the solution (10.72) for string oscillations in the stress tensor expression (10.64), we obtain the components of the stress tensor in Fourier space (10.71)

$$T_{11} = \mu \left(I_+ + I_- + 2I \cos 2\beta \right),$$

$$T_{22} = \mu \cos^2\frac{\phi}{2} \left(I_+ + I_- - 2I \cos 2\beta \right),$$

$$T_{33} = 2\mu \sin 2\beta \cos\frac{\phi}{2} I,$$

$$T_{13} = T_{23} = 0, \qquad (10.73)$$

where

$$I = \frac{1}{\pi} (-1)^n J_{n-1}(na) J_{n+1}(na),$$

$$I_{\pm} = \frac{\pi}{2} (-1)^n J_{n\pm 1}^2(na),$$

$$a = \left[1 - \sin^2\frac{\phi}{2} \sin^2(\hat{k} \cdot \hat{e}_1) \right]^{1/2},$$

$$\cos\beta = a^{-1} \cos(\hat{k} \cdot \hat{e}_1). \qquad (10.74)$$

The other components of the stress tensor can be evaluated using the stress tensor conservation equations, $\omega_n T_{0i} = -k^j T_{ij}$ and $\omega_n^2 T_{00} = k^i k^j T_{ij}$.

Using (10.73) in (10.70), the rate of energy loss per solid angle is given by

$$\frac{dE}{dt d\Omega} = \sum_{n=1}^{\infty} 8\pi G\mu^2 n^2 \left[J_n'(na) \right]^2 . \tag{10.75}$$

Integrating over angle and performing the sum numerically, we have n expression for the energy radiated by strings in gravitational waves given by

$$\frac{dE}{dt} = \gamma G\mu^2, \tag{10.76}$$

where $\gamma \simeq 50$.

References

1. B. Allen, The Stochastic gravity wave background: sources and detection. [arXiv:gr-qc/9604033 [gr-qc]]
2. B. Allen, J.D. Romano, Detecting a stochastic background of gravitational radiation: Signal processing strategies and sensitivities. Phys. Rev. D **59**, 102001 (1999). https://doi.org/10.1103/PhysRevD.59.102001. [arXiv:gr-qc/9710117 [gr-qc]]
3. E.E. Flanagan, The Sensitivity of the laser interferometer gravitational wave observatory (LIGO) to a stochastic background, and its dependence on the detector orientations. Phys. Rev. D **48**, 2389–2407 (1993). https://doi.org/10.1103/PhysRevD.48.2389. [arXiv:astro-ph/9305029 [astro-ph]]
4. M. Maggiore, Gravitational wave experiments and early universe cosmology. Phys. Rept. **331**, 283–367 (2000). [arXiv:gr-qc/9909001 [gr-qc]]
5. M. Maggiore, Stochastic backgrounds of gravitational waves. ICTP Lect. Notes Ser. **3**, 397–414 (2001). [arXiv:gr-qc/0008027 [gr-qc]]
6. S. Weinberg, *Gravitation and Cosmology* (Wiley, New York, 1972)
7. J. Chakrabortty, R. Maji, S.K. Patra, T. Srivastava, S. Mohanty, Roadmap of left-right models based on GUTs. Phys. Rev. D **97**(9), 095010 (2018). [arXiv:1711.11391 [hep-ph]]
8. E.J. Chun, L. Velasco-Sevilla, Tracking Down the Route to the SM with Inflation and Gravitational Waves. [arXiv:2112.14483 [hep-ph]]
9. R.N. Mohapatra, J.C. Pati, A natural left-right symmetry. Phys. Rev. D **11**, 2558 (1975). https://doi.org/10.1103/PhysRevD.11.2558
10. A. Vilenkin, E.P.S. Shellard, *Cosmic Strings and Other Topological Defects* (Cambridge University Press, Cambridge, 1994)
11. P. Sikivie, Of axions, domain walls and the early universe. Phys. Rev. Lett. **48**, 1156–1159 (1982)
12. T. Hiramatsu, M. Kawasaki, K. Saikawa, Evolution of string-wall networks and axionic domain wall problem. J. Cosmol. Astropart. Phys. **08**, 030 (2011). [arXiv:1012.4558 [astro-ph.CO]]
13. M. Hindmarsh, Analytic scaling solutions for cosmic domain walls. Phys. Rev. Lett. **77**, 4495–4498 (1996). https://doi.org/10.1103/PhysRevLett.77.4495. [arXiv:hep-ph/9605332 [hep-ph]]
14. T. Hiramatsu, M. Kawasaki, K. Saikawa, On the estimation of gravitational wave spectrum from cosmic domain walls. J. Cosmol. Astropart. Phys. **02**, 031 (2014). https://doi.org/10.1088/1475-7516/2014/02/031. [arXiv:1309.5001 [astro-ph.CO]]
15. Y.B. Zeldovich, I.Y. Kobzarev, L.B. Okun, Cosmological consequences of the spontaneous breakdown of discrete symmetry. Zh. Eksp. Teor. Fiz. **67**, 3–11 (1974)
16. M. Kawasaki, K. Saikawa, T. Sekiguchi, Axion dark matter from topological defects. Phys. Rev. D **91**(6), 065014 (2015). [arXiv:1412.0789 [hep-ph]]

17. K. Saikawa, A review of gravitational waves from cosmic domain walls. Universe **3**(2), 40 (2017). https://doi.org/10.3390/universe3020040. [arXiv:1703.02576 [hep-ph]]

18. M. Gleiser, R. Roberts, Gravitational waves from collapsing vacuum domains. Phys. Rev. Lett. **81**, 5497–5500 (1998). [arXiv:astro-ph/9807260 [astro-ph]]

19. T. Hiramatsu, M. Kawasaki, K. Saikawa, Gravitational waves from collapsing domain walls. J. Cosmol. Astropart. Phys. **05**, 032 (2010). https://doi.org/10.1088/1475-7516/2010/05/032. [arXiv:1002.1555 [astro-ph.CO]]

20. K. Kadota, M. Kawasaki, K. Saikawa, Gravitational waves from domain walls in the next-to-minimal supersymmetric standard model. J. Cosmol. Astropart. Phys. **10**, 041 (2015). https://doi.org/10.1088/1475-7516/2015/10/041. [arXiv:1503.06998 [hep-ph]]

21. T. Vachaspati, A. Vilenkin, Gravitational radiation from cosmic strings. Phys. Rev. D **31**(12), 3052–3058 (1985)

22. C.J. Burden, Gravitational radiation from a particular class of cosmic strings. Phys. Lett. B **164**, 277–281 (1985)

23. D. Garfinkle, T. Vachaspati, Radiation from kinky, cuspless cosmic loops. Phys. Rev. D **36**, 2229 (1987)

Inflation

11

Abstract

Inflation is the period of exponential expansion of the universe, which gives a natural explanation for the observed flatness and solves the problem of how regions of the universe which were not in causal contact, share the same anisotropy pattern of the CMB. Inflation also generates density fluctuations which seeds galaxy formation and generates primordial gravitational waves. In this chapter we give a survey inflation and discuss the tensor modes generated at different epochs. The primordial tensor modes are predicted to show up as B-mode polarisations of the CMB. Secondary gravitational waves generated by the scalars perturbation during inflation and during the scalar field oscillations in reheating may be seen as stochastic gravitational waves by Pulsar timing arrays. We study these issues in this chapter and predict the gravitational wave signals which can arise in inflationary era. We also study inflationary perturbations in the non-equilibrium in-in formalism which can answer questions like how primordial zero-point fluctuations become classical density perturbations.

11.1 Inflation

Inflation is the cosmological theory [1–6] where the universe goes through a period of exponential expansion. An e^N fold expansion of the scale factor with $N = 50-70$ can solve the puzzle of the observed homogeneity -why the Cosmic Microwave Background anisotropy is $\Delta T/T \sim 10^{-5}$ over regions in the present universe which were never in causal contact in the standard hot big bang era of the universe. The superhorizon anisotropy in the CMB was first observed by COBE [7] and confirmed with increasing precision by WMAP [8] and Planck [9] experiments.

Inflation also gives a natural explanation of the observed flatness $\Omega_k = 0.0007 \pm 0.0037(95\%\text{CL})$ [10] even though in the standard hot big bang cosmology, Ω_k

© The Author(s), under exclusive license to Springer Nature Switzerland AG 2023
S. Mohanty, *Gravitational Waves from a Quantum Field Theory Perspective*,
Lecture Notes in Physics 1013, https://doi.org/10.1007/978-3-031-23770-6_11

grows with the scale factor as $\Omega_k \propto a^2$ during the radiation era and $\Omega_k \propto a$ during the matter domination.

In addition inflation also provides an explanation of the observed large scale structure in the universe as it provides the primordial scale invariant perturbations [11] which under gravitational collapse to form galaxies and clusters of galaxies [12–15].

In the theory of inflation, the size of perturbations grow faster than the size of the horizon in the inflationary stage of the universe. After inflation ends there is a phase of reheating and then the standard hot big bang cosmology of radiation and matter dominated era follows. In the radiation and matter era the horizon grows faster than the length scale of perturbations and they *perturbations re-enter the horizon*. This explains the presence of superhorizon perturbations first observed by COBE. The amplitude of the (adiabatic) perturbations remain the same and the large scale perturbations we observe in the CMB are primordial—generated during inflation and unaffected by any sub-horizon causal physics.

To obtain perturbations of super horizon length scales, the wavelength of perturbations $\lambda = \lambda_0 a$ during an early era must become larger than the Horizon $R_h \simeq H^{-1} = a/\dot{a}$. This implies that in some early era we require

$$\frac{d}{dt}\left(\frac{\lambda_0 a}{H^{-1}}\right) > 0 \Rightarrow \ddot{a} > 0, \tag{11.1}$$

i.e. there must have been a period of positive acceleration. The Friedman equation $\ddot{a} = -(4\pi G/3)(\rho + 3p)$ then implies that in some period in the past the universe had a dominant component with negative pressure $p < -\frac{1}{3}\rho$. If $p/\rho \simeq -1$ then the universe goes through a period of exponential expansion $a \sim e^{Ht}$. One way to get a negative pressure is by having a scalar field potential. The energy momentum tensor of the scalar field is

$$T_{\mu\nu}(\phi) = \partial_\mu\phi\partial_\nu\phi - g_{\mu\nu}\left[\frac{1}{2}\partial^\rho\phi\partial_\rho\phi + V(\phi)\right]. \tag{11.2}$$

For the homogeneous background field, the energy momentum tensor takes the form of a perfect fluid with energy density and pressure for scalar fields given by

$$\rho_\phi = \frac{\dot{\phi}^2}{2} + V(\phi), \tag{11.3}$$

$$p_\phi = \frac{\dot{\phi}^2}{2} - V(\phi). \tag{11.4}$$

The resulting equation of state is

$$\omega_\phi \equiv \frac{p_\phi}{\rho_\phi} = \frac{\frac{\dot{\phi}^2}{2} - V(\phi)}{\frac{\dot{\phi}^2}{2} + V(\phi)}. \tag{11.5}$$

If the potential energy of the field dominates over its kinetic energy i.e. $\dot{\phi}^2 <<$ $V(\phi)$, then the above simple relation (11.5) implies that the scalar field can act as a negative pressure source i.e. $\omega_\phi < 0$ and can provide accelerated expansion i.e. $\omega_\phi < -\frac{1}{3}$. The Friedmann equation and the equation of motion of the scalar field are respectively

$$H^2 = \frac{8\pi G}{3} \left(\frac{\dot{\phi}^2}{2} + V(\phi) \right), \qquad (11.6)$$

$$\ddot{\phi} + 3H\dot{\phi} + V'(\phi) = 0. \qquad (11.7)$$

These equations determine the dynamics of the scale factor and scalar field in a FRW universe.

Slow-roll inflation occurs when $\dot{\phi}^2 \ll V(\phi)$. A second order differentiation of the condition $\dot{\phi}^2 \ll V(\phi)$ implies $\ddot{\phi} \ll V'(\phi)$. In the slow-roll approximation equation (11.6) becomes

$$3H^2 \simeq \frac{8\pi}{M_{pl}^2} V(\phi), \qquad (11.8)$$

and (11.7) becomes

$$3H\dot{\phi} \simeq -V'(\phi). \qquad (11.9)$$

By differentiating equation (11.8) w.r.t. time and combining the result with Eq. (11.9) we obtain

$$\dot{H} \simeq -\frac{4\pi}{M_{pl}^2} \dot{\phi}^2. \qquad (11.10)$$

The slow-roll conditions $\dot{\phi}^2 \ll V(\phi)$ and $\ddot{\phi} \ll V'(\phi)$ can be defined in terms of dimensionless parameters as

$$\epsilon = -\frac{\dot{H}}{H^2} \simeq \frac{4\pi}{M_{pl}^2} \left(\frac{\dot{\phi}}{H} \right)^2 \simeq \frac{M_{pl}^2}{16\pi} \left(\frac{V'(\phi)}{V(\phi)} \right)^2, \qquad (11.11)$$

$$\eta = -\frac{\ddot{\phi}}{H\dot{\phi}} \simeq \frac{M_{pl}^2}{8\pi} \left[\frac{V''(\phi)}{V(\phi)} - \frac{1}{2} \left(\frac{V'(\phi)}{V(\phi)} \right)^2 \right]. \qquad (11.12)$$

After the slow-roll phase $\epsilon \ll 1$, the period of Inflation ends when $\dot{\phi}^2 \approx V(\phi)$ and $\epsilon(\phi_e) \simeq 1$.

It is convenient to define the slow-roll parameters in terms of the potential

$$\epsilon_V \equiv \frac{M_{pl}^2}{16\pi}\left(\frac{V'}{V}\right)^2 = \epsilon,$$

$$\eta_V \equiv \frac{M_{pl}^2}{8\pi}\left(\frac{V''}{V}\right) = \eta + \epsilon. \tag{11.13}$$

During the slow-roll phase $V(\phi)$ is nearly constant and solving (11.8), we see that during this period scale factor evolves exponentially $a(t) \sim e^{Ht}$. The number of e-foldings before the inflation ends $\epsilon(\phi_e) = 1$ is given by

$$N(\phi) = \int_{t_i}^{t_e} H dt = \int_{\phi_i}^{\phi_e} \frac{H}{\dot{\phi}} d\phi$$

$$\simeq \frac{8\pi}{M_{pl}^2}\int_{\phi_e}^{\phi_i} \frac{V}{V'} d\phi = \frac{2\sqrt{\pi}}{M_{pl}}\int_{\phi_e}^{\phi_i} \frac{d\phi}{\sqrt{\epsilon}}, \tag{11.14}$$

where we used the slow-roll equations (11.9) and (11.8). To solve the horizon and flatness problems we required the number of e-foldings during inflation to exceed 60

$$N_{tot} \equiv \ln\frac{a_e}{a_i} \gtrsim 40 - 60. \tag{11.15}$$

The value of N_{tot} depends on the inflation potential. The slow-roll phase, should last a minimum $N = 40 - 60$ e-folds to solve the horizon and flatness problems (there is nonphenomenological limit on the maximum duration of inflation). During the slow-roll phase the quantum fluctuations in density and pressure of the scalar field called the inflaton are generated which are then imprinted on the CMB. In the CMB anisotropy measurements we probe the last 8 e-foldings before the end of inflation.

11.2   Perturbations of de-Sitter background

Linear order perturbations in the metric and field around the homogeneous background solutions of the field $\phi(t)$ and the metric $g_{\mu\nu}(t)$ can be written as

$$\delta\phi(t, \mathbf{x}) \equiv \phi(t, \mathbf{x}) - \bar{\phi}(t), \tag{11.16}$$

$$\delta g_{\mu\nu}(t, \mathbf{x}) \equiv g_{\mu\nu}(t, \mathbf{x}) - \bar{g}_{\mu\nu}(t). \tag{11.17}$$

The perturbed spatially flat FRW in the conformal Newtonian gauge can be written as

$$ds^2 = -(1 + 2\Psi)dt^2 + a(t)^2\left[(1 - 2\Phi)\delta_{ij} + h_{ij}\right]dx^i dx^j, \tag{11.18}$$

where Ψ, Φ are the scalar perturbations and h_{ij} are the tensor perturbations. In the linear order in perturbations, the scalar, vector and tensor perturbations are decoupled during inflation and thereafter evolve independently (*SVT decomposition theorem* [13]). If, however, higher order terms are retained then as we will study in this chapter scalar perturbations at second order can source tensor perturbations.

One can make use of the general coordinate transformation of GR to further restrict the dynamical d.o.f in the metric perturbations.

An important gauge-invariant observable is the comoving curvature perturbation introduced in [16]

$$\mathcal{R} \equiv \Phi - \frac{H}{\rho + p}\delta q. \tag{11.19}$$

The scalar momentum perturbation δq is the $0i$−component of the perturbed energy momentum tensor $\delta T_i^0 = \partial_i \delta q$. During inflation $\delta T_i^0 = -\dot{\phi}\partial_i \delta\phi$, comparing these two relations we see that $\delta q = -\dot{\phi}\delta\phi$. The background density and pressure of the scalar field obey the relation $\rho + p = \dot{\phi}^2$. Therefore the comoving curvature perturbations (11.19) during inflation becomes

$$\mathcal{R} \simeq \Phi + \frac{H}{\dot{\phi}}\delta\phi. \tag{11.20}$$

Geometrical interpretation of $\mathcal{R}$ is that it measures the spatial curvature of the comoving hypersurface where $\delta\phi = 0$, i.e.

$$\mathcal{R} = \Phi\Big|_{\delta\phi=0}. \tag{11.21}$$

Another important gauge-invariant quantity is curvature perturbations on constant energy density hypersurfaces defined as

$$\zeta \equiv \Phi + \frac{H}{\dot{\rho}}\delta\rho. \tag{11.22}$$

During slow-roll (from Eq. (11.3)) one has

$$\delta\rho = \dot{\phi}\delta\dot{\phi} + V'\delta\phi \simeq V'\delta\phi$$

and

$$\dot{\rho} = \dot{\phi}\ddot{\phi} + V'\dot{\phi} \simeq V'\dot{\phi}.$$

These equations imply

$$\frac{\delta\rho}{\dot{\rho}} \simeq \frac{\delta\phi}{\dot{\phi}}.$$

As a consequence, ζ becomes

$$\zeta \simeq \Phi + \frac{H}{\dot{\phi}} \delta\phi. \tag{11.23}$$

The geometrical interpretation of ζ is that it is the spatial curvature of the uniform density hypersurface, i.e.

$$\zeta = \Phi \Big|_{\delta\rho=0}. \tag{11.24}$$

In a single field slow-roll inflation models the perturbations produced are adiabatic. Non-adiabatic perturbations can arise in multi-field inflation models [17] and are defined as the isocurvature entropy perturbation

$$\Gamma \equiv \frac{\delta p}{\dot{p}} - \frac{\delta\rho}{\dot{\rho}}, \tag{11.25}$$

which is gauge invariant. The non-adiabatic part of the pressure perturbations are defined as

$$\delta p_{nad} \equiv \dot{p}\Gamma \equiv \delta p - \frac{\dot{p}}{\dot{\rho}}\delta\rho. \tag{11.26}$$

In inflationary models with more than one field, some perturbations can be non-adiabatic [17] and can have observable signatures in the CMB anisotropy. For example, the relative density perturbations (isocurvature or entropy perturbations) between photon and CDM that can be defined as

$$S_{m\gamma} \equiv \frac{\delta\rho_{cdm}}{\rho_{cdm}} - \frac{3}{4}\frac{\delta\rho_\gamma}{\rho_\gamma} \tag{11.27}$$

can have a observable signature in the CMB spectrum. CMB observations suggest that if the isocurvature perturbations are present, their amplitude is small compared to amplitude of the adiabatic (curvature) perturbations [9].

The time evolution of the gauge-invariant curvature perturbations is given by

$$\dot{\mathcal{R}} = -\frac{H}{\rho + p}\delta p_{nad} + \left(\frac{k}{aH}\right)^2 \left[\frac{H^2}{3(\rho + p)}\delta q\right], \tag{11.28}$$

therefore if there are no non-adiabatic matter perturbations $\delta p_{nad} = 0$ or no isocurvature perturbations $\Gamma = 0$, the curvature perturbations $\mathcal{R}$ are conserved on superhorizon scales $k \ll aH$. This ensures that when adiabatic perturbations exit the horizon during inflation and re-enter during the radiation or matter era, their amplitude remains the same.

11.3 Curvature Perturbation During Inflation

For a metric with small perturbations, the Einstein tensor $G_{\nu\mu} = R_{\mu\nu} - (1/2)g_{\mu\nu}R$ can be written as

$$G_{\nu\mu} = \bar{G}_{\nu\mu} + \delta G^{(1)}_{\nu\mu} + \delta G^{(2)}_{\nu\mu}....,$$

where $\delta G^{(1)}_{\nu\mu}$ and $\delta G^{(2)}_{\nu\mu}$ represent the terms to the linear metric quadratic orders respectively of perturbations $\delta g_{\mu\nu}$. The stress energy tensor T^{ν}_{μ} can be expanded in terms density and pressure perturbations and we can write perturbations in Einstein field equations as,

$$\delta G^{\nu}_{\mu} = 8\pi G\, \delta T^{\nu}_{\mu}. \tag{11.29}$$

In conformal Newtonian gauge the linear order perturbation of FRW line element is

$$ds^2 = -(1 + 2\Psi)dt^2 + a(t)^2\left[(1 - 2\Phi)\delta_{ij} + h_{ij}\right]dx^i dx^j, \tag{11.30}$$

and in this metric the components of the perturbed Einstein tensor are,

$$\delta G^0_0 = -2\nabla^2\Phi + 6H^2\Psi + 6H\dot{\Phi}, \tag{11.31}$$

$$\delta G^0_i = -2\partial_i(H\Psi + \dot{\Phi}), \tag{11.32}$$

$$\delta G^i_j = \partial^i\partial_j(\Phi - \Psi) + [\nabla^2(\Psi - \Phi) + 2\ddot{\Phi} + (4\dot{H} + 6H^2)\Psi + H(2\dot{\Psi} + 2\dot{\Phi})]\delta^i_j. \tag{11.33}$$

The stress-energy–momentum tensor for the scalar field ϕ as defined in Eq. (11.2), the components of the perturbed $T_{\mu\nu}$ are given by

$$\delta T^0_0 = -\delta\rho = \dot{\phi}^2\Psi - \dot{\phi}\delta\dot{\phi} - V'\delta\phi, \tag{11.34}$$

$$\delta T^0_i = \delta q = -\dot{\phi}\partial_i\delta\phi, \tag{11.35}$$

$$\delta T^i_j = \delta p = [-\dot{\phi}^2\Psi + \dot{\phi}\delta\dot{\phi} - V'\delta\phi]\delta^i_j, \tag{11.36}$$

where we have used the relation

$$\delta T^{\nu}_{\mu} = \delta(g^{\nu\tau}T_{\mu\tau}) = \delta g^{\nu\tau}T_{\mu\tau} + g^{\nu\tau}\delta T_{\mu\tau}.$$

To compute the curvature perturbation $\mathcal{R}$, one first considers the ij-component of the perturbed Einstein field Eq. (11.29). First taking the off-diagonal components, $i \neq j$, of the Eqs. (11.33) and (11.36), we have

$$\partial^i\partial_j(\Phi - \Psi) = 0 \quad \Longrightarrow \quad \Phi = \Psi. \tag{11.37}$$

Now consider the diagonal components, $i = j$, of Eqs. (11.33) and (11.36),

$$\ddot{\Phi} + 4H\dot{\Phi} + (2\dot{H} + 3H^2)\Phi = -\dot{\phi}^2\Psi + \dot{\phi}\dot{\delta\phi} - V'\delta\phi. \tag{11.38}$$

Since ϕ is background quantity, which is only time dependent, Eqs. (11.32) and (11.35) for $0i$-components give

$$\dot{\Phi} + H\Phi = 4\pi G \dot{\phi}\delta\phi = \epsilon H^2 \frac{\delta\phi}{\dot{\phi}}, \tag{11.39}$$

where $\epsilon = 4\pi G \frac{\dot{\phi}^2}{H^2}$ (slow-roll parameter). Similarly the Eqs. (11.31) and (11.34) for 00-component gives

$$\nabla^2\Phi - 3H\dot{\Phi} - 3H^2\Phi = 4\pi G(\dot{\phi}\dot{\delta\phi} - \dot{\phi}^2\Phi + V'\delta\phi). \tag{11.40}$$

To describe the spatial distribution of the perturbations it is convenient to work in terms of the Fourier decomposition of the metric and the field perturbations, which relates perturbation corresponding to a given comoving wavenumber k in Fourier space with corresponding comoving wavelength $\lambda = \frac{2\pi}{k}$. The physical wavelength is $\lambda_{phy} = a(t)\lambda$ and the physical wavenumber of the perturbations is $k_{phy} = k/a(t)$.

We can express the perturbations Ψ as a sum of Fourier modes as comoving wavenumber $\mathbf{k}$:

$$\Phi(t, \mathbf{x}) = \int \frac{d^3\mathbf{k}}{(2\pi)^{3/2}} \Phi_{\mathbf{k}}(t)e^{i\mathbf{k}.\mathbf{x}}. \tag{11.41}$$

With a similar expression for the Fourier transformation of $\delta\phi$.

We now add the Fourier transforms of Eqs. (11.38) and (11.40) to arrive at the equation of motion of gravitational potential Φ_k as

$$\ddot{\Phi}_k + \left(H - 2\frac{\ddot{\phi}}{\dot{\phi}}\right)\dot{\Phi}_k + 2\left(\dot{H} - H\frac{\ddot{\phi}}{\dot{\phi}}\right)\Phi_k + \frac{k^2}{a}\Phi_k = 0, \tag{11.42}$$

where we have used the background equation for scalar field $V' \simeq -3H\dot{\phi}$ and the relation $\dot{H} \simeq -4\pi G \dot{\phi}^2$. Using the slow-roll parameter relation

$$\delta = \eta - \epsilon = \frac{-\ddot{\phi}}{H\dot{\phi}},$$

the Eq. (11.42) can also be written as

$$\ddot{\Phi}_k + H\left(1 - 2\epsilon + 2\eta\right)\dot{\Phi}_k + 2H^2\left(\eta - 2\epsilon\right)\Phi_k + \frac{k^2}{a}\Phi_k = 0. \tag{11.43}$$

Since the slow-roll parameters satisfy $\epsilon \ll 1$ and $\eta \ll 1$, we see from Eq. (11.43) that on superhorizon scales $k \ll (aH)$,

$$\dot{\Phi}_k \simeq 2(2\epsilon - \eta)H\Psi_k \quad \Rightarrow \quad \dot{\Psi}_k \ll H\Psi_k \qquad (11.44)$$

which implies that on superhorizon scales the time variations of the perturbations Ψ_k can be neglected compared to $H\Psi_k$. This condition also holds true for field perturbations $\delta\dot{\phi}_k \ll H\delta\phi_k$. Therefore on superhorizon scales, from Eq. (11.39), we can relate the gravitational potential and field perturbations as

$$\Phi_k \simeq \epsilon H \frac{\delta\phi}{\dot{\phi}}. \qquad (11.45)$$

This can be used to compute the comoving curvature perturbation $\mathcal{R}_k$ on superhorizon scale (11.20) as

$$\mathcal{R}_k \simeq \Phi_k + \frac{H}{\dot{\phi}}\delta\phi_k \simeq (1+\epsilon)\frac{H}{\dot{\phi}}\delta\phi_k \approx \frac{H}{\dot{\phi}}\delta\phi_k .$$

The comoving curvature perturbations are related to the scalar field perturbations whose origin is quantum fluctuations as we will discuss in the following sections.

11.4 Power Spectrum of Curvature Perturbation

The power spectrum of comoving curvature perturbation $\mathcal{R}$ is defined as

$$\mathcal{P}_{\mathcal{R}}(k) = \frac{k^3}{2\pi^2}\langle|\mathcal{R}_k|^2\rangle , \qquad (11.46)$$

where $\langle|\mathcal{R}_k|^2\rangle$ denotes the statistical average of the quantity $|\mathcal{R}_k|^2$.

Using (11.3), the power spectrum of comoving curvature perturbation on superhorizon scale $k \ll (aH)$ becomes

$$\mathcal{P}_{\mathcal{R}}(k) \simeq \frac{k^3}{2\pi^2}\frac{H^2}{\dot{\phi}^2}\langle|\delta\phi_k|^2\rangle \qquad (11.47)$$

$$\simeq \frac{k^3}{4\pi^2\epsilon M_{pl}^2}\langle|\delta\phi_k|^2\rangle ,$$

where $\langle|\delta\phi_k|^2\rangle$ stands for the expectation value of the quantum fluctuations of the inflaton $|\delta\phi_k|^2$ in the vacuum quantum state $|0\rangle$.

11.5   Quantum Fluctuations of the Inflaton

We now calculate the scalar field perturbation mode amplitudes $\delta\phi_k$. Consider the perturbation of (11.7) for scalar field ϕ

$$\delta\ddot{\phi}_k + 3H\delta\dot{\phi}_k + \frac{k^2}{a^2}\delta\phi_k + V''\delta\phi_k = -2V'\Psi_k + 4\dot{\phi}\dot{\Psi}_k, \tag{11.48}$$

where we have used the background Eq. (11.9). Since on superhorizon scales $|2V'\Psi_k| \gg |4\dot{\phi}\dot{\Psi}_k|$ (which follows from the condition $\dot{\Psi}_k \ll H\Psi_k$ upon using the relation $V' \simeq -3H\dot{\phi}$). From Eq. (11.45) and (11.9), the e.o.m or the scalar field is

$$\delta\ddot{\phi}_k + 3H\delta\dot{\phi}_k + (V'' + 6\epsilon H^2)\delta\phi_k + \frac{k^2}{a^2}\delta\phi_k = 0. \tag{11.49}$$

Substituting $\delta\phi_k = \sigma_k/a$ and introducing the conformal time[1] $d\tau = dt/a$, we get

$$\sigma_k'' - \frac{1}{\tau^2}\left(v^2 - \frac{1}{4}\right)\sigma_k + k^2\delta\sigma_k = 0, \tag{11.50}$$

where prime denotes the derivatives w.r.t. conformal time τ and

$$v^2 = \frac{9}{4} + 9\epsilon_V - 3\eta_V, \tag{11.51}$$

and we have used the relation

$$\frac{a''}{a} = \frac{1}{\tau^2}\left(v^2 - \frac{1}{4}\right) \simeq \frac{1}{\tau^2}(2 + 3\epsilon).$$

To compute the zero-point fluctuation of the fields we treat $\sigma(k)$ as the mode functions of the quantum field

$$\hat{\sigma}(x, \tau) = \int \frac{d^3k}{(2\pi)^{3/2}}\left(\sigma_k a_k e^{-i\mathbf{k}\cdot\mathbf{x}} + \sigma_k^* a_k^\dagger e^{i\mathbf{k}\cdot\mathbf{x}}\right), \tag{11.52}$$

where $a_k^\dagger$ and a_k are creation and annihilation operators which satisfy commutation relation

$$[a_k, a_{k'}] = \delta^3(\mathbf{k} - \mathbf{k'}). \tag{11.53}$$

---

[1] In quasi de-Sitter expansion the Hubble rate is not a constant and $\dot{H} = -\epsilon H^2$, the scale factor goes as $a(\tau) = -\frac{1}{H\tau}\frac{1}{1-\epsilon}$.

The vacuum fluctuation of $\hat{\sigma}$ is given by

$$\langle 0|\hat{\sigma}(\tau,0)\hat{\sigma}^\dagger(\tau,0)|0\rangle = \int \frac{d^3k}{(2\pi)^{3/2}} \int \frac{d^3k'}{(2\pi)^{3/2}} \sigma_k \sigma_k^* \langle 0|[a_k, a_{k'}^\dagger]|0\rangle$$

$$= \int d\ln k \frac{k^3}{2\pi^2} |\sigma_k|^2. \tag{11.54}$$

The solution of scalar field perturbation Eq. (11.50) can be written as Hankel functions

$$\sigma_k = \sqrt{-\tau}\left[c_1(k)H_\nu^{(1)}(-k\tau) + c_2(k)H_\nu^{(2)}(-k\tau)\right], \tag{11.55}$$

where $H_\nu^{(1)}$ and $H_\nu^{(2)}$ are the Hankel's functions of the first and second kind, respectively. We follow the *Bunch-Davies initial condition* [18, 19] that in the ultraviolet regime, $k \gg aH$ $(-k\tau \gg 1)$ the solutions goes to the positive energy plane wave solution $\sigma_k(\tau) \sim e^{-ik\tau}/\sqrt{2k}$. In the limit $-k\tau \gg 1$ Hankel's functions are given by

$$H_\nu^{(1)}(-k\tau \gg 1) \sim \sqrt{\frac{2}{-k\tau\pi}} e^{i\left(-k\tau-\frac{\pi}{2}\nu-\frac{\pi}{4}\right)}, \tag{11.56}$$

$$H_\nu^{(2)}(-k\tau \gg 1) \sim \sqrt{\frac{2}{-k\tau\pi}} e^{i\left(-k\tau-\frac{\pi}{2}\nu-\frac{\pi}{4}\right)}. \tag{11.57}$$

Imposing the Bunch-Davies initial condition we have

$$c_1(k) = \frac{\sqrt{\pi}}{2} e^{i\left(\nu+\frac{1}{2}\right)\frac{\pi}{2}}, \qquad c_2(k) = 0,$$

from Eq. (11.55) we get the exact solution for $\delta\sigma_k$

$$\sigma_k = \frac{\sqrt{\pi}}{2} e^{i\left(\nu+\frac{1}{2}\right)\frac{\pi}{2}} \sqrt{-\tau} H_\nu^{(1)}(-k\tau). \tag{11.58}$$

As we are interested in the modes which have become superhorizon $k \ll aH$ $(-k\tau \ll 1)$ during inflation, knowing that in the limit $-k\tau \ll 1$ Hankel's function have solution

$$H_\nu^{(1)}(-k\tau \ll 1) \sim \sqrt{\frac{2}{\pi}} \frac{\Gamma(\nu)}{\Gamma(3/2)} 2^{\nu-\frac{3}{2}} e^{-i\frac{\pi}{2}} (-k\tau)^{-\nu}, \tag{11.59}$$

the solution (11.58) on superhorizon scales becomes

$$\sigma_k \simeq \frac{\Gamma(\nu)}{\Gamma(3/2)} 2^{\nu-\frac{3}{2}} e^{i\left(\nu+\frac{1}{2}\right)\frac{\pi}{2}} \frac{1}{\sqrt{2k}} (-k\tau)^{\frac{1}{2}-\nu}. \tag{11.60}$$

Since $\epsilon \ll 1$ and $\eta \ll 1$, we can set $\nu \sim \frac{3}{2}$ in the factors but will not do the same in the exponent because exponent term $(-k\tau)^{\frac{1}{2}-\nu}$ gives the small scale dependence of the power spectrum of perturbations. Going back to original variable $\delta\hat{\phi}_k = \hat{\sigma}_k/a$, we find the fluctuations on superhorizon scales in cosmic time

$$|\delta\hat{\phi}_k(t)|^2 \simeq \frac{H^2}{2k^3}\left(\frac{k}{aH}\right)^{3-2\nu}. \tag{11.61}$$

11.6   Scalar Power Spectrum

The power spectrum of scalar fluctuations (11.47) can therefore be written as,

$$\mathcal{P}_{\mathcal{R}}(k) \simeq \frac{1}{8\pi^2\epsilon}\frac{H^2}{M_p^2}\left(\frac{k}{aH}\right)^{3-2\nu} \tag{11.62}$$

$$\equiv \Delta_{\mathcal{R}}^2\left(\frac{k}{aH}\right)^{n_s-1},$$

where we have defined the spectral index n_s of the comoving curvature perturbations, which determines the tilt of the power spectrum as

$$n_s - 1 \equiv \frac{d\ln\mathcal{P}_{\mathcal{R}}}{d\ln k} \tag{11.63}$$

$$= 3 - 2\nu = -6\epsilon_V + 2\eta_V,$$

where we have used ν from Eq. (11.51). Since the slow-roll parameters ϵ_V and η_V are much smaller than unity, therefore $n_s - 1 \simeq 0$ which implies that in slow-roll inflation curvature perturbations are generated with an almost scale invariant spectrum. For comparison with the observations, the power spectrum (11.62) can be given as

$$\mathcal{P}_{\mathcal{R}}(k) = \Delta_{\mathcal{R}}^2(k_0)\left(\frac{k}{k_0}\right)^{n_s-1}, \tag{11.64}$$

where $k_0 = a_0H_0$ is the pivot scale. The pivot scale corresponds to a wavelength $\lambda_0 = 2\pi/k_0^{-1}$ at which the CMB anisotropy measurement has the maximum sensitivity. $\Delta_{\mathcal{R}}^2(k_0)$ is the amplitude of the power spectrum at the pivot scale k_0.

11.7 Tensor Power Spectrum

Along with density fluctuations (or scalar perturbations), inflation also predicts the existence of gravitational waves which are identified with the tensor perturbations in the metric [20–24]. The equation of motion for h_{ij} can be obtained from second order expansion of the Einstein–Hilbert action [12, 15]

$$S^{(2)} = \frac{M_p^2}{2} \int dx^4 \sqrt{-g} \frac{1}{2} \partial_\rho h_{ij} \partial^\rho h_{ij} \tag{11.65}$$

$$= \frac{M_p^2}{4} \int d\eta dx^3 \frac{a^2}{2} \left[(h'_{ij})^2 - (\partial_l h_{ij})^2 \right].$$

This is the same action as for the conformally coupled massless scalar field in FRW universe. We do a Fourier expansion

$$h_{ij}(\mathbf{x}, t) = \int \frac{d^3 k}{(2\pi)^{3/2}} \sum_{\lambda=+,\times} \left(h_{\mathbf{k},\lambda}(\tau) \, e_{ij}^\lambda(k) e^{i\mathbf{k}.\mathbf{x}} + h_{\mathbf{k},\lambda}^*(\tau) \, e_{ij}^\lambda(k) e^{-i\mathbf{k}.\mathbf{x}} \right),$$

$$\tag{11.66}$$

where e_{ij}^λ are the polarisation tensors which satisfy the properties

$$e_{ij}^\lambda(k) = e_{ji}^\lambda(k), \quad e_{ii}^\lambda(k) = 0, \quad k^i e_{ij}^\lambda(k) = 0, \quad e_{ij}^\lambda(k) e_{ij}^{\lambda'}(k) = 2\delta_{\lambda\lambda'}. \tag{11.67}$$

Using (11.66) and (11.67), the action (11.65) leads to the e.o.m. for the quantity $h_{\mathbf{k}}$

$$h''_{\mathbf{k},\lambda} + 2\frac{a'}{a} h'_{\mathbf{k},\lambda} + k^2 h_{\mathbf{k},\lambda} = 0. \tag{11.68}$$

Defining the canonically normalised field $\tilde{h}_{\mathbf{k}\lambda} \equiv \frac{1}{2} a h_{\mathbf{k},\lambda} M_p$, the e.o.m. (11.68) becomes

$$\tilde{h}''_{\mathbf{k}\lambda} + \left(k^2 - \frac{a''}{a} \right) \tilde{h}_{\mathbf{k}\lambda} = 0, \tag{11.69}$$

where

$$\frac{a''}{a} = \frac{2}{\tau^2} (2 + 3\epsilon)$$

during quasi de-Sitter epoch when $\dot{H} = -\epsilon H$.

Once again to determine the zero-point fluctuations of the canonically normalised graviton field $\tilde{h}_\lambda(\mathbf{x}, \tau)$ we treat it as a quantum field

$$\hat{\tilde{h}}_\lambda(\mathbf{x}, \tau) = \int \frac{d^3k}{(2\pi)^3} \sum_{\lambda=+,\times} \left(\tilde{h}_{k\lambda}(\tau) a_{k\lambda} e_{ij}^\lambda(k) e^{i\mathbf{k}\cdot\mathbf{x}} + \tilde{h}_{k\lambda}^*(\tau) a_{k\lambda}^\dagger e_{ij}^\lambda(k) e^{-i\mathbf{k}\cdot\mathbf{x}} \right),$$

$$(11.70)$$

where the graviton creation and annihilation operators obey the commutation relation

$$\left[a_{k\lambda}, a_{k'\lambda'}^\dagger \right] = \delta^3(\mathbf{k} - \mathbf{k}') \delta_{\lambda\lambda'}. \tag{11.71}$$

The zero-point fluctuation of the gravitons is obtained as

$$\langle 0 | \hat{\tilde{h}}_\lambda(\tau, 0) \hat{\tilde{h}}_{\lambda'}^\dagger(\tau, 0) | 0 \rangle = \int \frac{d^3k}{(2\pi)^3} \int \frac{d^3k'}{(2\pi)^3} \tilde{h}_{k\lambda} \tilde{h}_{k'\lambda'}^* \langle 0 | [a_{k\lambda}, a_{k'\lambda'}^\dagger] | 0 \rangle$$

$$= \int d\ln k \frac{k^3}{2\pi^2} |\tilde{h}_{k\lambda}|^2 \delta_{\lambda\lambda'}. \tag{11.72}$$

The solution of (11.68) can be written as (as in the scalar case (11.61)) as

$$|\tilde{h}_{k\lambda}|^2 = \frac{1}{M_p^2} \frac{a^2 H^2}{2k^3} \left(\frac{k}{aH} \right)^{3-2\nu_T}. \tag{11.73}$$

Using the relation

$$\frac{a''}{a} = \frac{1}{\tau^2} \left(\nu_T^2 - \frac{1}{4} \right) \simeq \frac{1}{\tau^2} (2 + 3\epsilon),$$

we obtain the tensor spectral index ν_T, given by

$$\nu_T \simeq \frac{3}{2} + \epsilon.$$

The power spectrum of tensor perturbations is

$$\mathcal{P}_T \equiv \frac{k^3}{2\pi^2} \sum_{\lambda=+,\times} |h_{\mathbf{k},\lambda}|^2 = 2 \times \frac{k^3}{2\pi^3} \frac{4|\tilde{h}_{\mathbf{k}\lambda}|^2}{a^2}, \tag{11.74}$$

where the factor of 2 is from two the polarisations. Substituting the solution (11.73), we get the amplitude of the tensor power spectrum

$$\mathcal{P}_T = \frac{2}{\pi^2} \frac{H^2}{M_p^2} \left(\frac{k}{aH} \right)^{3-2\nu_T} \tag{11.75}$$

$$\equiv \Delta_T^2 \left(\frac{k}{aH} \right)^{n_T} .$$

We define the *tensor spectral index* n_T as

$$n_T \equiv \frac{d \ln \mathcal{P}_T}{d \ln k} \tag{11.76}$$

$$= 3 - 2\nu_T = -2\epsilon .$$

As the action for tensor perturbations (11.65) is that of a massless scalar field, there will be no appearance of slow-roll parameter η in ν_T.

11.8 Tensor-to-Scalar Ratio and Energy Scale of Inflation

The *tensor-to-scalar ratio* r is defined as the ratio of the two amplitudes

$$r \equiv \frac{\Delta_T^2}{\Delta_{\mathcal{R}}^2} = 16\epsilon , \tag{11.77}$$

which determines the relative contribution of the tensor modes to mean squared low multipole CMB anisotropy. where in the last equality in (11.77), we have used the amplitude relations (11.62) and (11.75) for scalar and tensor perturbations. The scalar amplitude is fixed from the observations $\Delta_{\mathcal{R}}^2 \simeq 1.95 \times 10^{-9}$ and, from (11.75), amplitude of the tensor perturbations $\Delta_T^2 \propto H^2 \approx V(\phi)$, therefore the value of tensor-to-scalar ratio is a direct measure of the energy scale of inflation potential

$$V(\phi)^{1/4} \sim \left(\frac{r}{0.01} \right)^{1/4} 10^{16} \, GeV . \tag{11.78}$$

Therefore measurement of the tensor amplitudes will give us a direct handle on the energy scale of inflation potential.

11.9 Inflationary Observables in CMB

The physical quantities/observables are: the amplitude of the power spectrum of curvature perturbations $\Delta_{\mathcal{R}}^2$, spectral index n_s, running of spectral index α_s and tensor-to-scalar ratio r. These are obtained by measuring the temperature and E-mode polarisation anisotropy spectrum. The tensor-to-scalar ratio r is determined from the B-mode polarisation (after removing the B-mode polarisation from dust in the foreground and the B-model due to lensing of E-mode polarisation). The inflationary observables as given by PLANCK-2018 [9], from the combination of *PLANCK TT, TE, EE, low E +lensing* are, $\ln(10^{10}\Delta_{\mathcal{R}}^2) = 3.044 \pm 0.014$, 68% CL and $n_s = 0.9670 \pm 0.0037$, 68% CL

The value of amplitude and spectral index are given at 68% CL at the pivot scale $k = 0.05 \, \text{Mpc}^{-1}$. Whereas the upper bound on tensor-to-scalar ratio is determined at 95% CL at $k = 0.002 \, \text{Mpc}^{-1}$ [9].

The bound on the tensor-to-scalar ratio r (at 95% CL at the pivot scale $k = 0.002 \, \text{Mpc}^{-1}$) becomes more stringent by combining the PANCK data with that of BICEP and Keck (BKP) [9, 25] $r_{0.002} < 0.064$ 95% CL.

The running of the spectral index is consistent with zero [9],

$$\frac{dn_s}{d \ln k} = -0.0045 \pm 0.0067, \quad 68\% \text{ CL.} \tag{11.79}$$

The stringent bound on the n_s and r severely rules out many models of inflation.

11.10 Stochastic Gravity Waves from Inflation

CMB probes primordial gravitational waves whose frequency in the present epoch is around $f_{CMB} = (\frac{c}{2\pi})0.05 \, \text{Mpc}^{-1} = 1.94 \times 10^{-17} \, \text{Hz}$. PTA is sensitive to stochastic GW in the frequency range $f \sim 10^{-9} - 10^{-7}$ Hz while terrestrial detectors LIGO/VIRGO/KAGRA probe GW in the frequency range $f = (10 - 100) \, \text{Hz}$. The stochastic GW detectors are sensitive to the energy density of GW. The energy density can be related to the power spectrum of the tensor modes which we have discussed earlier as follows. The energy–momentum pseudo-tensor for GW is given by

$$t_{\mu\nu}^{GW} = \frac{1}{32\pi G} \left[\partial_\mu h^{\alpha\beta} \partial_\nu h_{\alpha\beta} - \frac{1}{2} \bar{g}_{\mu\nu} \partial_\sigma h^{\alpha\beta} \partial^\sigma h_{\alpha\beta} \right]. \tag{11.80}$$

In terms of the Fourier modes (11.66) the energy density from (11.80) can be written as

$$\rho_{GW} = \frac{1}{8\pi G} \int d \ln k \left(\frac{k^3}{16\pi^2} \right) \sum_{\lambda=+,\times} \langle \dot{h}_{\mathbf{k},\lambda} \dot{h}_{-\mathbf{k},\lambda} \rangle + \frac{k^2}{a^2} \langle h_{\mathbf{k},\lambda} h_{-\mathbf{k},\lambda} \rangle, \tag{11.81}$$

where the angular brackets denote time average over times longer than the period of the gravitational waves. For real fields the Fourier modes obey the condition $h_{\mathbf{k},\lambda}^* = h_{-\mathbf{k},\lambda}$. For free-gravitational waves propagating in FRW space

$$\dot{h}_{\mathbf{k},\lambda} = \frac{|\mathbf{k}|}{a} h_{\mathbf{k},\lambda}. \tag{11.82}$$

The fraction of energy density in GW as a function of frequency is given by

$$\Omega_{GW}(k) = \frac{1}{\rho_c}\frac{d\rho_{GW}}{d\ln k}, \tag{11.83}$$

where $\rho_c = 3H^2/(8\pi G)$. We can write the two point correlation of GW amplitudes in terms of the power spectrum as

$$\sum_{\lambda=+,\times} \langle h_{\mathbf{k},\lambda} h_{-\mathbf{k},\lambda} \rangle \equiv \frac{2\pi^2}{k^3} \mathcal{P}_T(k) \tag{11.84}$$

and the fractional energy density is then related to the tensor power spectrum as

$$\Omega_{GW}(k) = \frac{k^2}{12a^2 H^2} \mathcal{P}_T(k). \tag{11.85}$$

This can also be written in terms of the scalar power and scalar-tensor ration r as

$$\Omega_{GW}(f) = \left(\frac{r\Delta_{\mathcal{R}}^2}{24}\right)\left(\frac{f}{f_{CMB}}\right)^{n_t}\Omega_r. \tag{11.86}$$

The bound on Ω_{GW} of stochastic gravitational waves from LIGO/VIRGO/KAGRA observations [26] is $\Omega_{GW} < 5.8 \times 10^{-9}$ at 95% C.L. assuming a flat spectrum (Fig. 11.1).

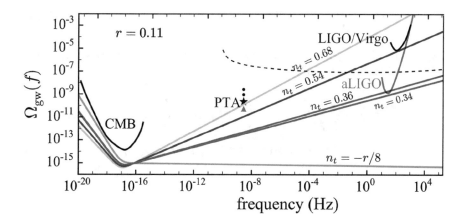

Fig. 11.1 First order gravitational waves generated during inflation. The tensor-scalar ratio taken is $r = 0.11$ with different values of n_t. The consistency relation for slow-roll inflation models is $n_t = -r/8$. Dashed line shows indirect bounds from BBN. Reprinted from [27]. Figure credit: ©The Author(s). Reproduced under CC-BY-3.0 license

11.11 The In-in Formalism for Evaluating Cosmological Correlators

During inflations perturbations the length scale of perturbations grows with the scale factor $\lambda = e^{Ht}\lambda_0$ and at some point they exit the horizon that is, λ becomes larger than the horizon size H^{-1}. In the past $t = -\infty$, the perturbation scales are very small compared to the horizon $ak \ll H$ and perturbations behave as free particles with mode function $u(x) = e^{ik\cdot x}$. The vacuum state $|0\rangle$ in the far past can then be defined as the zero-particle state w.r.t the Minkowski modes $u(x)$. The correlation function at any general time can be defined w.r.t the past vacuum only by taking the following time ordering of the operators

$$\langle O(\tau)\rangle = \langle 0|\bar{T}e^{i\int_{-\infty(1+i\epsilon)}^{\tau} dt' H_{\text{int}}(t')} O(\tau) T e^{-i\int_{-\infty(1-i\epsilon)}^{\tau} dt' H_{\text{int}}(t')}|0\rangle \qquad (11.87)$$

where $\bar{T}$ is the ant-time ordering operator and the $i\epsilon$ are included to ensure convergence in the limits at $t = -\infty$. This is the 'in-in' formalism [28–31] for computing correlations which is better suited for cosmological applications where the initial condition is the well defined Bunch-Davies vacuum which has zero positive frequency particles.

The correlation function (11.87) can computed using the non-perturbative path integrals or can be computed perturbatively by expanding the exponential factors to various orders. The expansion of (11.87) to first order in H_{int} is

$$\langle O(\tau)\rangle = \langle 0|\bar{T}\left(1 + i\int_{-\infty}^{\tau} dt' H_{\text{int}}\right) O(\tau)\, T\left(-i\int_{-\infty}^{\tau} dt' H_{\text{int}}\right)|0\rangle$$

$$= i\int_{-\infty}^{\tau} dt' \langle 0|[H_{\text{int}}(t'), O(\tau)]|0\rangle. \qquad (11.88)$$

The perturbative expansion of (11.87) to all orders can be written as nested commutators in the form [32],

$$\langle O(\tau)\rangle = \sum_{n=0}^{\infty} i^n \int_{-\infty}^{\tau} dt_n \int_{-\infty}^{\tau} dt_{(n-1)} \cdots \int_{-\infty}^{\tau} dt_1$$

$$\langle 0|[H_{\text{int}}(t_1), [H_{\text{int}}(t_2), \cdots [H_{\text{int}}(t_n), O(\tau)] \cdots]]|0\rangle. \qquad (11.89)$$

11.12 Non-Gaussianity in the In-in Formalism

As an application of the in-in formalism, we calculate the three-point correlation for the inflaton field with an interaction vertex

$$H_{int} = \mu \int d^3x \sqrt{-g}\varphi(\mathbf{x}, t)^3$$

$$= \mu a^4(\tau) \int \frac{d^3q_1}{(2\pi)^3} \frac{d^3q_2}{(2\pi)^3} \frac{d^3q_3}{(2\pi)^3} \varphi(\mathbf{q}_1, \tau)\varphi(\mathbf{q}_2, \tau)\varphi(\mathbf{q}_3, \tau)(2\pi)^3 \delta(\mathbf{q}_1 + \mathbf{q}_2 + \mathbf{q}_3),$$

$$(11.90)$$

where $a(\tau) = -1/(H\tau)$, $\tau \in (-\infty, 0)$. The massless inflaton solutions in momentum space are

$$\varphi(\mathbf{q}, \tau) = f_q(\tau)a_q + f_q^*(\tau)a_{-q}^\dagger,$$

$$f_q(\tau) = \frac{H}{\sqrt{2q^3}}(1 + iq\tau)e^{-iq\tau}, \quad q = |\mathbf{q}|.$$

$$(11.91)$$

We calculate the 3-point correlation or the bi-spectrum of φ using (11.89) and retain terms to the linear order in H_{int} to obtain

$$\langle \varphi(\mathbf{k}_1, \tau)\varphi(\mathbf{k}_2, \tau)\varphi(\mathbf{k}_3, \tau) \rangle = i \int_{-\infty}^{\tau} d\tau'[H_{int}(\tau'), \varphi(\mathbf{k}_1, \tau)\varphi(\mathbf{k}_2, \tau)\varphi(\mathbf{k}_3, \tau)].$$

$$(11.92)$$

For Hermitian H_{int} and O we can write

$$\langle [H_{int}, O] \rangle = \langle H_{int}O \rangle - \langle (H_{int}O)^\dagger \rangle = 2i \, \mathrm{Im} \, \langle H_{int}O \rangle.$$

$$(11.93)$$

Using this we can write the 3-point correlation as

$$\langle \varphi(\mathbf{k}_1, \tau)\varphi(\mathbf{k}_2, \tau)\varphi(\mathbf{k}_3, \tau) \rangle = -2\mu \, \mathrm{Im} \int_{-\infty}^{\tau} d\tau' a^4(\tau') \int \frac{d^3q_1}{(2\pi)^3} \frac{d^3q_2}{(2\pi)^3} \frac{d^3q_3}{(2\pi)^3}$$

$$(2\pi)^3 \delta(\mathbf{q}_1 + \mathbf{q}_2 + \mathbf{q}_3)\langle \varphi(\mathbf{q}_1, \tau')\varphi(\mathbf{q}_2, \tau')\varphi(\mathbf{q}_3, \tau')\varphi(\mathbf{k}_1, \tau)\varphi(\mathbf{k}_2, \tau)\varphi(\mathbf{k}_3, \tau) \rangle.$$

$$(11.94)$$

One can now write the correlations as a product of 2-point correlations

$$\langle \varphi(\mathbf{q}_a, \tau')\varphi(\mathbf{k}_b, \tau) \rangle = f_{q_a}(\tau')f_{k_b}^*(\tau)(2\pi)^3 \delta(\mathbf{q}_a + \mathbf{k}_b), \quad a, b = 1, 2, 3.$$

$$(11.95)$$

The 2-point correlations of $\langle \varphi(q_a, \tau')\varphi(q_b, \tau')\rangle$ and $\langle \varphi(k_a, \tau)\varphi(k_b, \tau)\rangle$ gives disconnected diagrams which get cancelled in the normalisation. Substituting in (11.94) and carrying out the d^3q_a using the delta functions we obtain

$$\langle \varphi(\mathbf{k}_1, \tau)\, \varphi(\mathbf{k}_2, \tau)\, \varphi(\mathbf{k}_3, \tau)\rangle = (2\pi)^3 \delta(\mathbf{k}_1 + \mathbf{k}_2 + \mathbf{k}_3) B_3(\mathbf{k}_1, \mathbf{k}_2, \mathbf{k}_3), \quad (11.96)$$

where

$$B_3(\mathbf{k}_1, \mathbf{k}_2, \mathbf{k}_3) = -2\mu\, 3!\, \mathrm{Im}\, \prod_{a=1}^{3} f_{k_a}^*(\tau) \int_{-\infty}^{\tau} d\tau'\, \frac{1}{H^4 \tau'^4} \prod_{a=1}^{3} f_{k_a}^*(\tau')$$

$$= -\frac{2}{3}\frac{\mu H^2}{(k_1 k_2 k_3)^3}\, \mathrm{Im}\, \prod_{a=1}^{3}(1 - i k_a \tau) \int_{-\infty}^{\tau} d\tau'\, \frac{1}{\tau'^4} e^{-iK(\tau'-\tau)} \prod_{b=1}^{3}(1 + i k_b \tau'),$$

$$(11.97)$$

where $K = k_1 + k_2 + k_3$. The time integration terms are of the form

$$\int_{-\infty}^{\tau} d\tau'\, \frac{1}{\tau'^n} e^{-iK\tau'}, \quad n = 1, 2, 3, 4. \quad (11.98)$$

The time integral can be performed by repeated integration by parts with the aim of reducing the integrand to the exponential integral form

$$Ei(x) = -\int_{x}^{\infty} \frac{e^{-t}}{t} dt. \quad (11.99)$$

The time integral gives [33]

$$\int_{-\infty}^{\tau} d\tau'\, \frac{1}{\tau'^4} e^{-iK(\tau'-\tau)} \prod_{b=1}^{3}(1 + i k_b \tau')$$

$$= -\frac{i}{3}\sum_{a=1}^{3} k_a^2 Ei(-i k_a \tau) - \frac{e^{-iK\tau}}{3\tau^3}\left[1 + iK\tau + \left(\sum_{a=1}^{3} k_a^2 - \sum_{a\neq b} k_a k_b\right)\tau^2\right].$$

$$(11.100)$$

We use the asymptotic limit of the exponential function in the $K\tau \to 0$ limit

$$Ei(-iK\tau) = \gamma_E + \log(K\tau) - \frac{i\pi}{2}, \quad \gamma_E = 0.577. \quad (11.101)$$

The 3-point function (11.97) in the $\tau \to 0$ limit then reduces to the form [33]

$$\langle \varphi(\mathbf{k}_1, \tau) \varphi(\mathbf{k}_2, \tau) \, \varphi(\mathbf{k}_3, \tau) \rangle = (2\pi)^3 \delta(\mathbf{k}_1 + \mathbf{k}_2 + \mathbf{k}_3) \frac{\mu H^2}{2(k_1 k_2 k_3)^3}$$

$$\times \left[\sum_a k_a^3 \left(\gamma_E - 1 + \ln(-K\tau) \right) + k_1 k_2 k_3 - \sum_{a \neq b} k_a^2 k_b \right].$$

$$(11.102)$$

11.13   Schwinger-Keldysh Treatment of Cosmological Correlations

In order to deal with dissipative effects during slow-roll and non-Bunch-Davies initial conditions like a thermal background, we need to doubling of variables and consequently we have four Green's function as discussed for gravitational waves in the Minkowski background in Chap. 6.

We double the fields from φ to φ_1 and φ_2. We evolve the fields along the contour shown in Fig. 6.2. The field φ_1 is time evolved from $t = -\infty$ to $t = \tau$ and the field φ_2 is evolved backward in time from $t = \tau$ to $t = -\infty$. At time τ the two fields must match $\varphi_1(\tau) = \varphi_2(\tau)$. The fields φ_1 and φ_2 have sources J_1 and J_2. The φ_2 fields can be evolved backward in time by choosing $\mathcal{L}(\varphi_2(t), \dot{\varphi}_2(t), t) = -\mathcal{L}(\varphi_1(t), \dot{\varphi}_1(t), t)$. The generating function this theory written as a path integral over all configurations of $\varphi_{1,2}$ is

$$Z(J) = e^{iW(J_1, J_2)} = \int \mathcal{D}\varphi_1 \mathcal{D}\varphi_2 \, \delta(\varphi_1(\tau, \mathbf{x}) - \varphi_2(\tau, \mathbf{x}))$$

$$- \int_{-\infty - i\epsilon}^{\tau} dt \int d^3 y \left(\frac{1}{2} \partial_\mu \varphi_2(y) \partial^\mu \varphi_2(y) - \frac{1}{2} m^2 \varphi_2^2(y) + J_2(y) \varphi_2(y) \right) \bigg\}.$$

$$(11.103)$$

We can write the delta function which equates φ_1 with φ_2 at $t = \tau$ as

$$\delta(\varphi_1(\tau) - \varphi_2(\tau)) = \lim_{\epsilon \to 0} \left(\frac{i}{\pi \epsilon} \right)^{1/2} e^{-\frac{1}{\epsilon}(\varphi_1(\tau) - \varphi_2(\tau))^2}. \qquad (11.104)$$

Integrating out φ_1 and φ_2 we obtain the generating function as a function of the sources and propagators, which can be written as

$$Z(J_1, J_2) = e^{iW(J_1, J_2)} = \exp\left[\frac{i}{2} \int d^4x \, d^4y \, J^a(x) G_{ab}(x - y) J^b(y) \right], \quad a, b = 1, 2.$$

$$(11.105)$$

The different propagators operating on the sources $J_a = (J_1(x), J_2(y))$ are written as 2×2 matrices

$$G_{ab}(x - y) = \begin{pmatrix} G_{11}(x - y) & G_{12}(x - y) \\ G_{21}(x - y) & G_{22}(x - y) \end{pmatrix}. \tag{11.106}$$

These propagators correspond to the following two point functions of the fields φ_1 and φ_2

$$G_{11}(x, x') = \langle T\varphi_1(x)\varphi_1(x') \rangle = \frac{-i\delta}{\delta J_1(x)} \frac{-i\delta}{\delta J_1(x')} Z(J_1, J_2) \Big|_{J_{a,b}=0}$$

$$G_{22}(x, x') = \langle \bar{T}\varphi_2(x)\varphi_2(x') \rangle = \frac{-i\delta}{\delta J_2(x)} \frac{-i\delta}{\delta J_2(x')} Z(J_1, J_2) \Big|_{J_{a,b}=0}$$

$$G_{12}(x, x') = \langle \varphi_1(x)\varphi_2(x') \rangle = \frac{-i\delta}{\delta J_1(x)} \frac{-i\delta}{\delta J_2(x')} Z(J_1, J_2) \Big|_{J_{a,b}=0}$$

$$G_{21}(x, x') = \langle \varphi_2(x)\varphi_1(x') \rangle = \frac{-i\delta}{\delta J_2(x)} \frac{-i\delta}{\delta J_1(x')} Z(J_1, J_2) \Big|_{J_{a,b}=0}. \tag{11.107}$$

The time ordering operator T means ordering along the contour shown in Fig. 6.2. In terms of time t the φ_1 fields are time ordered and the φ_2 fields are anti-time ordered by the operator $\bar{T}$. For the Wightman functions G_{12} and G_{21} no time ordering is needed as the φ_2 fields always occur 'later' than the φ_1 fields along the contour.

Doing a partial Fourier transform of the Greens functions

$$G_{ab}(k; t, t') = \int d^3\mathbf{x} d^3\mathbf{y} e^{-i\mathbf{k}\cdot(\mathbf{x}-\mathbf{y})} G_{ab}(t, \mathbf{x}; t'\mathbf{y}) \tag{11.108}$$

we can write the Greens functions in terms of the mode functions $u(\mathbf{x}, t)$ which solve the homogenous wave equation. The mode functions of the minimally coupled massless scalar in De-Sitter background are

$$\varphi(\mathbf{x}, \tau) = \int d^3x e^{i\mathbf{k}\cdot x} \left(f_k(\tau)a_k + f_k^*(\tau)a_{-k}^\dagger \right), \tag{11.109}$$

where $\tau = -1/(aH)$ is the conformal time and

$$f_k(\tau) = \frac{H}{\sqrt{2k^3}}(1 + ik\tau)e^{-ik\tau}, \quad k = |\mathbf{k}|. \tag{11.110}$$

We can express the different Greens functions in terms of the mode functions. The Wightman functions are

$$G_+(k; \tau, \tau') = f_k(\tau) f_k^*(\tau') = \frac{H^2}{2k^3}(1 + ik\tau)(1 - ik\tau')e^{-ik(\tau - \tau')},$$

$$G_-(k; \tau, \tau') = f_k^*(\tau) f_k(\tau') = \frac{H^2}{2k^3}(1 - ik\tau)(1 + ik\tau')e^{ik(\tau - \tau')}. \quad (11.111)$$

The Greens functions can be written in terms of these Wightmans functions as

$$G_{11}(k; t, t') = G_+(k; t, t')\theta(t - t') + G_-(k; t, t')\theta(t' - t),$$

$$G_{22}(k; t, t') = G_-(k; t, t')\theta(t - t') + G_+(k; t, t')\theta(t' - t),$$

$$G_{12}(k; t, t') = G_+(k; t, t'),$$

$$G_{21}(k; t, t') = G_-(k; t, t'). \quad (11.112)$$

These obey the complex conjugacy relations $G_{11}(k; t, t') = G_{22}^*(k; t, t')$ and $G_{12}(k; t, t') = G_{21}^*(k; t, t')$. The four Greens functions are not independent as they obey the relation

$$G_{11} + G_{22} = G_{12} + G_{21}. \quad (11.113)$$

This means that we can choose a basis where we can choose one of the elements of the 2×2 is zero. One such basis is the Keldysh representation.

In the Keldysh representation we take the following combinations of the fields and the corresponding sources [6]

$$\varphi_+ = (\varphi_1 + \varphi_2)/2, \quad \varphi_- = (\varphi_1 - \varphi_2),$$

$$J_+ = (J_1 + J_2)/2, \quad J_- = J_1 - J_2. \quad (11.114)$$

Classical fields are given by the average $\varphi_+ = (\varphi_1 + \varphi_2)/2$ and the quantum fluctuation is the difference $\varphi_- = (\varphi_1 - \varphi_2) = 0$.

The generating function in the (φ_1, φ_2) basis is

$$Z(J_+, J_-) = e^{iW(J_+, J_-)} = \exp\left[\frac{i}{2} \int d^4x\, d^4y\, J_A(x) G^{AB}(x - y) J_B(y)\right], \quad A, B = 1, 2.$$

$$(11.115)$$

The different propagators in the $J_A = (J_+(x), J_-(y))$ basis are

$$G^{AB}(x - y) = \begin{pmatrix} 0 & G_{adv}(x - y) \\ G_{ret}(x - y) & \frac{i}{2}G_H(x - y) \end{pmatrix}, \quad (11.116)$$

where G_{adv} is the advanced, G_{ret} is the retarded and G_H is the Hadamard propagator. These are related to the Wightman functions as follows. The commutator (or causal) and anti-commutator (Hadamard) two point functions are

$$G_c(x - x') = \langle [\varphi(x), \varphi(x')] \rangle = G_+(x - x') - G_-(x - x'),$$
$$G_H(x - x') = \langle \{\varphi(x), \varphi(x')\} \rangle = G_+(x - x') + G_-(x - x'). \quad (11.117)$$

The retarded and advanced Greens functions are

$$G_{ret}(x - x') = \theta(t - t') G_c(x - x') = \theta(t - t') \left(G_+(x - x') - G_-(x - x') \right),$$
$$G_{adv}(x - x') = \theta(t' - t) G_c(x - x') = \theta(t' - t) \left(G_+(x - x') - G_-(x - x') \right).$$
$$(11.118)$$

In the momentum space the retarded, advanced and Hadamard Greens functions for a massless scalar in de-Sitter background are

$$G_{ret}(k, \tau - \tau')$$
$$= \theta(t - t') \frac{H^2}{2k^3} \left[(1 + ik\tau)(1 - ik\tau') e^{-ik(\tau - \tau')} - (1 - ik\tau)(1 + ik\tau') e^{ik(\tau - \tau')} \right],$$
$$G_{adv}(k, \tau - \tau')$$
$$= -\theta(t' - t) \frac{H^2}{2k^3} \left[(1 + ik\tau)(1 - ik\tau') e^{-ik(\tau - \tau')} - (1 - ik\tau)(1 + ik\tau') e^{ik(\tau - \tau')} \right],$$
$$G_H(k, \tau - \tau')$$
$$= \frac{i}{2} \frac{H^2}{2k^3} \left[(1 + ik\tau)(1 - ik\tau') e^{-ik(\tau - \tau')} + (1 - ik\tau)(1 + ik\tau') e^{ik(\tau - \tau')} \right].$$
$$(11.119)$$

11.14 Tensor Propagators in de-Sitter Space

The propagators for gravitational waves in the de-Sitter space have the same form as the propagators for the minimally coupled massless scalars we have discussed in the last section, apart from the projection operator and a normalisation.

Tensor perturbations h_{ij} of the metric are independent of the gauge choice and for the de-Sitter spacetime the metric with tensor perturbations can be written as

$$ds^2 = a(\tau)^2 \left[-d\tau^2 + (\eta_{ij} + h_{ij}) dx^i dx^j \right], \quad (11.120)$$

where the scale factor goes as $a(\tau) = -1/H\tau$. The tensor perturbations obey the same wave equation as the minimal coupling massless scalars,

$$h_{ij}(k, \tau)'' + 2\frac{a'}{a}h_{ij}(k, \tau)' + k^2 h_{ij}(k, \tau) = 16\pi G\Lambda_{ij,kl}T_{kl}, \tag{11.121}$$

where $\Lambda_{ij,kl}$ is the transverse–traceless projection operator. The Hubble term can be removed by defining $v_{ij} = a(\tau)h_{ij}$, and the e.o.m for becomes

$$v_{ij}(k, \tau)'' + \left(k^2 - \frac{a''}{a}\right)v_{ij}(k, \tau) = 16\pi Ga(\tau)\Lambda_{ij,kl}T_{kl}. \tag{11.122}$$

In the de-Sitter background $\frac{a''}{a} = -\frac{2}{\tau^2}$ homogenous equation

$$v_{ij}(k, \tau)'' + \left(k^2 + \frac{2}{\tau^2}\right)v_{ij}(k, \tau) = 0 \tag{11.123}$$

with the Bunch-Davies initial condition has the solution

$$v_{ij}(k, \tau) = \frac{1}{\sqrt{2k^3}}(1 + k\tau)e^{-ik\tau}e_{ij}(k). \tag{11.124}$$

The Greens functions for the wave equation (11.123) can be constructed from the Wightmans functions

$$G_{+ij,kl}(k, \tau - \tau') = \Lambda_{kl,rs}v_{ij}(k, \tau)v_{rs}^*(k, \tau'),$$
$$G_{-ij,kl}(k, \tau - \tau') = \Lambda_{kl,rs}v_{ij}^*(k, \tau)v_{rs}(k, \tau'). \tag{11.125}$$

For example, the retarded Green's function is

$$G_{\mathrm{ret}ij,kl}(k, \tau, \tau') = \theta(\tau - \tau')\Big(G_{+ij,kl}(k, \tau - \tau') - G_{-ij,kl}(k, \tau - \tau')\Big)$$
$$= \frac{\Lambda_{ij,kl}}{2k^3}\theta(\tau - \tau')\Big((1 + k^2\tau\tau')\sin k(\tau - \tau') + k(\tau - \tau')\cos k(\tau - \tau')\Big). \tag{11.126}$$

The solution of the wave equation (11.121) for $h_{ij}(k, \tau)$ in terms of the source stress tensor $T(k, \tau')$ is then given by

$$h_{ij}(k, \tau) = 16\pi G\frac{1}{a(\tau)}\int d\tau' G_{\mathrm{ret}ij,kl}(k, \tau, \tau')a(\tau')T(k, \tau'). \tag{11.127}$$

11.15 Enhancement of Inflation Power Spectrum by Thermal Background

There is a possibility that the inflation era was preceded by a radiation era. The radiation era will transition to inflation when the radiation density ρ_r drops below the inflaton potential. During the radiation era the inflatons and gravitons can have a thermal distribution given by

$$n(k) = \frac{1}{e^{k_p/T_p} - 1}. \tag{11.128}$$

Here k_p and T_p are the physical wavenumber and temperature which are related to the comoving wavenumber k and comoving temperature T as $k_p(t) = k/a(t)$ and $T_p(t) = T/a(t)$. The distribution function $n(k)$ will be constant in time and given by

$$n(k) = \frac{1}{e^{k/T} - 1}. \tag{11.129}$$

As discussed in Sect. 6.10, the expectation value of the various combinations a_k, $a_k^\dagger$ operators for graviton and inflaton in thermal initial state are

$$\langle a_k \rangle_\beta = 0, \quad \langle a_k^\dagger \rangle_\beta = 0, \quad \langle a_k{}^\dagger a_{k'}^\dagger \rangle_\beta = 0, \quad \langle a_k a_{k'}' \rangle_\beta = 0,$$

$$\langle a_{k'}^\dagger a_k \rangle_\beta = n(k)\delta^3(\mathbf{k} - \mathbf{k}'), \quad \langle a_k a_{k'}^\dagger \rangle_\beta = (1 + n(k))\delta^3(\mathbf{k} - \mathbf{k}').$$

$$\tag{11.130}$$

Consider the mode functions of gravitons in the quasi-De-Sitter background

$$h_{ij}(x) = \sum_\lambda \int \frac{d^3k}{(2\pi)^3} \frac{H}{\sqrt{2k^{3/2}}} \left(\epsilon_{ij}(k) a_k e^{-ik\cdot x} + \epsilon_{ij}^*(k) a_k^\dagger e^{ik\cdot x} \right). \tag{11.131}$$

The equal-time two point function in the thermal background are then given by

$$\langle h_{ij}(0, \tau) h_{kl}(0, \tau)^\dagger \rangle = P_{ijkl} \int \frac{d^3k}{(2\pi)^3} \frac{H^2}{2k^3} \left(\langle a_k a_k^\dagger \rangle + \langle a_k^\dagger a_k \rangle \right)$$

$$= P_{ijkl} \int \frac{d^3k}{(2\pi)^3} \frac{H^2}{2k^3} ((1 + n_k) + n_k)$$

$$= P_{ijkl} \int \frac{d^3k}{(2\pi)^3} \frac{H^2}{2k^3} \coth\left(\frac{k}{2T}\right). \tag{11.132}$$

The thermal initial state therefore results in a factor of $\coth(k/2T)$ correction in the tensor power spectrum [34–37],

$$P_t(k, T) = A_t(k_0) \left(\frac{k}{k_0} \right)^{n_t} \coth \left(\frac{k}{2T} \right). \tag{11.133}$$

If the pre-inflation thermal bath has a thermal distribution of inflatons, then there will be the same factor of $\coth(k/2T)$ to the inflaton power spectrum [38].

For $k \gg T$, the $\coth(k/2T) \sim 1$. For $k \ll T$, $\coth(k/2T) \sim 2T/k$ and the power spectrum goes as $P_t = A_t k^{n_t-1}$. The thermal $\coth k/2T$ factor therefore enhances the superhorizon modes $k \sim H$. If there is a prior thermal distribution of gravitons, the B-mode polarisation of the CMB will be enhanced at low multipoles [35] as shown in Fig. 11.2 taken from reference [37]. Observation of the low-l enhancement of the B-mode polarisation will be a proof of the quantum nature of the gravitons as the thermal enhancement of the power spectrum is due to the phenomenon of stimulated emission of zero-point gravitons in the presence of thermal graviton background.

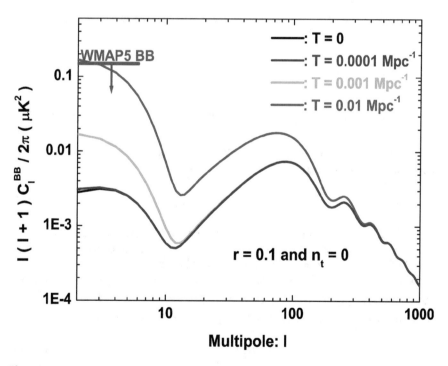

Fig. 11.2 Enhancement of the low-l B-mode polarisation due to thermal power spectrum of gravitons (11.133). Reprinted from [37]. Figure credit: ©The Author(s). Reproduced under CC-BY-3.0 license

11.16 Second Order Gravitational Waves from Scalar Perturbations

In the second order in inflaton and metric perturbations, there is a mixing between the tensor and the scalar modes. The second order scalar perturbations can therefore source the tensor perturbations [39–46].

The action for tensor perturbations h_{ij} to the second order in the metric perturbations $\Psi(\mathbf{x}, t)$, $\Phi(\mathbf{x}, t)$ (11.30) and the scalar field fluctuation over the homogenous background $\delta\phi(\mathbf{x}, t) = \phi(\mathbf{x}, t) - \bar{\phi}(t)$ is given by Domènech [45]

$$
S^{(2)} = \int d^3x dt \left[\frac{a^3}{8} \dot{h}_{ij}\dot{h}^{ij} - \frac{a^3}{8}\partial_i h_{kl}\partial^i h^{kl} - 2ah^{ij}\partial_i(\Phi + \Psi)\partial_j\Phi \right.
$$
$$
\left. + ah^{ij}\partial_i\Phi\partial_j\Phi + \frac{a}{2}h^{ij}\partial_i\delta\phi\partial_j\delta\phi \right] \tag{11.134}
$$

The e.o.m for h_{ij} from (11.134) is given by

$$
\ddot{h}_{ij} + 3H\dot{h}_{ij} - \frac{1}{a^2}\nabla^2 h_{ij} = \frac{1}{a^2}\Lambda_{ij}^{ab}\left\{ -8\partial_a(\Phi - \Psi)\partial_b\Phi + 4\partial_a\Phi\partial_b\Phi + 2\partial_a\delta\Phi\partial_b\delta\phi \right\}. \tag{11.135}
$$

In the absence of anisotropic stresses $\Phi - \Psi = 0$ and using (11.39) we can write

$$
\delta\phi = -\sqrt{\frac{2}{\epsilon}}\left(\Phi + \frac{\Phi'}{\mathcal{H}} \right) = v(\rho + P)^{1/2}, \tag{11.136}
$$

where Λ_{ij}^{ab} is the TT projection operator. We express the time dependence of $\Phi(k, \tau)$ using the transfer function

$$
\Phi(k, \tau) = T(k, \tau)\Phi_k, \tag{11.137}
$$

where Φ_k is the superhorizon perturbation generated during inflation and $T(k, \tau)$, (where τ is the conformal time) accounts for the time dependence after horizon entry in the radiation or matter era (or some-other non-standard cosmological era).

We operate on (11.135) with the polarisation tensor $e_\lambda^{ij}(k)$ and use the orthogonality property $e_\lambda^{ij}(k)e_{ij\lambda'}(k) = 2\delta_{\lambda\lambda'}$ and the relation $e_\lambda^{ij}\Lambda_{ij}^{ab} = e_\lambda^{ab}$. We assume that the scale factor at the time of horizon entry is $a \sim \tau^{1+b}$, where

$$
b = \frac{1 - 3\omega}{1 + 3\omega}. \tag{11.138}
$$

The ϵ parameter in terms of the equation of state during horizon entry is

$$
\epsilon = -\frac{\dot{H}}{H} = \frac{3}{2}(1 + \omega). \tag{11.139}
$$

The equation of motion (11.135) can then be written in the form from,

$$h''_{k,\lambda} + 2\mathcal{H}h'_{k,\lambda} + k^2 h_{k,\lambda} = S_{k,\lambda}, \tag{11.140}$$

where $\lambda = +, \times$ is the polarisation and the source function $S_{k,\lambda}$ is given by

$$S_{k,\lambda} = 4 \int \frac{d^3q}{(2\pi)^3} \, e_\lambda^{ij}(k) \, q_i q_j \, \Phi_q \Phi_{|k-q|} \, f(q, |k-q|, \tau), \tag{11.141}$$

where

$$f(q, |k-q|, \tau) = T(q, \tau)T(|k-q|, \tau)$$
$$+ \frac{1+b}{2+b}\left(T(q,\tau) + \frac{T'(q,\tau)}{\mathcal{H}}\right)\left(T(|k-q|,\tau) + \frac{T'(|k-q|,\tau)}{\mathcal{H}}\right). \tag{11.142}$$

In deriving (11.140)

The gravitational wave can be determined from (11.140) using the Green's function

$$h_{k,\lambda}(\tau) = \frac{1}{a(\tau)} \int_{\tau_i}^{\tau} d\tilde{\tau} \, G(k, \tau, \tilde{\tau}) \, a(\tilde{\tau}) S_{k,\lambda}(\tilde{\tau}). \tag{11.143}$$

The Green's function is the solution of the wave equation (11.140) with a delta function source

$$G''(k, \tau, \tilde{\tau}) + \left(k^2 - \frac{a''}{a}\right) G(k, \tau, \tilde{\tau}) = \delta(\tau - \tilde{\tau}). \tag{11.144}$$

The boundary conditions are $h_{k,\lambda}(\tau_i) = 0$, $h'_{k,\lambda}(\tau_i) = 0$.

In the radiation era $\omega = 1/3$, $b = 0$, $a'' = 0$ and the retarded Green's function in conformal time is given by

$$G(k, \tau, \tilde{\tau}) = \frac{4}{\pi^2 k}\left(\sin(k\tau)\cos(k\tilde{\tau}) - \cos(k\tau)\sin(k\tilde{\tau})\right)\Theta(\tau - \tilde{\tau}). \tag{11.145}$$

In the matter era, $\omega = 0$, $b = 1$, $a \sim \tau^2$ and $\frac{a''}{a} = \frac{2}{\tau^2}$. The retarded Greens function in the matter era is given by

$$G(k, \tau, \tilde{\tau}) = k\tau\tilde{\tau}\left(j_1(k\tau)y_1(k\tilde{\tau}) - y_1(k\tau)j_1(k\tilde{\tau})\right)\Theta(\tau - \tilde{\tau}), \tag{11.146}$$

where $j_1(x)$ and $y_1(x)$ are spherical Bessel functions.

11.17 Gravitational Wave from Pre-Heating

At the end of inflation there may be inhomogeneities in the of the inflaton at sub-horizon scales which can source gravitational waves [47]. The equation of motion of the tensor perturbations is

$$\ddot{h}_{ij} + 3H\dot{h}_{ij} - \frac{1}{a^2}\nabla^2 h_{ij} = 16\pi G \Pi_{ij}^{TT}, \tag{11.147}$$

where Π_{ij}^{TT} is the transverse traceless projection of the stress tensor

$$T_{\mu\nu} = \partial_\mu \phi \partial_\nu \phi - \bar{g}_{\mu\nu}\left(\frac{1}{2}\partial^\alpha \phi \partial_\alpha \phi - V(\phi)\right) \tag{11.148}$$

and is given by

$$\Pi_{ij}^{TT} = T_{ij} - p\bar{g}_{ij} = \partial_i \phi \partial_j \phi - \frac{1}{3}(\partial^k \phi \partial_k \phi). \tag{11.149}$$

As an example consider the inflaton potential [48,49]

$$V(\phi,\chi) = \lambda M_P^{4-p}|\phi|^p + \frac{1}{2}g^2\phi^2\chi^2, \tag{11.150}$$

where $p = 1, 2/3, 2/5$ as arising in some well-motivated models. With $\lambda = 3, 4, 5 \times 10^{-10}$ for $p = 1, 2/3, 2/5$ one can have inflation with N=50 e-foldings with scalar index $n_s = 0.970, 0.973, 0.976$ and scalar-tensor ratio $r = 0.08, 0.05, 0.03$ respectively.

In this cuspy set of potential oscillations of ϕ produces the light particles χ for reheating at the end of inflation [49]. In addition the oscillations of ϕ at the end of inflation produce gravitational waves very efficiently (Fig. 11.3).

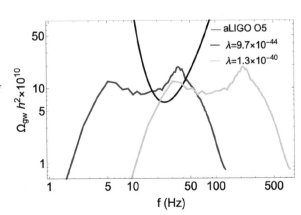

Fig. 11.3 Stochastic GW from pre-heating in a hybrid model with $aV = \lambda M_p^{4-p}|\phi|^p$, cuspy potential. Plots shown are for $p = 1$ with different values of λ. Reprinted from [48]. Figure credit: ©2018 American Physical Society. Reproduced with permissions. All rights reserved

References

1. A.H. Guth, Inflationary universe: a possible solution to the horizon and flatness problems. Phys. Rev. D **23**, 347 (1981)
2. A.A. Starobinsky, Spectrum of relict gravitational radiation and the early state of the universe. JETP Lett. **30**, 682 (1979)
3. D. Kazanas, Dynamics of the universe and spontaneous symmetry breaking. Astrophys. J. **241**, L59 (1980)
4. A.D. Linde, A new inflationary universe scenario: a possible solution of the horizon, flatness, homogeneity, isotropy and primordial monopole problems. Phys. Lett. B **108**, 389 (1982)
5. K. Sato, First order phase transition of a vacuum and expansion of the universe. Mon. Not. Roy. Astron. Soc. **195**, 467 (1981)
6. A. Albrecht, P.J. Steinhardt, Cosmology for grand unified theories with radiatively induced symmetry breaking. Phys. Rev. Lett. **48**, 1220 (1982)
7. G.F. Smoot, Summary of Results from COBE. astro-ph/9902027
8. C.L. Bennet et al., WMAP Coll., Nine-year Wilkinson Microwave Anisotropy Probe (WMAP) observations: final maps and results. Astrophys. J. Suppl. **208**, 20 (2013)
9. Y. Akrami et al. [Planck Collaboration], Planck 2018 results. X. Constraints on inflation. arXiv:1807.06211 [astro-ph.CO]
10. N. Aghanim et al. [Planck Collaboration], Planck 2018 results. VI. Cosmological parameters. arXiv:1807.06209
11. V.F. Mukhanov, H.A. Feldman, R.H. Brandenberger, Theory of cosmological perturbations. Part 1. Classical perturbations. Part 2. Quantum theory of perturbations. Part 3. Extensions. Phys. Rept. **215**, 203 (1992)
12. A. Riotto, Inflation and the theory of cosmological perturbations. ICTP Lect. Notes Ser. **14**, 317 (2003). [hep-ph/0210162]
13. S. Dodelson, *Modern Cosmology* (Academic Press, Amsterdam, 2003), 440p
14. V.F. Mukhanov, *Physical Foundations of Cosmology* (Cambridge University Press, Cambridge, 2005)
15. D. Baumann, TASI lectures on Inflation. arXiv:0907.5424
16. J.M. Bardeen, P.J. Steinhardt, M.S. Turner, Spontaneous creation of almost scale-free density perturbations in an inflationary universe. Phys. Rev. D **28**, 679 (1983)
17. C. Gordon et al., Adiabatic and entropy perturbations from inflation. Phys. Rev. D **63**, 023506 (2001)
18. T.S. Bunch, P.C.W. Davies, Quantum field theory in de Sitter space: renormalization by point splitting. Proc. Roy. Soc. Lond. A **360**, 117 (1978)
19. N.A. Chernikov, E.A. Tagirov, Quantum theory of scalar fields in de Sitter spacetime. Annales Poincare Phys. Theor. A **9**, 109 (1968)
20. L.F. Abbott, M. Wise, Constraints on generalized inflationary cosmologies. Nucl. Phys. B **244**, 541 (1984)
21. A.A. Starobinsky, Spectrum of relict gravitational radiation and the early state of the universe. JETP Lett. **30**, 682–685 (1979)
22. V. Sahni, Energy density of relic gravity waves from inflation. Phys. Rev. D **42**, 453 (1990)
23. A.G. Polnarev, Polarization and anisotropy induced in the microwave background by cosmological gravitational waves. Sov. Astron. **29**, 607–613 (1985)
24. U. Seljak, M. Zaldarriaga, Signature of gravity waves in polarization of the microwave background. Phys. Rev. Lett. **78**, 2054–2057 (1997). https://doi.org/10.1103/PhysRevLett.78.2054. [arXiv:astro-ph/9609169 [astro-ph]]
25. P.A.R. Ade et al., [BICEP2 and Planck Collaborations], Joint analysis of BICEP2/$KeckArray$ and $Planck$ data. Phys. Rev. Lett. **114**, 101301 (2015)
26. R. Abbott et al. [KAGRA, Virgo and LIGO Scientific], Upper limits on the isotropic gravitational-wave background from Advanced LIGO and Advanced Virgo's third observing run. Phys. Rev. D **104**(2), 022004 (2021). [arXiv:2101.12130 [gr-qc]]

27. P.D. Lasky, C.M.F. Mingarelli, T.L. Smith, J.T. Giblin, D.J. Reardon, R. Caldwell, M. Bailes, N.D.R. Bhat, S. Burke-Spolaor, W. Coles, et al., Gravitational-wave cosmology across 29 decades in frequency. Phys. Rev. X **6**(1), 011035 (2016). https://doi.org/10.1103/PhysRevX. 6.011035. [arXiv:1511.05994 [astro-ph.CO]]

28. P. Adshead, R. Easther, E.A. Lim, The 'in-in' Formalism and Cosmological Perturbations. Phys. Rev. D **80**, 083521 (2009). [arXiv:0904.4207 [hep-th]]

29. N. Arkani-Hamed, J. Maldacena, Cosmological Collider Physics. [arXiv:1503.08043 [hep-th]]

30. X. Chen, Y. Wang, Z.Z. Xianyu, Schwinger-Keldysh diagrammatics for primordial perturbations. J. Cosmol. Astropart. Phys. **12**, 006 (2017). [arXiv:1703.10166 [hep-th]]

31. P. Christeas, L. Thomas, Manifestly Unitary Cosmological Perturbation Theory. [arXiv:2205.15363 [gr-qc]]

32. S. Weinberg, Quantum contributions to cosmological correlations. Phys. Rev. D **72**, 043514 (2005) [arXiv:hep-th/0506236 [hep-th]]

33. E. Pajer, Field Theory in Cosmology. http://www.damtp.cam.ac.uk/user/ep551/FTC.html

34. M. Gasperini, M. Giovannini, G. Veneziano, Squeezed thermal vacuum and the maximum scale for inflation. Phys. Rev. D **48**, R439–R443 (1993). [arXiv:gr-qc/9306015 [gr-qc]]

35. K. Bhattacharya, S. Mohanty, A. Nautiyal, Enhanced polarization of CMB from thermal gravitational waves. Phys. Rev. Lett. **97**, 251301 (2006). [arXiv:astro-ph/0607049 [astro-ph]]

36. R.O. Ramos, L.A. da Silva, Power spectrum for inflation models with quantum and thermal noises. J. Cosmol. Astropart. Phys. **03**, 032 (2013). [arXiv:1302.3544 [astro-ph.CO]]

37. W. Zhao, D. Baskaran, P. Coles, Detecting relics of a thermal gravitational wave background in the early Universe. Phys. Lett. B **680**, 411–416 (2009). [arXiv:0907.4303 [gr-qc]]

38. K. Bhattacharya, S. Mohanty, R. Rangarajan, Temperature of the inflaton and duration of inflation from WMAP data. Phys. Rev. Lett. **96**, 121302 (2006). [arXiv:hep-ph/0508070 [hep-ph]]

39. S. Matarrese, S. Mollerach, M. Bruni, Second order perturbations of the Einstein-de Sitter universe. Phys. Rev. D **58**, 043504 (1998). [arXiv:astro-ph/9707278 [astro-ph]]

40. V. Acquaviva, N. Bartolo, S. Matarrese, A. Riotto, Second order cosmological perturbations from inflation. Nucl. Phys. B **667**, 119–148 (2003). [arXiv:astro-ph/0209156 [astro-ph]]

41. K.N. Ananda, C. Clarkson, D. Wands, The Cosmological gravitational wave background from primordial density perturbations. Phys. Rev. D **75**, 123518 (2007). https://doi.org/10.1103/ PhysRevD.75.123518. [arXiv:gr-qc/0612013 [gr-qc]]

42. D. Baumann, P.J. Steinhardt, K. Takahashi, K. Ichiki, Gravitational Wave Spectrum Induced by Primordial Scalar Perturbations. Phys. Rev. D **76**, 084019 (2007). https://doi.org/10.1103/ PhysRevD.76.084019. [arXiv:hep-th/0703290 [hep-th]]

43. L. Alabidi, K. Kohri, M. Sasaki, Y. Sendouda, Observable spectra of induced gravitational waves from inflation. J. Cosmol. Astropart. Phys. **09**, 017 (2012). https://doi.org/10.1088/1475-7516/2012/09/017. [arXiv:1203.4663 [astro-ph.CO]]

44. M.C. Guzzetti, N. Bartolo, M. Liguori, S. Matarrese, Gravitational waves from inflation. Riv. Nuovo Cim. **39**(9), 399–495 (2016). [arXiv:1605.01615 [astro-ph.CO]]

45. G. Domènech, Scalar induced gravitational waves review. Universe **7**(11), 398 (2021). [arXiv:2109.01398 [gr-qc]]

46. S. Bhattacharya, S. Mohanty, P. Parashari, Implications of the NANOGrav result on primordial gravitational waves in nonstandard cosmologies. Phys. Rev. D **103**(6), 063532 (2021). [arXiv:2010.05071 [astro-ph.CO]]

47. J.F. Dufaux, A. Bergman, G.N. Felder, L. Kofman, J.P. Uzan, Theory and numerics of gravitational waves from preheating after inflation. Phys. Rev. D **76**, 123517 (2007). [arXiv:0707.0875 [astro-ph]]

48. J. Liu, Z.K. Guo, R.G. Cai, G. Shiu, Gravitational waves from oscillons with cuspy potentials. Phys. Rev. Lett. **120**(3), 031301 (2018). [arXiv:1707.09841 [astro-ph.CO]]

49. H.B. Moghaddam, R. Brandenberger, Preheating with fractional powers. Mod. Phys. Lett. A **31**(39), 1650217 (2016). [arXiv:1502.06135 [hep-th]]

Index

Printed in the United States
by Baker & Taylor Publisher Services